Production Factor Mathematics

Martin Grötschel · Klaus Lucas · Volker Mehrmann
Editors

Production Factor Mathematics

 Springer

Editors

Martin Grötschel
Konrad-Zuse-Zentrum
für Informationstechnik Berlin (ZIB)
Takustr. 7
14195 Berlin
Germany
groetschel@zib.de

Klaus Lucas
RWTH Aachen
FB 4, Inst. Thermodynamik
Schinkelstr. 8
52062 Aachen
Germany
lucas@ltt.rwth-aachen.de

Volker Mehrmann
TU Berlin
Fak. II Mathematik & Naturwissenschaften
Institut für Mathematik
Straße des 17. Juni 136
10623 Berlin
Germany
mehrmann@math.tuberlin.de

Translation of the German edition *Produktionsfaktor Mathematik*,
Martin Grötschel, Klaus Lucas, Volker Mehmann (Eds.)
© acatech – Deutsche Akademie der Technikwissenschaften, Springer-Verlag Berlin Heidelberg 2009.
All rights reserved.

ISBN 978-3-642-11247-8   e-ISBN 978-3-642-11248-5
DOI 10.1007/978-3-642-11248-5
Springer Heidelberg Dordrecht London New York

Library of Congress Control Number: 2010933524

Mathematics Subject Classification (2010): 0-99, 00A69

Cover design: WMXDesign GmbH

Printed on acid-free paper

Springer is part of Springer Science+Business Media (www.springer.com)

Preface

In economics, the classical factors of production are *land, labor and capital*. At first glance, it may seem provocative to call mathematics a production factor as well. This provocation has a goal. It intends to bring the importance of mathematics for technology and the economy in general into focus, to help advance the visibility of the achievements of mathematics, and to uncover the benefits society has from mathematics. This volume is designed to contribute to this aim.

Management science differentiates production factors in greater detail and adds *entrepreneurship*, or, phrased in a more unassuming way, *planning* as further production factor. Its role is the best possible combination of the other factors. The goal of production in general is to maximize the benefit for all participating parties taking constraints on resources, technical possibilities, laws of physics and chemistry, budget limitations and also government requirements into account.

The terms *maximization, best possible combination, constraints* point already at the utilization of mathematics. The influence of mathematics is not restricted to the employment of optimization technology, though. Mathematics enters deeply into the design of products, the layout of production processes, and supply chains. Mathematics delivers the language to describe scientific, technological, and economic processes in an abstract way. It enables the modelling, simulation, and optimization of products and processes before their realization is started. Mathematics is not only a production factor for the improvement of products and services, it also proves to be a key technology in mastering complex technologies and a basic science for innovations. It is the goal of this book to substantiate these general statements by describing concrete cases where these roles of mathematics become apparent.

This volume, issued by *acatech*, the German Academy of Science and Engineering for the "Year of Mathematics" 2008 in Germany, focusses on the interplay between mathematics and engineering sciences: How can mathematics contribute to the improvement of technological processes and products? Where is this happening already? Are there deficits, and if so, where? What can we expect for the future? What can be the result of closer cooperations of engineers and mathematicians?

These are questions discussed in this volume of 19 articles on particularly relevant topics. The majority of them has been written by mixed teams of authors from engineering, industry, and mathematics, which has helped to avoid the overemphasis of discipline specific points of view. All authors were asked to write for a broad audience, and in most cases, this has been achieved. In a few cases technical details were necessary to describe the concrete applications or mathematical methods.

Mathematics is used with different intensity in different industries. In some branches, engineering progress is directly connected to top mathematical research. In other areas, well-known mathematical techniques are not used. Why this is so is described in several articles. The reasons are multifold. Language and terminology barriers play a role, also deficits in the education of engineers and mathematicians. Even regulatory reasons (such as monopolies), hierarchies in companies, or the psychology of decision making sometimes yield that mathematics is not utilized in an adequate way. All authors, however, are united in the opinion that an appropriate mastery of mathematics is a clear competitive advantage.

This volume with its diverse authorship is an authentic report on the interplay between mathematics and the engineering sciences. And thus, it also contains diverging opinions. One can find articles that contradict others in their suggestions for the treatment of engineering problems. Mathematicians for example require deeper theoretical understanding, while engineers who have great success in practice with ad hoc methods may question the usefulness of theory.

Germany has a very good starting position for a more intensive and deeper utilization of mathematics. The engineering sciences in Germany play leading roles worldwide, and the mathematics done in Germany is internationally recognized, in particular in those areas of mathematics that have actively sought contact with practitioners and addressed their problems. More and more countries are taking up this direction of progress and recognize the tremendous scientific and economic advantages that result from a closer cooperation of mathematics and industry.

Nevertheless there is a lot left to be done, and this is clearly visible in this volume. Some articles give concrete suggestions how to improve the cooperation between mathematics and engineering.

Our world is getting increasingly global. This leads to larger and tighter networks of production, trade, logistics, and transportation as well as to very involved processes and new forms of distributed and cooperative decision making. Thus, there will be growing demand for the understanding, planning, and control of complex systems. Mastering these tasks is impossible without mathematics.

Berlin   Martin Grötschel, Klaus Lucas and Volker Mehrmann

Contents

Networks

Medicine

Introduction

Martin Grötschel, Klaus Lucas,
and Volker Mehrmann

This book is the central contribution of the German Academy of Science and Engineering *acatech* to the "Year of Mathematics" in which the importance of mathematics and the many facets of this discipline were presented to the general public in Germany in 2008. This volume discusses the relationship of mathematics and the engineering sciences, it illuminates the contribution of mathematics to the industrial creation of value and the key position of mathematics in the handling of complex systems, in other words, the role of mathematics as a production factor and amplifier for innovations. The volume not only presents the state-of-the-art, it also points out strengths and weaknesses, it sketches future directions for research and development and it suggests activities to improve the interaction between mathematics and engineering as well as between academic and industrial research.

It is obvious that such a volume cannot discuss all parts of engineering and all aspects of mathematical research. The editors have, nevertheless, chosen a large bandwidth of topics, in which the tensions and the positive interactions between the two sciences become apparent. In this way a book has been created that is coherent and heterogeneous at the same time. The editors and authors hope that their analyses and suggestions will have influence that reaches beyond the concrete subjects that are discussed.

The coherence of the volume lies in the common structure of the 19 chapters. For every selected engineering field the relationship to mathematics is illuminated. It is examined in which way mathematics is successfully contributing to the field. Here

M. Grötschel (✉)
Zuse Institute Berlin, Takustraße 7, 14195, Berlin, Germany
e-mail: groetschel@zib.de

K. Lucas
RWTH Aachen, FB 4 Inst. Thermodynamik, Schinkelstr. 8, 52062 Aachen, Germany
e-mail: lucas@ltt.rwth-aachen.de

V. Mehrmann
TU Berlin, MA 4-5, Straße des 17. Juni 136, 10623 Berlin, Germany
e-mail: mehrmann@math.tu-berlin.de

M. Grötschel et al. (eds.), *Production Factor Mathematics*,
DOI 10.1007/978-3-642-11248-5_1, © Springer-Verlag Berlin Heidelberg 2010

scientific topics are discussed, but the role of mathematics as production factor in the corresponding industrial sector is analyzed as well. Every chapter begins with an executive summary that briefly summarizes the contributions of the paper. Then the main part starts with two to four success stories of a productive interaction between this field of engineering and mathematics, followed by a description of the state-of-the-art. From the analysis of strengths and weaknesses then implications and challenges for the engineering field and the associated mathematical disciplines are derived. This includes the necessary conditions for a successful cooperation between mathematics and engineering, the transfer to industry as well as the education at universities and the continuing education of people working in industry. Visions of potential developments and actions recommended to decision makers, industrial partners and funding agencies complete the picture.

The heterogeneity arises from the diversity of the teams of authors and their scientific fields. Several chapters are written jointly by mathematicians, engineers and industry representatives. In some chapters engineering is dominating and in others mathematics. There are major differences between the various technical fields, and the approaches taken vary considerably, e.g., in building services engineering, chemical engineering, signal processing, or logistics. These differences illuminate domains of interaction and/or conflict within the engineering sciences as well as between mathematics and engineering and they raise positive expectations for future developments. Many discussions have started, and we hope they will continue far beyond the group of authors of this volume.

The allocation of the 19 chapters of the book to the topics

- processes
- networks
- materials and mechanics
- energy and construction
- medicine

is rather coarse. Some chapters could have been assigned to several of these parts, since they do not deal with a clearly defined technical discipline only. In some of the papers the influence of a certain mathematical methodology that turned out to be very successful in one application area is described also for other applications (see, e.g., the chapters by S. Engell et al., G. Leugering et al., W. Dahmen and W. Marquardt, R. King et al.). These chapters nicely demonstrate the universal applicability of mathematics. Mathematics very often supports the transfer of knowledge between different scientific disciplines and industrial applications that are otherwise unrelated.

To advance mutual understanding, we also want to point in this introduction to the different interpretation of concepts and use of terminology. In particular, the terms modelling, simulation, and optimization call for explanation. These appear in almost every chapter but often with different meaning.

Mathematical models are used to describe observations of the real world, physical or social processes in the language of mathematics. They are constructed from mathematical building blocks. Modelling begins by defining the sets that are necessary to describe the potential space of solutions. In some cases, parameters, variables, and constants can take arbitrary real numbers as their values, occasionally complex numbers are needed, other values have to be integers, etc. Variables are introduced to describe potential ranges of reality. They operate on the chosen sets, and the relationships between the variables are given by equations and inequalities; in optimization problems, cost functions that need to be maximized or minimized come up in addition.

Anybody who deals with modelling knows that a mathematical model can never give an exact image of nature. The art of modelling is to filter out the important parameters and relationships that are necessary for the goals of the investigation, to describe these precisely and to leave out less important factors. This procedure implies that modelling cannot be done by mathematicians alone without the support from the application sciences and industry partners. Good mathematical modelling is an interactive process between several groups. Besides engineers and mathematicians, this could be computer scientists, biologists, chemists, physicists, economists, managers, or lawyers. The final product of the modelling process should be a mathematical model of which all the participating partners are convinced that it describes the "reality of the problem" in a reasonable way.

The created model, however, often has to satisfy further constraints. Sometimes one has a perfect model, but no algorithmic methods to solve the resulting equations. Models have to be simplified to make them "computable", see, e.g., the chapter by R. King et al. But it is also possible that the data for a good model are not at hand. Then the model has to be reduced in such a way that only available data are employed. Some theoreticians get nervous in such circumstances, but engineers are used to this and have learned to deal with the uncertainty of the real world.

At the end of the modelling process a mathematical model for the problem in question is established; but then further differences between mathematics and engineering arise. Mathematicians expect models in an analytic representation, expressed, e.g., by linear inequalities, ordinary or partial differential equations. But this is not always possible. Often, the model an engineer in practice comes up with, is a computer program, possibly containing many if-then-else-conditions. Such a model gives mathematicians a headache because it is difficult to treat such models with mathematical tools. Such models (given by codes) must be treated with methods that are uncommon in mathematics. Examples of this kind can be found in energy and building services engineering, see, e.g., the chapter by D. Hartmann et al. Similar observations hold for systems that are formed by a network of subsystems, see, e.g., the chapters by D. Abel or D. Müller, where there are appropriate models for each subcomponent, but the complete model is out of the current reach of mathematical analysis.

Nevertheless the intensive cooperation of mathematicians and engineers in the solution of concrete practical problems has led to enormous progress. Among other areas, this applies to the development and analysis of new materials, see, e.g., the chapters by W. Dreyer, and W. Ehlers and P. Wriggers, where material properties are predicted based on the mathematical model equations. Similar success stories can be observed in the development of production processes, see, e.g., the chapter by B. Denkena et al. However, not only the understanding of the technological processes has improved, but also the associated mathematics was developed further, new exciting mathematical questions have evolved and led to new solution algorithms, or known methods could be extremely accelerated, see, e.g., the chapter by R. Borndörfer et al. Without the close contact to the engineering practice, many of these questions would not have been considered by mathematicians. On the other hand, the contact to theory has made many new and efficient mathematical tools available for practical engineering.

Very often mathematical models and methods are not only useful for the problem for which they were developed. The theory and the solution algorithms can often be transferred to other applications with little effort. This is for example clearly visible in the area of networks, where methods of integer programming were developed for communication networks and are now used in logistics and traffic networks, and vice versa, see, e.g., the chapters by R. Borndörfer et al., R.H. Möhring and M. Schenk, J. Eberspächer et al., H. Boche and A. Eisenblätter. The same holds for multi-scale models and methods, which can be used in image compression and coding as well as in the process industry and in aerodynamics, see, e.g., the chapter by W. Dahmen and W. Marquardt.

The next step after creating the mathematical model is the numerical simulation. This is nothing else but a repeated solution of the model equations under variation of parameters. Simulation allows to check the model. Are the results as expected or are there surprising effects? Are modelling errors the reason for these unexpected results or wrong data, or has there—by chance—actually a new effect been discovered? Numerical simulation allows to gain confidence in the model, but it does not deliver a proof of correctness. It supports the "gradual understanding" of processes in a productive way. Besides experiment and theory, numerical simulation has established itself as an increasingly important third column of gaining insight.

When the numerical simulation is working well and efficiently, extensive simulation studies are usually conducted that may lead to good combinations of parameters for the real processes (e.g., in chemical engineering or the construction of a power station). In engineering practice you sometimes find the belief that it is possible to optimize via simulation studies. Mathematicians strongly disagree with this belief.

Optimization can only start when the model has proved itself, e.g., in simulation studies. Practitioners often speak of optimization, when merely an improvement has been achieved (e.g., with heuristic methods). Mathematicians view this more stringently. They want to find model-based solutions that are provably optimal. If this can't be achieved, they aim at providing quality criteria, for example in the form that the value of the cost function of the solution found is maximally 5% away from the optimal value. For many large scale problems, though, mathematicians have to resort to approximation methods as well, and they often employ heuristics that engineers use successfully in practice.

Often, "simple optimization" is not enough. Many practical problems do not have a unique cost function. This is for example the case when economical (maximization of profit), ecological (little negative impact on nature) and social (high employment rate) goals need to be combined. In such cases one speaks of multi-criteria optimization. In this field the state of the current mathematical research is not satisfactory, and in practical applications, nothing but scenario analysis and heuristic methods are used.

This leads to another question. How about the stability and robustness of computed optimal solutions? Do small variations in the parameters lead to completely different solutions? These types of questions nowadays arise frequently in all application areas. In the construction of systems or processes one is forced to incorporate potential disturbances already in the design phase and to make sure that they can be compensated at runtime in the real process in a reasonable way.

Random changes, expected values, or risk evaluations are currently hot topics in financial mathematics, but stochastic processes are also of great importance in medicine, bio technology and systems control. Disturbances come from outside (weather, accidents) but also data errors (faulty measuring instruments) or model errors can contribute to uncertainty. In the planning of traffic or logistic systems, see, e.g., the chapters by R. Borndörfer et al., R.H. Möhring and M. Schenk, one is for example interested in generating time tables or tour plans in such a way that local disturbances or traffic jams do not lead to global effects, and that delays can be compensated with little inconvenience for the customers. Mathematical stochastics deals with such questions and begins to deliver solutions in many application areas. But despite theoretical progress, simulation is still one of the major tools in the analysis of robustness and stability.

This leads to the area of control, which is intensively covered in the thematic part on processes, see, e.g., the chapters by L. Grüne et al., R. King et al., S. Engell et al. In closed-loop control one generates feedback controls that interact with the process or system with the goal to minimize resources and to keep the dynamic process stable. In open-loop control one determines a priori input parameters that lead (without further intervention) to a desired behavior of the system. There is almost no mathematical subfield that has not contributed to control theory. Similar methods also play an important role in modern signal processing, see, e.g., the chapter by H. Boche and A. Eisenblätter.

Almost everything that has been sketched so far is unthinkable without the existence of modern computers and information technology. The developments described in this book have only become possible by the utilization of fast and reliable computers. Mathematics not only has contributed in the design and implementation of efficient codes for the solution of a variety of problems, the development of modern computer chips is not possible without mathematical methods, see, e.g., the chapter by J. Koehl et al. Mathematical methods turn out to be similarly important in the analysis of economic data and the treatment of mass data sets, see, e.g., the chapter by J. Garcke et al.

Besides the described successes and perspectives in the interaction of mathematics and engineering, all health/medicine related areas such as, medical technology and drug design are increasingly dependent on mathematical methods, see, e.g., the chapters by P. Deuflhard et al., and G.J. Bauer et al.

In all the described fields the future of research and development lies in the integration of modelling, simulation, and optimization. This can only be achieved by an interdisciplinary cooperation of engineers, mathematicians (and others) and it requires that the different disciplines move closer to each other in terms of concepts and language. A major suggestion that can be found in almost all chapters is a reform of educational programs in engineering and mathematics. For example, discrete mathematics and optimization have become very important in recent years, but these topics are missing in many engineering programs; on the other hand, mathematical modelling is only marginally treated in mathematical curricula.

Another central suggestion that can be found in the book is the initiation of further research funding directed towards innovations in the interface between mathematics and applications. As a prime example of support of this type we want to mention the program "Mathematics for innovations in industry and service" funded by the German Federal Ministry of Education and Research (BMBF), which has been an effective catalyzer for the current developments and has funded many of the authors in this volume. The demand and the existing potential are so immense that such funding should definitely be intensified. This is the only way to achieve the scientific and technological progress that has been sketched in this book which requires a close cooperation between mathematics, engineering, and industry.

In 2007, the BMBF published a brochure "Der Schlüssel zur Hochtechnologie, Mathematik für Innovationen in Industrie und Dienstleistungen" (The key to key technologies, mathematics for innovations in industry and service) and the Federal Minister of Education and Research, Dr. Annette Schavan, writes in her address: "Even in view of concrete economic and social challenges mathematics plays a central role. Mathematics enables innovations in the industrial and service sector, that lead to more jobs and an increasing competitiveness of Germany". And she continues: "In this way an old science accompanies us into the future, with a potential that has not nearly been exhausted". The authors and editors of this book happily agree with the minister's analysis. They combine this with the hope that this will have consequences in industry, politics, and funding organizations. Mathematics can contribute much more to product design, product safety, generation of value, saving of resources, etc. This volume bears witness to this and at the same time suggests appropriate measures to improve the situation.

Processes

Predictive Planning and Systematic Action—On the Control of Technical Processes

Lars Grüne, Sebastian Sager, Frank Allgöwer,
Hans Georg Bock, and Moritz Diehl

1 Executive Summary

Since the beginning of the industrial revolution control engineering has been a key technology in many technical fields. James Watt's centrifugal governor for steam engines is one of the early examples of an extremely successful controller concept, of which at the end of the 1860s approximately 75 000 devices were in use only in England [2, p. 24]. Around this time, motivated by the increasing complexity of the plants that had to be controlled, engineers started to investigate systematically the theoretical foundations of control theory. The dynamic behavior of controlled systems, however, can only be understood and advanced with the help of mathematics, or as Werner von Siemens formulated: "Without mathematics you are always in the dark."

L. Grüne (✉)
Universität Bayreuth, Mathematisches Institut, 95440 Bayreuth, Germany
e-mail: lars.gruene@uni-bayreuth.de

S. Sager
Interdisziplinäres Zentrum für Wissenschaftliches Rechnen, Der Ruprecht-Karls-Universität Heidelberg, Im Neuenheimer Feld 368, 69120 Heidelberg, Germany
e-mail: Sebastian.Sager@iwr.uniheidelberg.de

F. Allgöwer
Institute for Systems Theory and Automatic Control, University of Stuttgart, Pfaffenwaldring 9, 70550 Stuttgart, Germany
e-mail: allgower@ist.uni-stuttgart.de

H.G. Bock
Interdisciplinary Center for Scientific Computing, 69120 Heidelberg, Germany
e-mail: Bock@IWR.Uni-Heidelberg.de

M. Diehl
Optimization in Engineering Center (OPTEC) and Electrical Engineering Department ESAT, Division SCD, KU Leuven, Kasteelpark Arenberg 10, 3001 Leuven-Heverlee, Belgium
e-mail: Moritz.Diehl@esat.kuleuven.be

M. Grötschel et al. (eds.), *Production Factor Mathematics*,
DOI 10.1007/978-3-642-11248-5_2, © Springer-Verlag Berlin Heidelberg 2010

Therefore, in control engineering the question is not—today like more than one hundred years ago—*if*, but rather *which kind of* mathematics has to be used. In fact, from algebra via geometry and the theory of dynamic systems to optimization and numerics there is hardly any mathematical field that has not contributed significantly to control engineering and its neighboring mathematical disciplines systems and control theory.

In this article we want to point out two aspects. On the one hand, by means of two examples, we show the impact mathematics had in the long history of control engineering and which factors have been fundamental for the success of control methods. On the other hand—after a brief overview of the state-of-the-art in control engineering—we introduce a modern control methodology in more detail, namely *model predictive control* (MPC). We will briefly explain underlying mathematical concepts from systems theory, numerics, and optimization, and sketch some future challenges for mathematics.

In its linear form MPC has already become a standard tool in industrial applications, in particular in process engineering. In its nonlinear form (NMPC), on which we focus here, it is generally considered as one of the most promising modern methods for the control of complex technical processes. NMPC keeps finding its way into new application areas, due to the rapid progress on the theoretical as well as on the algorithmic side. Moreover, it is a prime example for a method that has reached its present state of development only by the interdisciplinary collaboration in the fields of control engineering, mathematical systems theory, numerics, and optimization, and whose future success will depend on an interdisciplinary approach, as well. Fortunately, such approaches are more and more supported by a variety of universities in clusters of excellence, research centers, or graduate schools. These joint interdisciplinary efforts in research and teaching will be an essential factor to tackle the future challenges and exploit the full potential of the production factor mathematics in control engineering.

2 A Long Success Story

Mathematical methods have played an essential role in control theory since the very beginning of this research field. We revise two prominent examples that illustrate this strong impact: the stability criterion of Hurwitz and Pontryagin's maximum principle. We will also discuss which factors made the mathematical developments this successful and are in a sense prototypical for mathematical impact in control engineering.

2.1 The Stability Criterion of Hurwitz

The stability criterion of Hurwitz was developed in 1893–1894 by the mathematician Adolf Hurwitz, an employed professor at the Polytechnikum in Zurich (today

ETH Zurich) [30]. The cause for the development was not so much Hurwitz' mathematical curiosity, but rather a concrete request of his Zurichian colleague Aurel Stodola, a mechanical engineer, who had been engaged in the development of regulators for hydraulic turbines. Stodola's problem is best illustrated by means of an everyday problem. Consider the control of a heater in which the temperature of the room needs to be adjusted to a desired value. An obvious strategy is to regulate the valve for the hot water flow in the heater depending on the current measured temperature. If the current temperature is lower than the desired one, the valve is opened, otherwise the valve is closed. Obviously, one does not need any mathematics to see that this approach will eventually lead to the desired temperature.

However, this only holds in the ideal case. The case gets more complicated if the influence of the control input (in the example the flow valve) on the value (i.e., the temperature) is less direct, e.g., due to delays in the system. Any user of a shower whose hot water supply reacts only delayed on the opening or closing of the valves knows this effect: instead of the desired temperature one constantly gets values alternating between "too hot" and "too cold". Only with much effort and fine tuning one eventually succeeds in adjusting the right temperature. This is a classical example for an unstable behavior of a control circuit. The effects in the mechanical turbine systems that Stodola considered are quite similar, caused by a multitude of mechanical couplings. Also in this case it is not an easy task to design a control that keeps the system stable on a given reference value.

Stodola knew and applied mathematical results that had been discovered approximately two decades before [62]. They allow to deduce a condition for the stability from a model of the control circuit. This condition says basically that a polynomial, i.e., a mathematical function of the form

$$P(x) = a_0 + a_1 x + a_2 x^2 + \cdots + a_n x^n$$

has only roots (i.e., complex numbers x with $P(x) = 0$) with negative real parts. If such a polynomial has only few summands, then this condition is easy to check. However, polynomials with few summands correspond to very simple control systems for which this theory is not necessary to ensure stability—an experienced engineer would have this knowledge anyway. Relevant for practice are polynomials with many terms. The main problem is that the roots cannot simply be computed due to algebraic reasons that are known since the work of Évariste Galois in the early 19th century.

Thus, at the suggestion of Stodola, Hurwitz looked for a while into this problem and finally found a criterion that allows to determine if the roots have negative real parts, without actually having to compute them. Mathematically, his solution consisted of combining the coefficients $a_0, a_1, \ldots, a_n$ of the polynomial in n different matrices of different sizes. Then it is sufficient to compute the determinants of these matrices in order to respond to the original question of the sign of the real parts of the eigenvalues.

Stodola was enthusiastic about this solution. In a letter to Hurwitz he wrote: "I am exceedingly indebted to you for the ingenious solution of the root problem that has bothered me so much." [5]. But also Hurwitz mentioned in a footnote to his corresponding publication in the "Mathematische Annalen", not entirely without pride, the successful practical application of his criterion: "Mr. Stodola uses my result in his paper [...] whose results have been used in the turbines system in the bath Davos with splendorous success." Remarkably in this context is the fact that the English mathematician Edward John Routh developed a similar solution already 20 years before Hurwitz, therefore today Hurwitz' criterion is mostly referred to as the method of Routh-Hurwitz. However, Hurwitz' result has found the way into applications earlier. For decades Routh's result was known only to a small academic circle, but hardly to any engineer working in practice [2, pp. 81f].

2.2 Pontryagin's Maximum Principle

In spite of its name, referring to the Russian mathematician Lew S. Pontryagin, Pontryagin's maximum principle has several authors. On the one hand, Pontryagin published and proved the principle released in 1956 not alone, but with essential contributions of the rarely mentioned mathematicians V.G. Boltyansky and R.L. Gamkrelidze [8] (see also [51]). On the other hand, cf. [49, pp. 55f], the principle can be found already in publications of the American mathematician M.R. Hestenes from 1950 and—even in more general but not completely proven form—of the American mathematician R.P. Isaacs from 1954/1955. Important fundamental ideas for the maximum principle can be found even earlier in Carathéodory's book from 1935, as stated in [50].

The maximum principle addresses the issue of how the motion of a system—such as the trajectory of a rocket or the movement of the arm of an industrial robot—can be optimally planned. Thereby, "optimal" is always related to a pre-defined criterion with corresponding constraints. For example: how does a rocket get with minimum energy consumption (criterion) in a given time (constraint) from Earth to a particular orbit, or how can the arm of an industrial robot be steered as fast as possible (criterion) and with constant energy expense (constraint) from a given position to another one? Here, the example of the rocket is not chosen at random, since space travel as well as military rocket technology was one of the main driving forces for the development of this kind of mathematics in the beginning of the cold war between the USSR and the USA. The improvement in efficiency and stability of industrial robots, surely more significant for the industrial development, can therefore—like the famous Teflon-pan—be seen as a by-product of space research.

Optimization problems of similar form had already been solved about 250 years before Pontryagin and the other developers of the maximum principle with the help of the so-called calculus of variations. A comprehensive description of this method surely exceeds the scope of this article, but at least a brief outline of the principle's functionality shall be given here. For a detailed description see, e.g., [49]. Starting point for the approach is a model describing a motion via mathematical equations. Usually, this is done by means of a system of differential equations. For the solution of the problem the position on the optimal path has to be computed at every time instant. The problem is that the choices of values at different time points depend on one another. For example, higher acceleration in the beginning requires stronger braking at the end and vice versa. The calculus of variations solves these complex dependencies by setting up an extended system of differential equations (the Euler-Lagrange equations), whose first part is solved "normally" forwards, while the second part is solved backwards in time—a so-called boundary value problem. Then, for each time point the optimal path can be determined from the solution of this system.

The main problem in the application of the calculus of variations to practical problems is that here the optimal trajectories are computed, rather than the actually relevant optimal control input which has to be transmitted to the rocket engines or to the motors of the robot arms. Since these input values do not occur in the computation, it is not possible to provide physically meaningful ranges in which these control values have to lie. In other words, the calculus of variations can indeed determine optimal trajectories, but it is impossible to exclude that these trajectories require, e.g., a thrust of the engines that is way beyond the physical possibilities. This is quite obviously a deficiency which causes serious practical problems not only in space travel but in virtually every imaginable industrial application.

Here, the maximum principle provides a remedy. Although conceptually quite similar, as also boundary value problems need to be solved, the focus is not on the computation of the optimal trajectory, but on the direct computation of the corresponding controls. This allows to directly include physical or economic restrictions on the control values in the computation. At the same time, the maximum principle yields a rather intuitive criterion, since the optimal control values in every time point are nothing but the solutions of new "small" optimization problems, with parameters resulting from the boundary value problem. Since such problems can be solved numerically, the principle is not only useful for theoretical analysis, but also as a basis for powerful algorithms.

2.3 Conclusion

The aforementioned mathematical concepts for the analysis and solution of control problems are only two of many possible examples. Yet they have a number of properties that are more or less typical for successful mathematical concepts:

- They are applicable without knowledge of the underlying mathematical theories that were necessary for their derivation.
- They yielded considerable progress for real applications and thus enabled new industrial developments.
- They are constructive in the sense that they are easy to formulate as algorithms and therefore also easy to implement with the availability of digital computers.

Despite the evident advantages that these mathematical concepts had for industrial applications, their development was fostered by yet another factor. The control engineers were aware of the necessity of mathematical methods in the first place, and accordingly had a solid mathematical basic education. Without Stodola's knowledge of the interpretation of polynomial roots, Hurwitz would probably never have thought of developing his criterion. And without the availability of convenient models of differential equations the development of the maximum principle would have been impossible from the very beginning.

3 State-of-the-Art and Current Developments: The Example "Model Predictive Control"

3.1 A Short Introduction to Control Engineering

The objective of control engineering is the manipulation of dynamical systems such that their behavior has one or several desired characteristics. Here the dynamic systems are influenced by the so-called *control input* or *control variables u*. In the dynamical system "car" these are, e.g., the turning angle of the steering wheel or the position of the gas pedal. Then, by suitable changing of the control variable over time, the system dynamics can be influenced in the desired manner. The behavior of interest of the controlled system is usually summarized in the so-called *output variables y*. Figure 1 shows schematically such a *controlled system*.

Two important control concepts are *open-loop* and *closed-loop control*. As shown in Fig. 2, in open-loop control the dynamic behavior is given in form of an open functional chain. The desired behavior of the output variable *y* is given by the *reference variable w* that provides the reference signal for the open-loop control. From this signal the control input is generated which influences the dynamical system such that the output follows the reference signal as accurately as possible.

In industrial practice open-loop controls are very important and in many cases they perform their task very well. Examples are robots, machine tools and production facilities. The basic assumption for the proper function of an open-loop control is that the real-world behavior of the dynamical system can be predicted sufficiently precise by means of the given model; if this is not the case, the actual behavior can significantly deviate from the theoretical forecast used for the computation of the control variable *u*.

This can happen if the dynamic system is affected by large external perturbations, as, e.g., strong side wind in automobiles. Often the behavior of the dynamic system is not precisely known, because the mathematical model does not represent all aspects of the real system, hence *uncertainty* is an issue. The effects of these uncertainties are particularly dramatic when the system behavior is *unstable*. In the shower

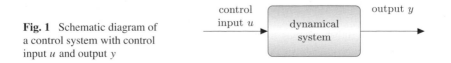

Fig. 1 Schematic diagram of a control system with control input *u* and output *y*

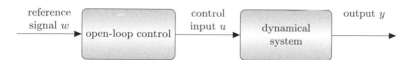

Fig. 2 The principle of open-loop control

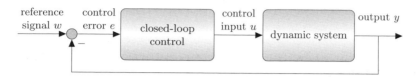

Fig. 3 Principle of a closed-loop control with feedback and reference/output value comparison

example in Sect. 2.1 instability would mean small changes in the control variable (valve) lead to large changes in the controlled variable (water temperature). In these cases *closed-loop control* has to be used in order to ensure stability of the system, i.e. in our example to ensure a temperature that stays close to the reference value.

In contrast to open-loop control, a closed-loop control makes use of a feedback structure as shown in Fig. 3. The controlled variable is measured, fed back and compared to the desired value, i.e., the reference signal. The difference between reference and current value, also called the *control error*, is provided to the controller. Due to this closed-loop structure, the influence of external perturbations and uncertainties can be explicitly detected and adjusted before they lead to larger deviations from the reference value. Therefore, closed-loop controls are also applicable in the case of perturbations, uncertainties and unstable control systems.

Certainly, the feedback structure alone does not automatically lead to stability of the closed-loop system: It is of crucial importance by which rule the control input u is computed from the control error e. As already exemplified in Sect. 2, the dynamic behavior of a controlled system can be analyzed with the mathematical methods of control theory. The most important analysis methods examine stability and the so-called *robustness of stability* of the controlled system. The analysis of the robustness of stability addresses the issue if stability is preserved when the current dynamic system differs from the assumptions made in the model. This property plays a central role for the use of control methods in practical applications.

Mathematics plays a vital role in the *design* of open-loop and closed-loop controls, i.e., in the derivation of the rule that tells how the control input is computed from the reference variable or the control error. Today, modern controllers are designed *model-based*, i.e., it is assumed that a mathematical model of the system to be controlled is given, in most cases in the form of differential equations. In addition, the control objectives are formulated in mathematical terms, too. Besides the aforementioned stability these objectives often include the so-called *control performance*, typically formulated in form of an optimality criterion. Different classes of controller design methods differ in both the assumptions for the models of the control circuit and the control objectives. Regarding the model assumptions the most important distinction is between linear and nonlinear systems. Linear systems are characterized by—slightly simplified—a proportional relation between the control input u and the output y. This means that if, for instance, the control input u is doubled then the output y will be twice as large, too (mathematically one says that the system fulfills the superposition principle). Linear systems are described by linear differential equations, usually in the so-called state space form

$$\dot{x} = Ax + Bu, \tag{1a}$$

$$y = Cx + Du. \tag{1b}$$

The design of controllers for linear systems is well-understood and there is a large number of extremely powerful linear controller design methods available. At this point we want to mention exemplary the LQ methods (LQR, LQG, LQG/LTR, etc.) [23, 36, 42, 47], in which for linear systems of the form (1) a controller is designed, such that the closed-loop system has optimal behavior with regard to the minimization of a quadratic integral criterion

$$J = \int_0^\infty x^T(\tau) Q x(\tau) + u^T(\tau) R u(\tau) \, d\tau.$$

However, no external perturbations such as fluctuating side wind are considered in LQ methods. In the more recent H_∞-controller design [24, 64] controllers are designed such that external disturbances are optimally rejected, i.e., their negative effects on the control system is minimized. This minimization can be carried out by different mathematical optimization criteria, which apart from the H_∞-methods also lead to, e.g., L_1-methods [14].

Even if hardly any system occurring in industrial practice is actually linear, linear models can nevertheless often be used. The reason for this is that most systems "almost" behave like linear systems when only a small domain for the values of the controlled variable is considered: If the temperature in a room is to be increased from 20° to 22° Celsius, then the valve of the heater has to be opened approximately twice as far as if the temperature is only to be increased to 21°—the relationship between valve opening and temperature increase is linear. For larger temperatures this does not hold anymore, simply because further heating is not possible once the maximal temperature is reached, regardless of how the valve is set.

If the controlled system in the relevant domain does not exhibit linear behavior, which is frequently the case in practice, then nonlinear controller design methods have to be used. Here, the dynamic system forming the basis of the design is described by a nonlinear differential equation of the form

$$\dot{x} = f(x, u),$$

$$y = h(x, u).$$

In the last 20 years huge progress has been made in the field of nonlinear control and many current research projects deal with this subject. As an example, we mention differential geometric methods of exact linearization [33], flatness-based control

[22, 58], passivity-based control [9, 55], and backstepping methods [37, 41]. In the simplest case[1] these methods yield a static function

$$u = k(x),\tag{2}$$

i.e., a mathematical formula $k(x)$ that depends on the current system state x and determines how the control variable u has to be chosen.

In the nonlinear case, the primary focus often lies on the stabilization problem. However, certainly the control performance is of interest, too. The nonlinear equivalent to the previously mentioned LQ methods is optimal control, in which again an optimization criterion and constraints are provided—as already described in Sect. 2.2. However, in the case of the nonlinear optimal control the computation of an explicit formula (2) in feedback form for u is often impossible even with high-performance computers. On the other hand, the computation of nonlinear optimal open-loop controls (i.e., a control input u depending on time) with computer based, e.g., on Pontryagin's maximum principle, cf. Sect. 2.2, or the more modern direct methods, cf. Sect. 3.5, is relatively easy.

Therefore, a new class of methods for the optimal control of nonlinear systems has been developed in the last years, in which no explicit formula of the form (2) is computed, but the feedback law $k(x)$ is computed online in real time from open loop optimal control signals. Although such an online-computation looks more complicated at first sight, this approach has considerable advantages and is appropriate for the solution of large scale nonlinear control problems in practice. The most prominent representative of this new class of control methods is model predictive control which will be considered and discussed in more detail in the following sections.

3.2 The Principle of Model Predictive Control

If a controller is designed, e.g., in order to keep the temperature in a house on a constant value, then the easiest idea is to measure the current temperature and subsequently to increase or decrease the hot water flow rate appropriately—as described in Sect. 2.1—, such that by well-chosen control parameters a stable controlled system can be reached that more or less compensates for the external variations in temperature caused by the weather. As the heater cannot supply heat arbitrary fast and also the cooling needs some time, the room temperature controlled this way will typically always slightly oscillate.

But what would happen if we used the weather forecast? Then we could slightly pre-heat the apartment if a cold front approaches, or, if warm weather is expected, the heater could be shut off already some hours in advance to have the house cool down. It is intuitively clear that the use of this "predictive" principle allows to better maintain the control objective "as small deviation from the reference value as possible", taking into account our limited heating or cooling capabilities.

---

[1]Often the controller is not given by a static equation, but is itself given by a differential equation.

Fig. 4 Anticipatory
driving—as it is supposed to
be …

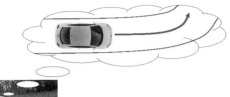

When we are driving a car, it is utterly indispensable that we drive "anticipatory",
i.e., "predictive". If we drove into a bend without braking in time, we would skid
off the road. That this does not happen is due to the advanced control technique in
our brain that has already anticipated the drive through the bend a long time before
we learn about the centrifugal force in the bend the hard way, see Fig. 4.

Model predictive control (MPC) realizes exactly this principle for the automatic
control of technical systems, cf., e.g., [11, 46]. We explain this in more detail using
the example of a typical stabilization problem: the system state is to be controlled to
a reference value and kept there by repeatedly measuring the state and, if necessary,
adjusting detected deviations by suitable adaptation of the control input.

Based on the current state x_0 of the system to be controlled and on the current predictions of the external influences, a model predictive controller computes
an (here for simplicity piecewise constant) optimal control for the near future, the
"prediction horizon" of length T. Typically, the employed optimality criterion minimizes the distance to the reference value, such that the computed optimal solution
has a distance from the reference value as small as possible. Then, only the first part
u_0 of the control is applied to the plant and used in order to steer the system for a
short time—the sampling time δ. After this short time the new current state is determined, the prediction horizon is shifted forward by δ, and a new optimal control
on this horizon is computed from which again only the first short part is used at the
plant. In this way, the online computed optimal open-loop controls are "composed"
to a closed-loop control. Figure 5 illustrates one step in this procedure.

The two essential advantages of model predictive control methods are the ability
to make use of look-ahead information and to include optimality criteria and restrictions. These advantages become apparent, e.g., in energy-intensive or time-critical
applications, since goals as "with minimal energy expense" or "within a given time"
can be directly included in the optimization criterion or in the constraint.

The two most important questions are:

- How can stability of the MPC controlled system be guaranteed?
- How can the optimal control problems be solved fast and reliably on the prediction horizon?

Both questions are of fundamentally mathematical nature. The first one belongs to
mathematical systems theory, the second one to numerical mathematics. We want to

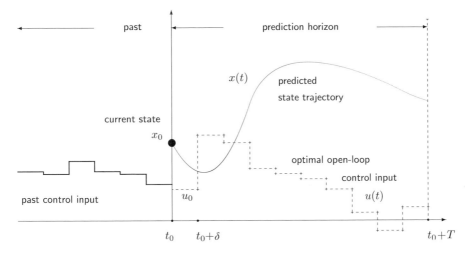

Fig. 5 Schematic diagram of model predictive control

discuss these questions—along with a presentation of applications of MPC—in the following sections.

3.3 Stability of MPC

As we want to avoid mathematical technicalities, here "stability" will simply denote the fact that the control algorithm fulfills its goal, i.e., that the system state is steered to the reference value. Intuitively, it seems plausible that by minimizing the distance to the reference value a stable control strategy is obtained. However, this may not be the case, a we illustrate with a small example, cf. Fig. 6. In this figure the position of a car (system state, in the beginning at location B) shall be steered to location A (reference value), while maintaining the velocity restrictions on the road. As there are mountains in the way, the road does not lead directly from A to B, but takes a detour via location C.

If you consider the air-line distance of the car from location A as optimization criterion, then this distance has to be increased along the road from B to A before it can eventually be decreased. An MPC strategy with a prediction horizon that covers, e.g., only the part of the road up to point C in the bend will not notice this, since every motion of the car within this horizon would only increase the distance. Therefore it is optimal to simply stay in point B and consequently the MPC controlled solution will stay forever in point B and never reach point A: the control algorithm is not stable.

Of course, this problem is known for a long time and different possibilities to find a remedy have been proposed. One approach often considered in the literature is to allow only for such solutions in the optimization on the horizon T that end up in the desired reference value. This method works theoretically (what has rigorously been

Fig. 6 Example of a control
problem

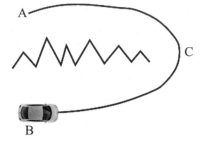

proven mathematically, cf. [38]) and is also occasionally employed in practice, but
it has the disadvantage that a further restriction is added to the optimization problem
that has to be solved and in addition in general it requires a very large horizon T
to ensure that under the given restriction (i.e., for instance the speed limit in our
road example) it is possible at all to reach the reference value at the time T. Either
can make the solution of the optimization problem considerably more expensive.
A further approach is the use of corrector terms in the optimization that compensate
the effect of too short horizons T, see, e.g., [13]. The computation of these terms,
however, is in general quite tedious, which is why this method is rarely applied in
practice.

The approach most frequently used in practice is the most apparent: the length of
the horizon T is simply increased until the algorithm becomes stable. Interestingly
enough, this solution is *not* considered in the major part of the theoretical MPC literature. Actually, for nonlinear MPC methods only recently it has been mathematically rigorously proven under relatively general assumptions that it indeed works
[26, 34]. The disadvantage of this method is that in general it is not clear in advance
how large T has to be chosen—and as T grows larger, more time is needed for the
solution of the optimization problem that has to be carried out online during the runtime of the system. Current research approaches [27, 28] therefore try to estimate
the required horizon T from the system properties and to use such estimates in order
to determine how the optimality criterion has to be chosen in order to obtain stability
with T as small as possible. At present, one can only speculate about the practicality
of this approach for industrially relevant processes. For our road example, however,
this approach yields a natural as well as efficient solution which perhaps already
came to the mind of some readers: if you measure the distance from location A to
B not by air-line, but via the length of the road, then for arbitrary short horizons
T it is always better to move towards A, since this will decrease the distance in
any case. If you optimize over this distance, then the method is stable for arbitrary
horizons T.

3.4 Application Fields

Model predictive control has originally been developed in chemical engineering,
where it is used, e.g., to control large and slow distillation columns—these sys-

tcms are extremely energy-intensive and the slow time scale leaves much time for the computation of optimal controls. Chemical engineering is even today the main application area of MPC: Dittmar and Pfeiffer [20] have counted 9456 MPC applications in a survey in Germany in 2005 (but estimate a considerably higher number in reality), of which more than 80% lie in the field of chemical process control. Primarily, MPC is used in the control of continuous reactors in which the state of the process (temperature, concentration of chemicals . . .) has to be kept close to a reference value for an arbitrary long time. Less often, MPC is used in so-called batch reactors in which the reaction does not occur continuously, but in a given time frame and in which the control problem consists of following a predetermined path of the process states.

Although there are examples for successful applications of nonlinear MPC (NMPC) in chemical engineering [52], here mostly linear MPC methods are applied, i.e., methods in which the underlying differential equation model is linear. This has the great advantage that the resulting optimization problems are linear, such that faster and more reliable solution algorithms are available. Although a comparative study on an industrial batch reactor [48] revealed that NMPC methods may well lead to a better control performance, so far the improvements are not as significant as to economically justify the time-consuming implementation of the method—it remains to be seen if this will change in the future with short running resources and more expensive energy. A further obstacle for the application of NMPC is often the lack of suitable nonlinear models in chemical engineering [52]. However, in this field recently considerable progress has been made with the aid of sophisticated mathematical methods, see Sect. 4.1.

Due to more efficient algorithms and theoretical advances, recently more and more application fields beyond chemical engineering have become accessible for MPC and in particular NMPC methods. Examples are the control of water gates in canal systems, the control of heating and climate systems in large buildings, or the optimal control of seasonal heat storage devices. The latter use the heat in the summer to heat up the ground under a building and at the same time to cool down the building, and reuse this heat in winter with the aid of a heat pump to save heating costs. Only predictive control allows to decide when the heat pump and additional heating can be used efficiently based on the seasonal weather forecast. Increasingly faster optimization algorithms allow even faster applications of model predictive control. Today, to-the-point control of train runs is possible and experimentally tested [25], as well as the control of the amount of fuel injection and air in automotive engine at sampling intervals of 50 milli-seconds [21]. Also, NMPC control of robot arms that need to carry out complex maneuvers as fast as possible and in constantly changing environments is within reach, see Fig. 7. One of the most visionary possible future application of NMPC is the automatic control of flying kites that supply wind energy from high altitudes by flying on periodic paths under changing wind conditions [12, 32] in a stable way, see Fig. 8. Another application of NMPC that is recently studied is the automatic control of insulin supply for diabetics.

Fig. 7 Robots at the university of Leuven currently running experiments with MPC in milli-second intervals

Fig. 8 Automatically controlled kite for energy production [12]

3.5 Direct Optimal Control Methods

In this section we want to give some insight into the current state-of-the-art of solution methods for the computationally most expensive step in model predictive control, the repeated solution of optimal control problems on finite time horizons. In what follows we concentrate on the more challenging case of nonlinear systems.

In Sect. 2.2, the maximum principle was addressed. So-called *indirect* methods based on this mathematical theorem are one possibility to compute optimal solutions for control problems. Even today, these methods are indeed applied. In particular, this holds for aerospace technology, see for example [10], but also for the analysis of problems in chemical engineering, see [59]. The maximum principle still plays an important role in the analysis of properties of optimal control problems. But, for the construction of algorithms that can be transferred to a computer these indirect methods have some disadvantages:

- the significant analytical effort even for small changes in the model or in the parameters,
- the complicated treatment of path and control restrictions which lead to state-dependent jumps in the differential variables,
- the boundary value problems with extremely small domains of convergence which are often very difficult to solve numerically, frequently only by the use of homotopies.

Because of these disadvantages the *direct* methods that are also known under the key word "first discretize, then optimize" have established themselves for the computation of control strategies in practice, see also the discussion in [4].

A further important question is the evolution of the computed optimal control functions over time. In the maximum principle the optimal control is a continuous function, structural restrictions on its shape cannot be made. From a practical point of view this does not necessarily make sense. For instance, you may think of a temperature controller that due to technical reasons cannot be adjusted continuously but only in fixed intervals from one constant value to another. Such a controller cannot exactly follow any arbitrary control function, but only approximate it. Such an approximation is the main principle of direct methods in the first place.

In these methods the control functions are approximated by functions that are determined by a finite number of control variables. In the simplest case these are piecewise constant functions, i.e., they take a constant value q_i on a given time interval $[t_i, t_{i+1}]$. In Fig. 9 this is illustrated.

This transformation to a mathematical optimization problem that now only depends on a finite number of degrees of freedom (namely exactly the values q_i) allows the use of sophisticated methods of nonlinear optimization (nonlinear programming). Here, mainly two types of algorithms compete: interior point and active set-based methods. While in the latter, the set of active inequalities (denoted as the active set) is carried along and modified from iteration to iteration, the interior point methods move within the admissible domain towards the solution on the so-called central path with the aid of logarithmic barrier functions.

Both types of algorithms have certain assets and drawbacks which lead to the fact that they determine the optimal solution faster for certain classes of optimization problems. As a rule of thumb, interior point methods often perform better for problems with many inequality constraints (since they come from the inside and do

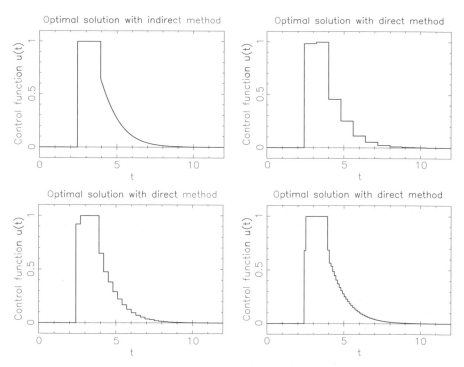

Fig. 9 *Top left*: control computed with indirect method. *The other diagrams* show approximations with the direct method on different discretization grids

not "notice" the many intersections of the inequality constraints that for active set-based methods have to be "visited" sequentially), while active set-based methods allow more efficient warm starts in the solution of related optimization problems. This is due to the fact that the sets of active inequalities often differ only marginally, e.g., when a slightly altered optimization problem results from newly measured data. Then, starting in the already computed solution of the original problem often leads to the result in a few iterations.

In the discretization of optimal control problems with direct methods commonly three types of algorithms are distinguished, see [4]: direct *single shooting*, direct *multiple shooting* and direct *collocation*. *Single shooting* was developed in the 1970s, see for example [29] and [54]. The basic idea is to consider the differential variables as dependent quantities of the independent control variables and to solve an initial value problem in each iteration of the optimization algorithm in order to compute the objective function and the constraints as well as the derivatives with respect to the control variables. In other words: we have an outer optimization algorithm in which our degrees of freedom are optimized, and an inner one, in which we "shoot" for fixed solutions, i.e., we solve the differential equation by integration in order to compute the value of the states at all time points. Figure 10 visualizes this basic idea.

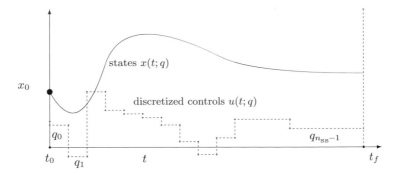

Fig. 10 Schematic illustration of the direct single shooting method. The controls are given by piecewise constant functions q_i, the corresponding states are determined by integration. The lengths of the intervals do not necessarily have to be equidistant, but might also be longer on the last intervals, as indicated

In optimization problems often the state at the end of the time horizon is of particular interest. If you think of determining the controls to fly a plane, then it becomes clear how important the fine tuning of all control variables is, if a reference in a long distance is to be reached exactly. To distribute these difficulties along the time horizon, the concept of single shooting has been extended to the direct *multiple shooting*. This method has been published in 1981 for the first time [6]. Since then, the method has constantly been improved or newly implemented, see for example [43].

As in the case of single shooting, the controls are discretized on a time grid. But, in order to better coordinate the flight of the plane over a long distance, checkpoints in form of additional variables are incorporated. Now, the plane only has to be "flown" to the next checkpoint, similar as in a human chain for extinguishing a fire. Since nonlinearities and possibly instabilities of the problem have smaller effects on a short time horizon than on a long one, it is now considerably easier to reach the focused target. This approach is a good example for the strategy to separate a hard problem into several small ones. This principle *divide et impera*, divide and conquer, can be found in many mathematical algorithms.

As new "checkpoints", multiple shooting variables $s_i^x \in \mathbb{R}^{n_x}$ with $0 \leq i < n_{ms}$ are introduced for the states on a given time grid $t_0 \leq t_1 \leq \cdots \leq t_{n_{ms}} = t_f$ that serve as initial values in the decoupled integration on the time intervals $[t_i, t_{i+1}]$. As shown in Fig. 11, such a composed trajectory is only continuous on the whole time horizon if the conditions

$$s_{i+1}^x = x(t_{i+1}; s_i^x, q_i, p) \tag{3}$$

hold. The left-hand side of (3) corresponds to the initial value on the time interval $[t_{i+1}, t_{i+2}]$, the right-hand side to the integrated value of the final state on the previous time interval $[t_i, t_{i+1}]$, depending on the initial value s_i^x and the control variables q_i and p. These equations are added to the optimization problem and have to be fulfilled by an optimal solution. For this reason, the direct multiple shooting method is

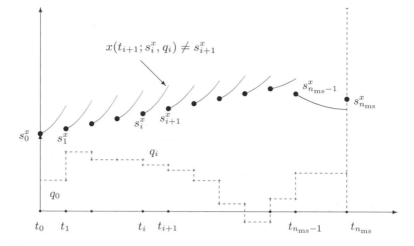

Fig. 11 Schematic illustration of direct multiple shooting. The controls are given by piecewise constant functions q_i, the corresponding states are determined by piecewise integration. The connection conditions are not yet fulfilled in this example, the resulting trajectory is still discontinuous

often denoted as *all-at-once* approach, since simulation and optimization are carried out at the same time in contrast to the sequential approach of single shooting.

Graphically spoken, each checkpoint has to make sure that his flight reaches the respective successor so that the plane is eventually controlled from the start to the end point. The decisions that need to be made between every two checkpoints, our controls q_i, of course also have to be made in the single shooting. But now, at first everyone takes care prior about his own time period and about how the plane can be brought to the neighbor, and in the same time nevertheless behaves optimally. This "minding one's own business" is one of the biggest strengths of this approach. Besides the gain in stability, it leads to unfolding of structures in the optimization problem that can algorithmically be used. The matrix that results from the analysis of the necessary optimality conditions has block form, whereby the blocks include all variables that belong to a time interval.

Because of the parameterizations of the state space, *direct collocation* and direct multiple shooting are quite similar. Collocation goes back to [61] and has been expanded amongst others in [3, 7, 56]. Here, the basic idea is not to use any independent integrators for the solution of the differential equation, but to discretize all equations and incorporate them in the optimization problem. Collocation and multiple shooting share a number of advantages in comparison to single shooting. For instance

- it is possible to use previous knowledge about the process (that is given, e.g., in form of measured data) for the initialization of the state variables,
- it is guaranteed that the arising systems of differential equations are actually solvable also for nonlinear systems, while single shooting can run into a singularity for badly chosen controls,

- also unstable systems can be solved, if they are well-posed, since perturbations do not propagate over the complete time horizon, but are damped by the tolerances in the parameterization variables,
- a considerably better convergence behavior in higher dimensional space is observed in practice.

The resulting optimization problems are larger than in the case of single shooting. But, this is compensated by specific structure exploiting algorithms. In the case of collocation these are solution algorithms that use the particular band structure of the sparse matrices. For direct multiple shooting condensation algorithms are used to reduce the size of the quadratic programs that need to be solved, while for the approximation of the Hessian so-called high-rank updates are used. These accommodate the special block structure given by the shooting-intervals and thus accumulate more curvature information in each iteration than general quasi-Newton methods.

In summary, from our point of view direct multiple shooting and collocation are the methods of choice to solve nonlinear optimal control problems efficiently. Furthermore, these methods can be adapted efficiently to the special structure of NMPC applications: in particular, the fact that here iteratively a series of similar optimal control problems is solved has been utilized in recent years to develop variants of these methods that are particularly adapted to NMPC [18, 44, 57] and for which, also with inaccurate numerical solutions, the stability of the controlled system can be verified [17, 19, 63].

4 Challenges

We hope that so far we have convinced the reader of the significant role of mathematics as a key factor for technical control processes and that we have given some—of course subjective—insight into the current state-of-the-art and the variety of applications. In this chapter, we want to address some of the tasks and challenges that lie ahead. For this purpose in Sect. 4.1 we consider aspects that arise in the modeling of the considered processes and see to what extent mathematics can contribute significantly also at this level. In Sect. 4.2, we discuss the robustness of MPC controls and in the concluding Sect. 4.3 we highlight some further current topics.

4.1 Modeling

The methods for the model-based control of processes described in the preceding sections have an essential assumption: *for the considered process an adequate mathematical description in form of a system of differential equations can be found.*

The keywords in this sentence suggest where the difficulties lie.

- "For the considered process" indicates that it is a matter of problem-specific modeling. From one plant to another several components may slightly differ—but for this reason the new process may require a completely different mathematical model.
- "Adequate" implicates at least three facts. First, that the model is correct, i.e., the essential properties that are of interest in the particular context are captured. Second, that the model only includes what is absolutely necessary in order to avoid excessive computing times. And third, that the resulting model is suited for numerical simulation and optimization.
- "Mathematical description" refers to the underlying principles as well as to the estimation of the specific problem parameters, both based on the comparison of simulated with measured data.
- The phrase "can be found" suggests a certain randomness. The question arises, how such "finding processes" can be organized systematically.

Mathematical modeling addresses the problem of finding a mathematical model for a real process. This is based on knowledge from the respective field of application. It is known at least since Newton that the acceleration is the second derivative of the (time dependent) position, which can therefore in turn be computed from the fixed initial position and velocity and given acceleration. Another example are conservation laws which often form the basis of physical or chemical models. All such models are of course only approximations of reality and simplify matters here and there. From the consideration of friction forces to the interaction on atomic level such a model can be refined arbitrarily. An important task of the modeler is to find the right compromise between simplicity of the mathematical model and a sufficiently close description of reality.

Unfortunately, for many problems in practice the considered processes are often still too complex, or not sufficiently understood. This is particularly true for systems biology, which has started to systematically make use of scientific computing methods only very recently. A different situation prevails in robotics, mechanics, or chemical engineering, where already for some decades methods of modeling, simulation and also optimization are applied. Here, the basic principles are often so-called *first principles*, i.e., basic and well-understood conservation laws of natural sciences. Nevertheless, a successful modeling—in particular in the field of nonlinear models—can only be achieved by very labor-intensive efforts based on all available expertise.

Therefore, for many relevant technical processes the assumptions mentioned in the beginning of this section are in fact a challenge for the future. Yet, there are encouraging results towards an algorithmic modeling that we want to highlight in the following—not without emphasizing that also this shifting away from the tedious trial-and-error based expert modeling towards standardized and verifiable methods would not be possible without underlying mathematical methods.

Parameter Estimation

Even though the basic principles are known, often certain model parameters are unknown. One may think for example of the masses of components in multi-body systems, of activation energies, or of reaction velocities. For the determination of these parameters measurements are required, in which frequently the parameters cannot be directly measured; instead only states of the system, or functions of these states are available. If you want to know certain physical properties of your heating, you will most probably need to measure the temperature in the room under different conditions and determine the intrinsic heating parameters from this experimental data and your mathematical model. The parameter estimation aims to minimize the distance between a model response and collected measured data by *fitting* of the unknown parameters. The method of least squares has been developed by Carl-Friedrich Gauss and Adrien-Marie Legendre already in the beginning of the 19th century. However, it experienced a significant advancement in the last decades, in particular concerning efficient numerics, restrictions on the parameter estimation problem, and robust parameter estimation by use of objective functionals that allow for outliers, see e.g., [40].

Optimum Experimental Design

The parameter estimation yields parameter values for which a simulation of the system of differential equations has a minimal distance to measured values. In addition, a statistical analysis allows for the computation of the covariance matrix, which is a measure for the accuracy with which the parameters have been determined. In simplified terms, this is analogous to the error bars which are frequently plotted in diagrams for measured data.

As a matter of fact, different experimental setups lead to different accuracies. The goal of optimum experimental design is to determine experimental setups that lead to confidence regions (error bars) as small as possible. Experimental setup means everything that can be influenced in carrying out an experiment. This includes control inputs as well as the decision at which time instances (or at which positions in spatially distributed processes, respectively) measurements are to be made.

While optimum experimental design has been examined in statistics for many years, there was hardly any transfer to technical applications. This is due to the fact that optimum experimental design has been examined for a long time as a theoretical construct, and concrete computations have been restricted to academic example problems—a comparable situation to the one of Hurwitz mentioned earlier. Only in recent years researchers started working on efficient numerical implementations and on a generalization to the nonlinear, dynamic case; pushed, of course, by a concrete demand from engineers in academia and industry. The achieved breakthroughs in the advancement of statistical analysis now allow the application even within industrial practice, where nonlinear dynamic systems with up to hundred differential states, and more and more also spatially distributed processes are considered.

In particular, for nonlinear processes the computed optimal experimental designs are often anything but intuitive. Yet they can yield significant improvements and insight. For example, methods developed at the Interdisciplinary Center for Scientific Computing in Heidelberg are by now regularly applied at the industrial partner BASF in Ludwigshafen: empirical data from more than two dozen industrial projects has shown that compared to traditional experimental setups designed by experts the number of experiments could be reduced by more that 80% while maintaining or even improving statistical quality [1]. In [39] an optimal experimental design is described with only 2 experiments that determines the parameters that have to be estimated accurately to 1%, while a test plan with 15 experiments designed by an expert for comparison determines a parameter merely up to plus/minus 30%. In the highly competitive field of chemical engineering this paves the way for potential savings in terms of costs, environment pollution, and time-to-market.

Further important research fields are *model discrimination* which focuses on the automatic design of experiment designs in order to set a boundary as significant as possible between competing models, as well as *model reduction* that deals with the simplification of an existing model without losing the relevant properties. One approach for this is the spectral analysis of the modes that separates the slow components from the fast ones, which often play only a subordinate role for the control.

4.2 Robustness of Solutions, Uncertainties

While (N)MPC is a concept for the control of processes that implicitly allows for uncertainties, in particular in the methods for offline-optimization introduced in Sect. 3.5, it is assumed that the process to be optimized behaves deterministically according to the mathematical model. In practice, this will not be the case as different kinds of uncertainties occur:

- *Model uncertainties.* Despite the efforts to obtain an adequate model for a considered process presented in Sect. 4.1, there will always be model errors that lead to (hopefully only marginal) deviations of the simulation results from the real process.
- *External perturbations.* Random perturbations (noise) cause influences on the system that are unknown at the time of optimization. Examples are the side wind for automobiles, rainfall fluctuations in sewerage plants, consumer behavior in economy, or external temperature variations.
- *Different scenarios.* This kind of uncertainty arises due to the fact that certain parameters of the examined process are open a priori. Turbines blades or aircrafts, for example, should have good aerodynamic properties for a whole bandwidth of upstream flow angles. Frequently, one further distinguishes between a discrete number of scenarios, a probability distribution in a continuous domain, and a worst-case scenario.

All three types of uncertainties pose challenges both to the user as well as to the mathematician.

Optimal solutions often have a property that limits their applicability in practice: they are very sensitive against uncertainties. This is due to the fact that the constraints and nonlinear effects are exploited as much as possible. Typically, there are one or more restrictions that are active in an optimal solution, i.e., they will be violated whenever a little something changes. If you want to drive in optimal time from A to B, then at some places you will drive exactly at the maximally allowed velocity. If now the model is inaccurate, or external perturbations occur (think of head or tail wind), then these restrictions will possibly be violated. In case of a speed limit small violations may be acceptable, but this is not the case in so-called *runaway* processes that cannot be reversed once a certain threshold is crossed. Who wants to be reliable for achieving maximal energy efficiency in a nuclear power plant at the expense of constantly being on the brink of an irreversible chain reaction?

From a practical point of view a solution that is *almost* as good as the optimal, but less sensitive against perturbations would be preferred. Approaches in this direction work with (a) feedback, in the same way as MPC does by including measured data in a realtime-context, or (b) with safety margins that are added to the restrictions, or (c) with weighted sums, multi-criteria optimization or game-theoretical approaches (here, there are first approaches in the context of NMPC [45] that are however algorithmically considerably more complex as the "normal" NMPC), or as well (d) by the integration of higher derivatives in the optimization routine. The sensitivity of the solution against uncertain parameters can be formally computed as a derivative. Modern approaches in robust optimization therefore often extend the mathematical model by terms with higher derivatives of the objective function or the restrictions to these parameters, respectively. Despite encouraging advances in this field, uncertainties and the determination of robust solutions will remain a big challenge further on.

4.3 Further Challenges

While theory and algorithmics of optimal control have long since been developed for finite dimensional dynamic systems, the door for the treatment of *spatial effects* has opened due to algorithmic advances and the availability of faster hardware obeying Moore's Law. For instance, this means that the temperature in a room should not be controlled by only using a representative sample value at a specific position, but rather by taking into account the temperature at each position in space.

While this will be hardly ever necessary for simple heaters in a room, there are many processes where the spatial component is important. In the production of steel beams, for example, you need to make sure the temperature variations within the beam do not become too large in the cooling phase to avoid subsequent cracks [60]. In aerodynamics the fluid behavior plays a decisive role. The spatial positions of air turbulence and shock fronts directly influences the flight properties [35].

The optimization of processes described by partial differential equations has gained more and more attention in recent years, for example due to a Priority Research Program of the Deutsche Forschungsgemeinschaft [15]. Nevertheless, it appears that the gap between mathematical fundamental research and practical application is still rather large at this point, not only in the field of algorithms for the solution of optimal control problems, but also in the system-theoretical fundamentals of MPC methodology.

A challenge of quite different nature is the combination of continuous and discrete events. Here, continuous means a connected domain from which a variable or a control can take a value. You can think for example of the velocity of a vehicle or of the position of the gas pedal. Discrete means a non-connected domain that is mathematically composed of a finite set of possible values. Here, the gearshift is a good example for a discrete control, while "the walking robot has contact with the ground or not" is a state in which either one applies. In the same way, the digital couplings in linked systems that are composed of a variety of smaller physical subsystems lead to discrete state components. Systems that combine both types of states and/or controls are denoted as *hybrid systems*. Frequently, hybrid systems arise due to a multi-scale modeling. Here, fast transient transitions are assumed to be instantaneous. A good example are valves that are assumed to be either open or closed in the model, while in reality they have to be transferred from one state to the other.

Mathematically, the integral or discrete variables that arise in the modeling pose a substantial challenge for the optimization. This may surprise on first sight, as only a restricted number of possibilities are available that theoretically can all be enumerated. However, the number of possibilities grows rapidly if the number of variables is increased. This is the case for hybrid systems, in particular to decide when and how often should be switched. A treatment of these variables by trial-and-error is impractical because of the immense number of possibilities. An overview of possible methods for discrete control functions and further literature on hybrid systems can be found for example in [53].

Also this research field turns out to be extremely active, and is for example supported by the European research network HYCON [31] and—for the special case of digitally networked systems—by a DFG Priority Research Program [16]. In particular, the development of reliable and efficient algorithms and a stability analysis in the sense of Sect. 3.3 will require much attention in the next years to be able to exploit the full optimization potential in hybrid systems.

5 Visions and Recommendations

In our opinion, MPC and in particular NMPC belong to the most promising methods for the solution of complex nonlinear control problems, which is why they will gain further importance in the future. Their ability not "only" to solve control problems but also to include technical and economical restrictions via optimality criteria and constraints yields—in particular in times of short running resources—distinct advantages over other methods. Moreover, the rapid advance in the field of optimization algorithms allows continuously new applications with increasingly faster dynamics.

Just like the success factors of the classical methods of Hurwitz and Pontryagin, MPC has the distinct advantage to be intuitively easy to understand and therefore to be applicable without deeper understanding of the underlying mathematics—be it the systems theoretical analysis of the control behavior or the basics of the used algorithms stemming from numerics and optimization.

Therefore, we want to restrict the following comments to this control method, although surely analogous statements are also valid for other modern methods in control engineering.

5.1 Theory and Praxis

From a theoretical point of view, the biggest conceptual advantage of (N)MPC namely the explicit inclusion of a mathematical model for the prediction of the system behavior—is at the same time one of the essential disadvantages in practice. MPC inevitably requires a sufficiently exact mathematical model for the complex process to be controlled, which is not always available. As described in Sect. 4.1, recently promising mathematical modeling methods have become available also for the nonlinear case. Nevertheless, the design of suitable models will most likely remain a major challenge. The necessary investments may only be made by industry if the achievable advantages for the optimization of production processes are supported by concrete examples. Therefore, many of the emerging applications sketched in Sect. 3.4 play a pioneering role and their success or failure (and naturally also the communication of corresponding achievements) will decide over the acceptance of NMPC in industrial application in the long run.

But also on the theoretical side there are a number of open questions. In particular, the assumptions in the major part of theoretical MPC literature differ distinctively from the methods used in practice, as presented in Sect. 3.3. Here, it seems desirable to close this gap between theory and praxis. Especially, theoretical work that aims at concrete improvements of MPC methods—such as improved stability, faster algorithms, better control quality, higher robustness—should try to achieve their results and improvements under realistic assumptions.

Finally, on the algorithmic side further advance is expected, mainly by intensified cooperations of system and control theorists on the one and optimizers and numerical analysts on the other hand. As an example, only a systems theoretical stability analysis of the overall control algorithm can give information in what sense the integration of higher derivatives in the optimization routine explained in Sect. 4.2 really increases the robustness—and how the optimality criterion should look like. On the other hand, e.g., the systems theoretical analysis of MPC algorithms should always consider realistic properties of the optimization routines (including the manifold "tricks" used in practice) in order to be able to make realistic and reliable statements about the resulting control method. Thus, in both examples the competence from both fields is necessary.

5.2 Interdisciplinary in Education

From our point of view, all these goals are only to be achieved if—just as in the cooperation of Hurwitz and Stodola—already in the education the cooperation between mathematicians and engineers and between the different fields of applied mathematics is emphasized. Only this way mathematicians can discover the methodical challenges of practical problems, only in this way engineers can profit from the newest mathematical methods, and only in this way methods and techniques from different mathematical areas can complement each other in a meaningful way.

Fortunately, there are a quite a number of very promising current interdisciplinary teaching initiatives, e.g.,

- the International Doctorate Program "Identification, Optimization and Control with Applications in Modern Technologies" of the Universities Bayreuth, Erlangen-Nürnberg and Würzburg within the Elite Network Bayern
- the Graduate School "Mathematical and Computational Methods for the Sciences" of the University Heidelberg
- the Center of Excellence "Optimization in Engineering" (OPTEC) at the University Leuven
- the Cluster of Excellence "Simulation Technology" (SimTech) of the University Stuttgart with postgraduate school as well as bachelor and master degree programs

to mention only those activities in which the authors of this article are involved. The formal framework for further advances is currently quite good, especially in the field of control engineering. It remains to hope that these efforts are continued such that many of the developed mathematical methods find their way into industrial applications and that the opportunities of the production factor mathematics are also taken in the future.

References

1. BASF, S.E.: BASF und Universität Heidelberg entwickeln gemeinsam neue Mathematik-Software für die Forschung. Press release P-08-308, 16 June 2008
2. Bennett, S.: A History of Control Engineering 1800–1930. Peter Peregrinus Ltd., London (1979). Paperback reprint 1986
3. Biegler, L.T.: Solution of dynamic optimization problems by successive quadratic programming and orthogonal collocation. Comput. Chem. Eng. **8**, 243–248 (1984)
4. Binder, T., Blank, L., Bock, H.G., Bulirsch, R., Dahmen, W., Diehl, M., Kronseder, T., Marquardt, W., Schlöder, J.P., Stryk, O.V.: Introduction to model based optimization of chemical processes on moving horizons. In: Grötschel, M., Krumke, S.O., Rambau, J. (eds.) Online Optimization of Large Scale Systems: State of the Art, pp. 295–340. Springer, Berlin (2001)
5. Bissell, C.C.: Stodola, Hurwitz and the genesis of the stability criterion. Int. J. Control **50**, 2313–2332 (1989)
6. Bock, H.G.: Numerical treatment of inverse problems in chemical reaction kinetics. In: Ebert, K.H., Deuflh. d, P., Jäger, W. (eds.) Modelling of Chemical Reaction Systems. Springer Series in Chemical Physics, vol. 18, pp. 102–125. Springer, Heidelberg (1981)

7. Bock, H.G.: Recent advances in parameter identification techniques for ODE. In: Deuflhard, P., Hairer, E. (eds.) Numerical Treatment of Inverse Problems in Differential and Integral Equations, pp. 95–121. Birkhäuser, Boston (1983)
8. Boltyanski, V.G., Gamkrelidze, R.V., Pontryagin, L.S.: On the theory of optimal processes. Dokl. Akad. Nauk SSSR **110**, 7–10 (1956) (in Russian)
9. Byrnes, C.I., Isidori, A., Willems, J.C.: Passivity, feedback equivalence, and the global stabilization of minimum phase nonlinear systems. IEEE Trans. Automat. Control **36**(11), 1228–1240 (1991)
10. Caillau, J.-B., Gergaud, J., Haberkorn, T., Martinon, P., Noailles, J.: Numerical optimal control and orbital transfers. In: Proceedings of the Workshop Optimal Control, Sonderforschungsbericht 255: Transatmosphärische Flugsysteme, Heronymus München, pp. 39–49. Greifswald, Germany (2002). ISBN 3-8979-316-X
11. Camacho, E.F., Bordons, C.: Model Predictive Control, 2nd edn. Springer, London (2004)
12. Canale, M., Fagiano, L., Ippolito, M., Milanese, M.: Control of tethered airfoils for a new class of wind energy generators. In: Proceedings of the 45th IEEE Conference on Decision and Control, San Diego, CA, pp. 4020–4026 (2006)
13. Chen, H., Allgöwer, F.: A quasi-infinite horizon nonlinear model predictive control scheme with guaranteed stability. Automatica **34**(10), 1205–1217 (1998)
14. Dahleh, M.A., Diaz-Bobillo, I.J.: Control of Uncertain Systems: A Linear Programming Approach. Prentice Hall, Englewood Cliffs (1995)
15. DFG Schwerpunktprogramm 1253: Optimization with partial differential equations. http://www.am.uni-erlangen.de/home/spp1253/
16. DFG Schwerpunktprogramm 1305: Regelungstheorie digital vernetzter dynamischer Systeme. http://spp-1305.atp.rub.de/
17. Diehl, M., Bock, H.G., Schlöder, J.P.: A real-time iteration scheme for nonlinear optimization in optimal feedback control. SIAM J. Control Optim. **43**(5), 1714–1736 (2005)
18. Diehl, M., Bock, H.G., Schloder, J.P., Findeisen, R., Nagy, Z., Allgöwer, F.: Real-time optimization and nonlinear model predictive control of processes governed by differential-algebraic equations. J. Process Control **12**(4), 577–585 (2002)
19. Diehl, M., Findeisen, R., Allgöwer, F., Bock, H.G., Schlöder, J.P.: Nominal stability of the real time iteration scheme for nonlinear model predictive control. IEE Proc. Control Theory Appl. **152**(3), 296–308 (2005)
20. Dittmar, R., Pfeiffer, B.-M.: Modellbasierte prädiktive Regelung in der industriellen Praxis. at—Automatisierungstechnik **54**, 590–601 (2006)
21. Ferreau, H.J., Ortner, P., Langthaler, P., del Re, L., Diehl, M.: Predictive control of a real-world diesel engine using an extended online active set strategy. Annu. Rev. Control **31**(2), 293–301 (2007)
22. Fliess, M., Lévine, J., Martin, P., Rouchon, P.: Flatness and defect of non-linear systems: introductory theory and examples. Int. J. Control **61**(6), 1327–1361 (1995)
23. Föllinger, O.: Optimierung dynamischer Systeme: Eine Einführung für Ingenieure, 2nd edn. Oldenbourg, München (1988)
24. Francis, B., Helton, J., Zames, G.: H_∞-optimal feedback controllers for linear multivariable systems. IEEE Trans. Automat. Control **29**(10), 888–900 (1984)
25. Franke, R., Meyer, M., Terwiesch, P.: Optimal control of the driving of trains. at—Automatisierungstechnik **50**(12), 606–614 (2002)
26. Grimm, G., Messina, M.J., Tuna, S.E., Teel, A.R.: Model predictive control: for want of a local control Lyapunov function, all is not lost. IEEE Trans. Automat. Control **50**(5), 546–558 (2005)
27. Grüne, L., Rantzer, A.: On the infinite horizon performance of receding horizon controllers. IEEE Trans. Automat. Control **53**(9), 2100–2111 (2008)
28. Grüne, L.: Analysis and design of unconstrained nonlinear MPC schemes for finite and infinite dimensional systems. SIAM J. Control Optim. **48**(2), 1206–1228 (2009)
29. Hicks, G.A., Ray, W.H.: Approximation methods for optimal control systems. Can. J. Chem. Eng. **49**, 522–528 (1971)

30. Hurwitz, A.: Über die Bedingungen, unter welchen eine Gleichung nur Wurzeln mit negativen reellen Theilen besitzt. Math. Ann. **46**, 273–284 (1895). Nachgedruckt in: Jeltsch R. et al. (eds.) Stability Theory. Birkhäuser, Basel, pp. 239–249 (1996)

31. HYCON: European Network of Excellence on Hybrid Control. http://www.ist-hycon.org/

32. Ilzhoefer, A., Houska, B., Diehl, M.: Nonlinear MPC of kites under varying wind conditions for a new class of large scale wind power generators. Int. J. Robust Nonlin. Control **17**(17), 1590–1599 (2007)

33. Isidori, A.: Nonlinear Control Systems, vol. 1, 3rd edn. Springer, Berlin (2002)

34. Jadbabaie, A., Hauser, J.: On the stability of receding horizon control with a general terminal cost. IEEE Trans. Automat. Control **50**(5), 674–678 (2005)

35. Jameson, A.: Aerodynamics. In: Stein, E., De Borst, R., Hughes, T.J.R. (eds.) Encyclopedia of Computational Mechanics, vol. 3, pp. 325–406. Wiley, New York (2004)

36. Kalman, R.E.: When is a linear control system optimal? Trans. ASME, Ser. D, J. Basic Eng. **86**, 51–60 (1964)

37. Kanellakopoulos, I., Kokotovic, P.V., Morse, A.S.: Systematic design of adaptive controllers for feedback linearizable systems. IEEE Trans. Automat. Control **36**(11), 1241–1253 (1991)

38. Keerthy, S.S., Gilbert, E.G.: Optimal infinite horizon feedback laws for a general class of constrained discrete-time systems: stability and moving horizon approximations. J. Optim. Theory Appl. **57**, 265–293 (1988)

39. Körkel, S.: Numerische Methoden für Optimale Versuchsplanungsprobleme bei nichtlinearen DAE-Modellen. PhD thesis, Universität Heidelberg, Heidelberg (2002)

40. Körkel, S., Kostina, E., Bock, H.G., Schlöder, J.P.: Numerical methods for optimal control problems in design of robust optimal experiments for nonlinear dynamic processes. Optim. Methods Softw. **19**, 327–338 (2004)

41. Krstic, M., Kanellakopoulos, I., Kokotovic, P.V.: Nonlinear and Adaptive Control Design. Wiley, New York (1995)

42. Kwakernaak, H., Sivan, R.: Linear Optimal Control Systems. Wiley, New York (1972)

43. Leineweber, D.B., Bauer, I., Schäfer, A.A.S., Bock, H.G., Schlöder, J.P.: An efficient multiple shooting based reduced SQP strategy for large-scale dynamic process optimization (Parts I and II). Comput. Chem. Eng. **27**, 157–174 (2003)

44. Li, W.C., Biegler, L.T., Economou, C.G., Morari, M.: A constrained pseudo-Newton control strategy for nonlinear systems. Comput. Chem. Eng. **14**(4/5), 451–468 (1990)

45. Limon, D., Alamo, T., Salas, F., Camacho, E.F.: Input to state stability of min-max MPC controllers for nonlinear systems with bounded uncertainties. Automatica **42**(5), 797–803 (2006)

46. Mayne, D.Q., Rawlings, J.B., Rao, C.V., Scokaert, P.O.M.: Constrained model predictive control: stability and optimality. Automatica **36**, 789–814 (2000)

47. Mehrmann, V.L.: The Autonomous Linear Quadratic Control Problem. Theory and Numerical Solution. Lecture Notes in Control and Information Sciences, vol. 163. Springer, Berlin (1991)

48. Nagy, Z., Mahn, B., Franke, R., Allgöwer, F.: Evaluation study of an efficient output feedback nonlinear model predictive control for temperature tracking in an industrial batch reactor. Control Eng. Pract. **15**, 839–850 (2007)

49. Pesch, H.J.: Schlüsseltechnologie Mathematik. Teubner, Stuttgart (2002)

50. Pesch, H.J., Bulirsch, R.: The maximum principle, Bellman's equation and Caratheodory's work. J. Optim. Theory Appl. **80**(2), 203–229 (1994)

51. Pontryagin, L.S., Boltyanski, V.G., Gamkrelidze, R.V., Miscenko, E.F.: The Mathematical Theory of Optimal Processes. Wiley, Chichester (1962)

52. Qin, S.J., Badgwell, T.A.: An overview of nonlinear model predictive control applications. In: Cantor, J.C., Garcia, C.E., Carnahan, B. (eds.) Nonlinear Model Predictive Control. Birkhäuser, Basel (2000)

53. Sager, S., Reinelt, G., Bock, H.G.: Direct methods with maximal lower bound for mixed-integer optimal control problems. Math. Program. **118**(1), 109–149 (2009)

54. Sargent, R.W.H., Sullivan, G.R.: The development of an efficient optimal control package. In: Stoer, J. (ed.) Proceedings of the 8th IFIP Conference on Optimization Techniques (1977), Part 2. Springer, Heidelberg (1978)

55. van der Schaft, A.J.: L_2-gain and Passivity Techniques in Nonlinear Control, 2nd edn. Springer, London (2000)
56. Schulz, V.H.: Solving discretized optimization problems by partially reduced SQP methods. Comput. Vis. Sci. **1**, 83–96 (1998)
57. Shimizu, Y., Ohtsuka, T., Diehl, M.: A real-time algorithm for nonlinear receding horizon control using multiple shooting and continuation/Krylov method. Int. J. Robust Nonlin. Control **19**(8), 919–936 (2009)
58. Sira-Ramírez, H., Agrawal, S.K.: Differentially Flat Systems. Marcel Dekker, New York (2004)
59. Srinivasan, B., Palanki, S., Bonvin, D.: Dynamic optimization of batch processes: I. Characterization of the nominal solution. Comput. Chem. Eng. **27**, 1–26 (2003)
60. Tröltzsch, F., Unger, A.: Fast solution of optimal control problems in the selective cooling of steel. Z. Angew. Math. Mech., 447–456 (2001)
61. Tsang, T.H., Himmelblau, D.M., Edgar, T.F.: Optimal control via collocation and non-linear programming. Int. J. Control **21**, 763–768 (1975)
62. Wischnegradski, J.: Sur la théorie générale des régulateurs. C. R. Acad. Sci. Paris **83**, 318–321 (1876)
63. Zavala, V.M., Biegler, L.T.: The advanced-step NMPC controller: optimality, stability and robustness. Automatica **45**(1), 86–93 (2009)
64. Zhou, K., Doyle, J.C., Glover, K.: Robust and Optimal Control. Prentice Hall, Upper Saddle River (1996)

Data Compression, Process Optimization, Aerodynamics: A Tour Through the Scales

Wolfgang Dahmen and Wolfgang Marquardt

1 Executive Summary

Increasing computational power facilitates a more and more precise analysis of complex systems in science and engineering through computer simulation. Mathematics as the interface between the real and the digital world provides on the one hand the foundations for the formulation of necessarily simplified models of reality. On the other hand it also presents the basic methodological principles for designing efficient algorithms, which can be used to gain quantitative information from such models. The fact, that real systems are mostly driven by phenomena covering a large range of time and length scales, constitutes the major challenge. Therefore, the development of mathematical methods explicitly dealing with the multiscale nature is of great importance. In this article, this issue will be explained and illustrated on the basis of most recent developments in different fields of application to demonstrate how many "birds" can be caught with a single "mathematical stone". In particular, fundamental algorithmic concepts relying on wavelet decomposition will at first be explained in a tutorial manner in the context of image compression and coding. Then it will be demonstrated that these concepts carry over to different applications such as analysis of measurement data in process plants, large-scale optimization in the process industries as well as complex fluid dynamics problems in aerodynamics. Stable decomposition in different time and length scales facilitates the implementation of adaptive concepts, which are capable of placing computational resources

W. Dahmen (✉)
Institut für Geometrie und Praktische Mathematik, RWTH Aachen, Templergraben 55, 52056 Aachen, Germany
e-mail: dahmen@igpm.rwth-aachen.de

W. Marquardt
Aachener Verfahrenstechnik–Prozesstechnik, RWTH Aachen, Turmstr. 46, 52064 Aachen, Germany
e-mail: wolfgang.marquardt@avt.rwth-aachen.de

M. Grötschel et al. (eds.), *Production Factor Mathematics*,
DOI 10.1007/978-3-642-11248-5_3, © Springer-Verlag Berlin Heidelberg 2010

automatically where needed to realize the desired quality of the solution, e.g. in terms of error tolerances, at the lowest computational effort.

2 Some Guiding Ideas

The role of mathematics in a modern technology-based society can be appraised from different angles. We wish to emphasize in this article the ability of mathematical concepts to unveil sometimes unexpected *shortcuts* as a guiding theme. Already from an economical point of view any sort of shortcut is crucial for an increasingly better understanding of complex real world phenomena through a virtual digital world.

Mathematics as an Interface Between Real and Virtual Worlds

The impact of digital information technology is nowadays obvious. A superficial consideration might even give rise to the impression that the mere increase of computing power will solve "all" problems in science and engineering. However, no matter how many wind mills are installed there will be no electricity without wind. The most powerful engines cannot create complex buildings without a good construction plan and no computer can solve a problem without being told by an *algorithm* which operations in which order are to be carried out. Algorithms in modern large-scale applications often give rise to computational tasks of enormous complexity. Especially, when dealing with challenging multiscale phenomena, realizing a substantial complexity reduction already on the level of designing an algorithm is an inherently mathematical task. The real world in its entire complexity can never be fully described. An increasingly better understanding of the basic mechanisms can be acquired only through simplified models. The formulation of such models already has to reflect prioritizing specific phenomena of interest. The key is to find a highest possible simplification without spoiling the target phenomena. Mathematics offers a framework to formulate such models and to quantify their components so as to find a good balance.[1] A typical example is modeling flow processes in nearly every area of technology such as aerodynamics in aircraft and automotive design, reactive and disperse flows in chemical engineering, or porous media flows when analyzing spreading of pollutants. Continuum mechanics serves as the model framework for those applications where the quantities of interest such as velocity, density or pressure fields can be regarded as functions of continuous temporal and spatial variables, neglecting the particulate character of the involved media. Classical balance equations for mass, momentum and energy allow one to interrelate rates of changes of the involved quantities typically providing systems of *differential or integral equations*.

---

[1] This is a central theme in the just started Priority Program 1324 of the German Research Foundation "Mathematical Methods for Extracting Quantifiable Information from Complex Systems".

Questions concerning the solvability of such equations and the properties of their solutions have always been a major source of mathematical developments— needless to stress their relevance for applications relying on such models.

Now aside from the natural curiosity of scientists, it is the tremendous continuing increase of computing capacities that encourages them to formulate more and more realistic models which hence become more and more complex posing further challenges to mathematics. It is then often necessary to leave the scope of continuum mechanics as, for instance, in modern semiconductor technology or in quantum chemistry to couple models for different scale ranges in order to properly capture the influence of unresolved scales on a effective macroscopic level.

A Universe of Scales

The above comments already indicate that an important parameter for the complexity of a model is the range of length scales that are relevant for the understanding of the real process. A systematic treatment of multiscale structures is therefore a central conceptual principle offered by a mathematical approach. A key objective is to disclose shortcuts through the universe of scales by developing strategies for identifying only the relevant components. This issue is currently driving several different directions of mathematical research. We shall concentrate here mainly on one specific aspect that, in particular, accounts for the fact that even within the scope of a single model frame like continuum mechanics one may well be confronted with a range of relevant length scales, e.g. in connection with turbulence, which in the near future will not be tractable with the aid of current methodologies by just increasing computer power. This hints at another important point namely that mathematics is not confined to the formulation of suitable models but that the insight gained through the underlying analysis feeds directly into the design of efficient algorithms in order to help reducing computational complexity already on the level of mathematical algorithmic foundations.

One important paradigm of this type centers upon *adaptive solution techniques*. Their objective is to assign in the course of the solution process computational resources (e.g. in terms of locally more accurate discretizations) in such a way that a desired solution quality, e.g. in terms of accuracy tolerances, is achieved at the expense of a possibly small number of degrees of freedom, viz, a possibly small size of a resulting system of equations. This role at the interface between real and digital world and some of its consequences are to be brought out in this article.

In this dynamic process of science evolution vast areas of mathematical research have changed significantly. They have become much more interdisciplinary. In fact, the full potential can unfold only in a direct synergetic cooperation with computer scientists, engineers and other scientists. This is mirrored also by the *nature* of success stories. There are certainly numerous examples where certain problems in applications are simply handed over to mathematicians who return, after some time, a solution package without much intermediate interaction. But even in the process of science evolution one discovers "multiscale effects". Perhaps less apt to simple concise description but in the long range the more important are those methodological advances whose manifestation in industrial applications becomes evident sometimes only after years or even decades after their principal discovery. This is simply due to the fact that the application context is often very complex, or that a single good idea is far from sufficient and multifaceted interdisciplinary input is required that makes it impossible to single out a clear cut solution component.

This article is therefore also a plea for not always insisting on short term results but accepting the necessity of long term interdisciplinary developments, quite in accordance with an appreciation of the foundational objectives of university research. In fact, these should complement modern industrial developments confined by ever shorter cycle times. The strength of methodological progress is last but not least reflected by another attribute of mathematics, namely the occasional ability of catching several birds with one stone, in our opinion without any doubt a relevant aspect from an economical point of view.

3 Success Stories

3.1 Information Retrieval from Huge Datasets

We shall demonstrate in subsequent discussions that one and the same basic principle will bring to bear in quite different application areas, namely signal processing, optimal control, and numerical solution of partial differential equations. It should be stressed though that it is not the same algorithm that is used in these contexts but rather the same common short cutting mathematical concept that drives the algorithms tailored to the specific application.

3.2 Information Retrieval from Huge Datasets

Today in industrial plants for material and energy conversion (such as power plants, refineries or chemical plants) data archives are established for the storage of measured plant data collected during operation. On the one hand, archiving of plant data is required by law to inspect if the plants are safely operated according to legal regulations. On the other hand, data archives are routinely used by the plant operators themselves to improve plant performance with respect to availability, capacity, product quality and production cost. Furthermore, plant operators analyze the data in real time to detect and diagnose faults, to facilitate preventive maintenance or to adjust operating conditions in real time such that an economically optimal operation of the plant can be guaranteed. The complexity of the problem becomes obvious if industrial practice is considered: several ten or even hundred thousands of measurements are sampled every second over a period of several years resulting in extremely large data sets.

As part of the archiving of time series sampled in the plant, data compression is conducted at first. Today, data compression is not aiming for a reduction of memory requirements because storage media for long-term data archiving are not expensive anymore. In fact, one is primarily interested in separating the coarse-scale signal trends from the superimposed fine-scale detail signal to create a solid foundation for signal interpretation. The coarse-scale trends reflect those dynamic phenomena, which determine plant behavior, while the superimposed details, e.g. the trend disturbances, result from measurement and transmission errors as well as from process noise. Many different heuristic methods are currently used for data compression and trend analysis. These methods not only lack unifying theoretical foundation, but they can only be used profitably by very experienced engineers, they require substantial maintenance effort and most often do not lead to satisfying results.

The key question arising in the described problem is the decomposition of the multivariate time series in suitable components reflecting the different time scales present in the signal. Recent development of wavelet-based multiscale signal processing methods [37] provide a distinguished foundation for the development of a collection of theoretically sound methods for compression and interpretation of multivariate plant data.

Methods for automatic tuning of heuristic data compression algorithms [39] routinely used in industry have been developed recently. They show excellent compression performance for time series with widely differing signal characteristics. These methods facilitate reliable tuning of the data compression algorithms at minimal engineering effort. At the same time, they guarantee that the archived measurement data after compression still capture the significant information content to reflect process performance. Moreover, these methods can be easily applied to very large data sets [3].

Novel multi-scale methods have been recently reported to identify time intervals in multivariate data sets, where every univariate time series is restricted to a tube like region around a polynomial trend with a given diameter. Only the consequent exploitation of the distinct properties of multiscale methods [24] allows the reliable and efficient determination of the start and end point of all time intervals showing a given polynomial trend in the signal. This way, a segmentation of the data set in a sequence of time intervals with qualitatively different behavior can be identified. This modern method has recently been validated in industrial applications [2] showing excellent results compared to established methods [14]. Hence, multiscale methods form an excellent foundation for the development of new software tools for process data analysis with particular emphasis on denoising, compression and trend detection. A schematic architecture of such a tool is depicted in Fig. 1. Due to their theoretical foundation, these methods yield more dependable and more accurate trend segmentations of large process data sets at drastically reduced engineering effort.

3.3 Adaptively Optimized

Everybody knows optimization from daily life. One always tries to generate the optimum out of given alternatives—the maximum earning from shares, the maximum

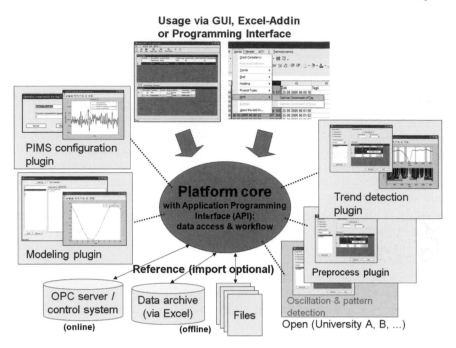

Fig. 1 Alanda, a modular platform for data analysis in offline and online applications

distance to travel with a tankful of gasoline, or the maximum profit from a manufacturing site. Some of these problems belong to the class of optimal control problems which have been studied in mathematics for a long time. The infinite-dimensional decision variables, e.g. often functions of time, are chosen to minimize a certain optimization objective subject to a mathematical model and additional point or path constraints. A simple example is driving a defined distance with a car in minimum time without violating speed limits. The car is supposed to be at rest at the beginning and at the end of the trip. Furthermore, acceleration and velocity of the car are constrained. The time-varying acceleration is the only degree of freedom of the optimal control problem. The numerical solution requires adequate discretization. Figure 2 gives an illustration of this example, where the car has to travel 250 m in the shortest possible time. The diagram on the left shows the optimal trajectory of the acceleration and the right diagram shows velocity and distance trajectories. Minimum traveling time can be achieved through proper selection of the control variable, which is the acceleration in this case. In the first phase, the car is accelerated at its maximum for 10 s, then the car runs at maximum speed for another 25 s. Finally, it is decelerated as much as possible for 35 s to come to rest at the finish line.

If this kind of problem is considered in the context of industrial plants in the process industry operated in a transient mode, decision variables are associated with energy and raw material supply rates which vary with time. Constraints are reflecting equipment limitations, safety or quality requirements. Obviously, the actuating

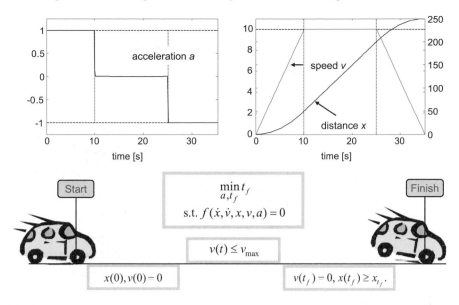

Fig. 2 Illustrating example for the solution of optimum control problems

variables—energy and material supply—can influence process behavior, economics, and environmental impact. Due to the large production capacity and hence the large energy and raw material demand, it is crucial to choose these actuating variables in an optimal manner to minimize resource consumption and to maximize plant availability. An adequate mathematical model of the process and an efficient numerical optimization method embedded in an efficient software implementation are essential premises of a successful application. The enormous size of the models required to adequately represent process dynamics and the very long time horizons used in optimization constitute the major technical challenges.

The formulation and solution of the optimal control problems dealing either with design or operational problems in the process industries have been considered a seminal research area since the 1970s [25]. First papers on this topic, for example [38] or [4], revealed the complexity of these problems, because the optimization problem has to be numerically solved in real time. Large-scale models, the prevailing nonlinearity and the repetitive solution at the sampling times of the control system require adaptive methods which exploit the problem structure [6]. In a joint research project of the authors in the mid 1990s a radically new algorithmic concept has been suggested. In contrast to established methods employing a fixed time-discretization of the decision variables, the new algorithmic concept aims for multiscale representation of the state and control variables [7]. This algorithmic framework facilitates the adaptive refinement of the resolution of all optimization variables in the multiscale bases to allow for the solutions of very large state estimation and control problems in real time. The basic idea of the novel multiscale optimization method (7) is to represent the measurement data, to discretize the state variables and to parameterize the control variables by means of a unifying math-

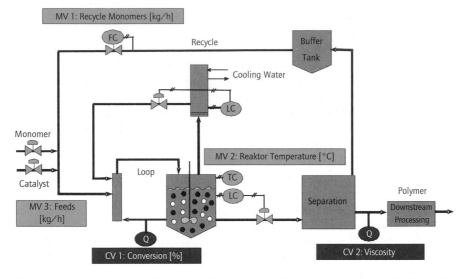

Fig. 3 A process flowsheet of an industrial case study which was done jointly with Bayer AG, Germany

ematical framework given by function representation using bi-orthogonal wavelet bases and by an adaptive and context-dependent refinement. State estimation in uncertain dynamic models, also known as dynamic data reconciliation, was chosen as a challenging problem class. The joint research resulted in completely new algorithmic approaches, which impressively demonstrate the advantages of adaptive multiscale methods [8, 9]. In particular, it could be shown that multiscale discretization is very suited to achieve the necessary compromise between bias and variance in ill-posed inverse data reconciliation problems exploiting adaptively adjusted discretization levels [10].

Despite this success, a series of algorithmic problems have been identified for the chosen adaptive simultaneous discretization of state and control variables using wavelet bases, the solution of which demands substantial fundamental research. At the same time, alternative methods were explored to find a practical but maybe conceptually not as elegant solution on a shorter time scale. These alternative solution concepts only discretize adaptively the control variables by wavelet bases and deal with the state variables by available error-controlled integration methods in the sense of single shooting [5]. This idea proved as promising and was therefore worked out successfully in multiple development cycles [36]. Its performance was demonstrated on medium-sized problems. For example, load and grade change strategies were optimized for an industrial case study shown in Fig. 3 [23]. The considered optimal control problem included about 1500 differential-algebraic constraints, several path and endpoint constraints as well as four control variables. The automatic detection of the switching structure resulting from the necessary conditions of optimality was another break through [34], because it enables a minimal parametrization of the control variables. Switching structure detection has also been combined with adaptive control parametrization [33].

In more recent research, this adaptive method for the solution of optimal control problems was continuously improved. Problems of a size which could not be dealt with before were solved after improving the adaption strategy, implementing a more efficient method for gradient calculation and a partial parallelization of the code. For example, a load change problem for an industrial chemical plant was solved in cooperation with an industrial partner. This optimal control problem comprised a nonlinear economical objective, a set of 15 000 differential-algebraic equations to model the plant dynamics, multiple path constraints and four controls. A differential-algebraic system of about 50 million equations would have had to be solved in every major iteration of the optimization algorithm to achieve the necessary solution accuracy. Since this is computationally prohibitive, the solution of this problem has been addressed by means of our adaptation strategy which resulted in a reduction to about 4 million equations [27]. In comparison to the established manual operation every optimized load change leads to a cost saving of several ten thousands of Euros.

3.4 Aerodynamics under a Mathematical Microscope

A seemingly rather different application area: Of course, aerodynamics has always played an important role in the design of aircrafts. Experiment and numerical simulation complement each other in the design process. Again the significance of numerical simulations depends on the validity of the underlying model. A widely accepted high quality model in aerodynamics is given by the compressible Navier-Stokes equations that are known to pose enormous challenges to theory, algorithmic development, and computing power. Therefore simulations are usually confined to cutouts or simplified scenarios and hence need to be validated and complemented by experiments.

Fluid-Structure-Interaction

The development of wide-bodied aircraft is an example where the scope of experimental analysis becomes more and more limited from a physical as well as economical point of view. Hence it becomes highly desirable and even necessary to expand the range of reliable and efficient simulations. At the same time with increasing size of the aircraft the aspect of fluid-structure-interaction, namely the effect of fluid forces on the deformation of wings which in turn changes flight performance, becomes increasingly important. The fact that a wing tip could deflect by 11 meters between a position at rest and a high lift phase indicates which forces are acting on the wing structure. In view of the effect of wing deformation on flight performance it is of paramount interest to reliably predict it in order to find a good balance between structure stiffness and weight reduction to reach an overall economical optimum. The numerical simulation of such fluid-structure-interaction is to provide among other things answers to the following questions: How should the wing geometry be designed in order to warrant possibly economical flight conditions? How to configure the structural components so that the dynamical forces on the structure will neither induce flutter nor unstable limit states that would cause material fatigue and eventually failure.

In order to numerically simulate these complex interactions between structure and fluid one has to face the following challenges: two media are coupled now in the mathematical model, namely the wing structure and the fluid flow around it. When coupling the flow and elasticity model it is for instance essential to formulate the coupling conditions in such a way that the overall problem is well posed in the sense that it admits a unique solution that depends continuously on the data. Otherwise any algorithm would fail to converge to a physically meaningful solution [29–31]. The involved processes on the fluid as well as on the structure side are inherently non-stationary. The coupling of the two media gives rise to a dynamical process driven by strongly varying time scales and is therefore stiff. Inexpensive explicit time integration schemes are thus inappropriate. Instead one has to resort to implicit schemes that require though to solve in each time step a nonlinear system of equations whose size is determined by the underlying spatial discretization. Thus it could involve millions of unknowns whose number results among other things from the enormous range of relevant spatial length scales governing the fluid phase. One usually avoids to design an algorithm for the whole coupled problem that would have to be developed largely from scratch. Instead one would rather prefer to couple already available solvers for the flow and structural components by alternatingly computing in each of the two media, using the results of the previous computation as boundary data in the next computation for the other medium, see e.g. [29–31]. Mathematically this can be interpreted as a fixed point iteration. One therefore has to ensure that sufficiently many inner iterations are performed to guarantee a sufficiently strong coupling before advancing the time step. Overall one has to deal with an enormously complex problem that is typically not tractable by off-the-shelf commercial software. Algorithms need to be tailored closely enough to the particular situation in order to take advantage of any possible shortcut.

Wake Turbulence

A second problem area concerns wake turbulence caused by airplanes during take-off. It has gained economical importance especially through modern wide bodied aircraft since wake turbulence could significantly delay takeoff frequencies since smaller airplanes have to wait until the wake turbulence has sufficiently decayed. Potential economical advantages of wide bodied airplanes could therefore well be (at least partially) gambled away. This problem has been present for a long time and has triggered numerous investigations towards the objective of finding ways to accelerate the decay of wake turbulence. The progress achieved primarily by experimental means has been modest. To gain better insight through numerical simulations

Fig. 4 Wing model in
HiReNASD

is likewise very limited by the enormous complexity of the problem. One would have to compute the whole vortex field over a range of roughly fifty wing lengths with sufficient accuracy, while current methods can cover a few wing widths at best. Again an essential obstruction lies in the enormous range of length scales that need to be resolved by a discretization in order to understand the key mechanisms.

The Collaborative Research Initiative (SFB) 401 "Modulation of Flow and Fluid-Structure-Interaction at Airplane Wings" is concerned with both problem areas. In a combination of several subprojects engineers and mathematicians developed a new flow solver concept QUADFLOW based on two main pillars, namely an adaptive multiscale discretization and a tailor-made grid generation method [13]. The latter one makes crucial use of concepts from Computer Aided Geometric Design (CAGD). These methods played, in particular, a pivotal role for the free form surface generation on which the numerically controlled milling process for the wind tunnel model was based upon. This wing model is currently used in the experiment project "High Reynolds Number Aerostructural Dynamics (HiReNASD)" in the European Transsonic Windtunnel (ETW). Figure 4 shows the wing model that was manufactured with the aid of the CAD model displayed in Fig. 5.

The methods developed in this cooperation allow us now to significantly advance the simulation frontiers in this area. QUADFLOW will soon be completely parallelized which offers a much more effective starting point for computing wake turbulence in a much larger range. Moreover, we shall be in a position to make reliable fluid-structure-interaction predictions based on the aeroelastic simulations accompanying the wind tunnel experiments. This can certainly be regarded as one of the highlights of the work in the SFB 401.

However, when setting up the experiment another obstruction arose. The available commercial CAGD tools turned out to be inappropriate for modeling the wing

Fig. 5 CAGD model

cap. In particular, it was important to use as few parametric surface patches so as to still merge them in a sufficiently smooth fashion to guarantee accurate milling results. The solution of these problems in [11, 12] benefited greatly from the close connection between the grid generation and surface modeling concepts. This led finally to a clean and accurate CAD model for variable winglet caps (see Fig. 16 on p. 69).

4 Status Quo

4.1 Multiscale-Decomposition: Wavelets

In all the above mentioned applications a central challenge lies in developing numerical techniques that take proper account of all length scales that are relevant for the respective area while not spending more computational work than necessary. In this section we would like to bring out the mathematical conceptual foundations.

The common mathematical framework resides on viewing the objects in question such as time series, control variables, fluid quantities as *functions*. Instead of approximating these functions on a *given* mesh, thereby discarding some possibly important information, one can seek suitable *representations* of functions that first keep all the information.

In principle this resembles the representation of numbers. The Roman system is less suitable for arithmetics than the Arabic one that is based on expansions into powers of ten (or in fact, any other integer). It takes infinitely many terms in such an expansion to represent the number π exactly. But a specific application might require only a crude accuracy and finitely many terms would suffice to meet such accuracy requirements. Of course, it is less clear what kind of "building blocks" would be appropriate for functions. Classical examples are the trigonometric system for representing periodic objects. The coefficients—"amplitudes"—weighting the harmonics reflect the strength of the various frequency components. This leads into classical harmonic analysis. Orthogonal polynomials are the algebraic counterpart. Which system to use will depend in a much stronger way on the particular context and the structure of the objects to decompose and analyze. Significant recent advances of modern harmonic analysis have centered around creating new systems of building blocks taking shortcomings of the classical systems into account. For instance, Fourier expansions offer a perfect localization in the frequency domain— they would recognize if only a single frequency was involved—but does not permit any localization of spatial phenomena such as the jumps of a piecewise constant—an object with very little information content. In fact, the Fourier expansion of a piecewise constant involves a huge range of significant frequencies with slowly decaying amplitudes. Modern computational harmonic analysis therefore aims at contriving systems of building blocks that, for instance, realize a better compromise between spatial and frequency localization (complete simultaneous localization in both domains not being possible). Therefore the coefficients in corresponding expansions convey much more information concerning the separation of spatially varying length scales. *Wavelet bases* are a prominent example in this direction that have triggered enormous developments over the past two decades. We shall indicate next, in a rather simplified way, some of the basic ideas in the context of *image compression* in order to be able to take them up later again in connection with other application areas.

Images are Just Functions

A digital image is a discrete object. In fact, sticking for simplicity with black and white images one deals with a matrix with integer entries which assign a grey value to each pixel quantifying just brightness. Now, storing in a naive way each pixel value the transmission of such images would soon clutter the Internet and extremely impose on the patience of the recipient. Only very few pictures could be stored in a digital camera. The key to much more economical ways of storing images lies in alternate representations of images that hinge on a remarkable step, namely to view the image as a *function*, more precisely, as a piecewise constant—a histogram—that assumes over each pixel area the corresponding grey value.

For simplicity, let us make this more precise for a piecewise constant on the unit interval $[0, 1]$ which corresponds for instance to the cross section indicated by the red line in the figure on the left. Defining the box-function

$$\phi(x) := \chi_{[0,1)}(x) := \begin{cases} 1, & x \in [0, 1); \\ 0, & \text{else,} \end{cases}$$

along with its shifted and scaled versions $\phi_{j,k}(x) := \phi(2^j x - k)$, and assuming that the pixel width is just 2^{-J}, one obtains the "single scale representation"

$$f(x) = \sum_{k=0}^{2^J-1} g_{J,k} \phi_{J,k}(x)$$

of the histogram.

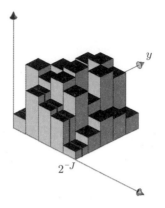

Now, in order to transmit the "image" f it first has to be encoded by a bitstream, a sequence of zeros and ones, that identify f. Since the grey values $g_{J,k}$ are integers this could be done by simply concatenating the binary representations of the $g_{J,k}$ in a prescribed order. However, regardless of small scale variations in the image or larger areas of almost equal light intensity, in this representation all grey values $g_{J,k}$ are equally important and would have to be stored. We shall explain next an alternate representation that will be seen to facilitate a much more economic storage and hence faster transmission. It allows one to exploit that image regions with little detail may have much less information content and will therefore require fewer bits to encode. To this end, consider again the cross section marked by the red line in the two-dimensional histogram shown above. Now, when replacing two neighboring constant pieces on the small intervals by their mean value on the union of these intervals, one can represent the resulting error as a multiple of the "fluctuation profile" $\psi_{J-1,k}(x) = \psi(2^J x - k)$, where

$$\psi(x) = \phi(2x) - \phi(2x - 1) = \begin{cases} 1, & 0 \leq x < 1/2; \\ -1, & 1/2 \leq x < 1; \end{cases} \tag{1}$$

is the so called *Haar-Wavelet*.

One easily derives from (1) that the original grey values $g_{J,k}$, $k = 0, \ldots, 2^J - 1$, are related to the coarser means $g_{J-1,k}$, $k = 0, \ldots, 2^{J-1} - 1$, and the split off fluctuations- or detail amplitudes $d_{J-1,k}$, $k = 0, \ldots, 2^{J-1} - 1$, by the following "two-scale relation"

$$g_{J-1,k} = \frac{1}{2}(g_{J,2k} + g_{J,2k+1}), \qquad d_{J-1,k} = \frac{1}{2}(g_{J,2k} - g_{J,2k+1}). \tag{2}$$

The so-called *fast wavelet transform* results from a repeated application of this splitting procedure

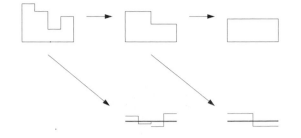

$$\sum_{k=0}^{2^J-1} p_{J,k}\phi_{J,k}(x) = \sum_{k=0}^{2^{J-1}-1} p_{J-1,k}\phi_{J-1,k}(x) + \sum_{k=0}^{2^{J-1}-1} d_{J-1,k}\psi_{J-1,k}(x).$$

The following diagram illustrates this schematically

(3)

The information carried by the original array $\mathbf{g}_J$ of grey values is still completely contained in the arrays of detail coefficients $\mathbf{d}_j$, $j = 0, \ldots, J-1$, and of the coarsest mean $\mathbf{g}_0$. So far the above process describes only a change of bases from the box-functions into the Haar-wavelets. However, the new representation has the following essential advantage. If two neighboring grey values are nearly equal, i.e. the picture exhibits locally little variation, the corresponding detail coefficient will be very small. One may expect that discarding this small value would hardly change the image, a fact that will indeed be confirmed in a few moments. This is the main point in wavelet based image compression.

Functions are Just Sequences

Before substantiating the above remarks let us pause to emphasize a further principal advantage of the above wavelet representation. It is not restricted to a fixed image resolution but could progressively be extended by including details on even

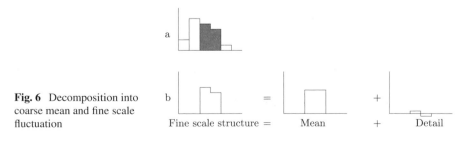

Fig. 6 Decomposition into coarse mean and fine scale fluctuation

smaller scales. This could in fact be driven to infinity from which one deduces that every square integrable function can be expanded in a wavelet basis for the whole infinite-dimensional function space L_2. More precisely, this space consists of all (measurable) functions whose L_2-norm

$$\|f\|_{L_2} := \left(\int_0^1 |f(x)|^2 dx \right)^{1/2}$$

is finite. Normalizing $\psi_{j,k}$ by the factor $2^{j/2}$, so that $\|\psi_{j,k}\|_{L_2} = 1$, and noticing that

$$\langle \psi_{j,k}, \psi_{l,m} \rangle := \int_0^1 \psi_{j,k}(x)\psi_{l,m}(x)dx = \delta_{(j,k),(l,m)}, \qquad (4)$$

one obtains the *wavelet representation*

$$f(x) = \langle f, \phi_{0,0} \rangle + \sum_{j=0}^{\infty} \sum_{k=0}^{2^j-1} \langle f, \psi_{j,k} \rangle \psi_{j,k} \qquad (5)$$

of f. It assigns to any function f a sequence of "digits" $\mathbf{d} = \{d_{j,k}\}_{j=-1,k=0}^{\infty,2^j-1}$ with $d_{j,k} = \langle f, \psi_{j,k} \rangle$ $(d_{-1,0} := \langle f, \phi \rangle)$, in analogy to representations of numbers. The particular suitability of this representation for various purposes rests on the following facts. First, because of the orthonormality of the basis functions (4) one has

$$\|f\|_{L_2} = \left\| \{d_{j,k}\}_{j,k} \right\|_{\ell_2} := \left(\sum_{j=-1}^{\infty} \sum_{k=0}^{2^j-1} |d_{j,k}|^2 \right)^{1/2}. \qquad (6)$$

Hence the Euclidean norm of the coefficient sequence *equals* the L_2-norm of the corresponding function. Moreover, for an arbitrary finite partial sum $P_\Lambda f := \sum_{(j,k)\in\Lambda} d_{j,k}\psi_{j,k}$ this means that

$$\|f - P_\Lambda f\|_{L_2} = \|\{d_{j,k}\}\|_{(j,k)\notin\Lambda} = \left(\sum_{(j,k)\notin\Lambda} |d_{j,k}|^2 \right)^{1/2}. \qquad (7)$$

Thus, the smaller the Euclidean norm of the discarded part of the sequence $\|\{d_{j,k}\}\|_{(j,k)\notin\Lambda}$ is, the better is the approximation $P_\Lambda f$ to the infinite expansion f. Small perturbations of the coefficient sequence cause therefore only small perturbations of the function—the image.

Now one should ask which properties of f cause some of the fluctuation coefficients $d_{j,k} = \langle f, \psi_{j,k} \rangle$ to be small? To see this one uses the fact that the wavelets $\psi_{j,k}$ are *orthogonal* to constant functions, more precisely to any function that is constant on the support of $\psi_{j,k}$, i.e. $\langle c, \psi_{j,k} \rangle = c \int_0^1 \psi_{j,k}(x)dx = 0$. One says that $\psi_{j,k}$ has a *vanishing moment*. Denoting by $I_{j,k} := [k2^{-j}, (k+1)2^{-j})$ the support of $\psi_{j,k}$, one obtains

$$|d_{j,k}| = |\langle f, \psi_{j,k}\rangle| = \inf_{c\in\mathbb{R}} |\langle f - c, \psi_{j,k}\rangle| \le \inf_{c\in\mathbb{R}} \|f - c\|_{L_2(I_{j,k})}$$

$$\le 2^{-j}\|f'\|_{L_2(I_{j,k})}, \tag{8}$$

Thus, whenever f has a small derivative on $I_{j,k}$, i.e. is smooth on the support of the wavelet, then the corresponding wavelet coefficient $d_{j,k}$ is small in absolute value. In this sense wavelets filter out regions on each dyadic length scale where the function exhibits only little variations relative to that length scale.

Roughly speaking, image compression works by transforming the image—the histogram—into a wavelet representation to discard then all coefficients below a given threshold value. The exact procedure in image compression/encoding as in JPEG Standard 2000 is of course much more involved although the basic idea is the same. In order to generate bit streams all coefficients are *quantized*. That means with increasing accuracy demands progressively more bits are assigned to each co-efficient above the current resolution level. A mathematically rigorous assessment of the performance of such encoders is typically based on a suitable mathematical model for the class of images of interest. For certain such model classes these en-coding schemes can be shown to be in some sense optimal, i.e. encode an image with a number of bits that for any required accuracy is comparable to the smallest possible one which in turn is identified by the metric entropy of the model class, see [19].

Denoising images, that is removing data perturbations, is one of many further tasks in image processing that can be tackled by similar techniques such as "soft thresholding" [22].

The Haar wavelet is just the simplest example. With considerably more effort one can construct wavelets that are even several times differentiable and have a higher number of vanishing moments which is responsible for better compression prop-erties although higher vanishing moments necessarily come with larger supports. More precisely, vanishing moments of order $m + 1$, which means orthogonality to all polynomials of degree at most m, imply that the wavelet coefficients $d_{j,k}$ of a function f are only of the order $2^{-(m+1)j}$ provided that f has derivatives up to the order $m + 1$ on the support of the corresponding wavelet. Moreover, the construction of such wavelets even with additional desirable symmetry properties becomes eas-ier when relaxing orthogonality to *biorthogonality*. The strict equality of norms in (6) is then weakened to inequalities where each norm can be bounded by a constant multiple of the other norm. These constants are independent of f so that one still retains the tight interrelation between function and coefficient norms that justify the thresholding concept. Such bases are called *Riesz bases* and always come in pairs of two biorthogonal bases, see [17] for more details.

All these constructions deal so far with functions of a single variable. Wavelets of two or more variables can easily be built from products of univariate wavelets. In d space dimensions one has then $2^d - 1$ types of fluctuation profiles.

In summary, the commonly used wavelet bases all share the following essential properties:

- *Locality:* the diameters of the wavelet supports scale like 2^{-j} for wavelets on the dyadic level j;
- *Vanishing moments of order m:* imply a fast decay of coefficients in regions where f is smooth;
- *Norm equivalence:* The Euclidean norm of the array of wavelet coefficients can be bounded from below and above by a constant multiple of the function norm so that small perturbations of the function cause only small perturbations in the coefficients and vice versa. Thus functions can be identified even quantitatively with sequences noting that the latter objects are much more amenable to computer processing.

The above attributes allow one in all application areas (such as image and signal processing, adaptive optimization techniques, or adaptive solution techniques for partial differential equations) to effectively separate components representing features on different length scales.

4.2 Multiscale Methods for Process Data Analysis

Measurement data compression through wavelet based decomposition is closely related to image compression as introduced in the previous section. The multivariate data-set is two-dimensional with one dimension comprising the measured process quantities while the second dimension refers to a discrete time. Wavelet-based process data compression methods have shown to be superior to established heuristic methods [39]. Despite these findings, these methods have not entered commercial process information systems because of the reluctance of software providers to enhance their codes. The commercially provided methods for data compression rely on a piecewise linear approximation of time series using either the swinging-door or the boxcar-backslope algorithm. These algorithms can lead to very good results, if the parameters of the algorithm are manually tuned for every time series. To reduce the tremendous tuning effort a multiscale method has been developed to largely automate the selection of tuning parameters to match the properties of the time series. The method described in detail elsewhere [3] calculates the parameters of these algorithms after a multiscale decomposition of the signal using bi-orthogonal spline wavelets of order $(3, 3)$ [17]. The standard deviation of the detail signal, which is the most important factor for automatic tuning, is estimated after denoising of the time series with the wave-shrink method [22].

The improvements achieved with the new wavelet-based tuning method are illustrated by two examples provided by chemical industry. A temperature measurement which is compressed too much with the original setting of the tuning parameters is shown in Fig. 7 on the upper left. At a medium absolute deviation of 1.24 K the transient is not expressed very well. The oscillations are captured much better with the revised compression tuning at a medium absolute deviation of 0.28 K and a compression factor of 3 (see Fig. 7, upper right). On the lower left of Fig. 7, a delusive downward trend of a temperature measurement results after compression with the

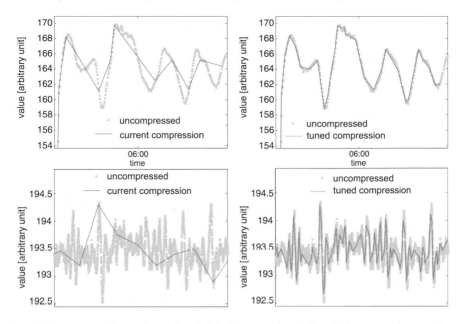

Fig. 7 Temperature time series with established compression (*left*) and after automatic correction of the tuning parameters of the compression algorithm (*right*)

original tuning. After the correction of the tuning with the wavelet-based method the oscillations present in the time series truly appear (see Fig. 7, lower right). The engineering effort is reduced significantly by the new tuning method and the quality of the time series at comparable compression rates can be improved. The compressed time series can therewith be interpreted with a high dependability.

The identification of trends is an important building block for the interpretation of multivariate time series. A multivariate measurement signal y_m is said to follow a trend of order n, if the measured trend error

$$\epsilon_m(t_0, \Delta t) := \frac{1}{\Delta t} \int_{t_0}^{t_0 + \Delta t} \left\| \frac{d y_m(t)}{dt} \right\|_2^2 dt. \tag{9}$$

is smaller than a given bound, which reflects the unavoidable noise in the signal. A stationary trend resulting for $n = 1$ is very important for the analysis of operating conditions in continuous chemical processes. Higher values of n point to polynomial trends. Linear ($n = 2$) and quadratic ($n = 3$) trends are of practical importance to identify drifts in selected process quantities. A suitable trend error can be deduced from the application context or from an analysis of the variation in the signals. The quantity Δt refers to the time interval in which the multivariate trend is observed.

For the identification of trends of order n on a given time interval (defined through a starting value and its length), the scalar signals $y_{m,i}$ are consecutively decomposed by means of wavelet bases, then they are denoised and differentiated [1] in the wavelet representation. Then, the trend error is computed for every sin-

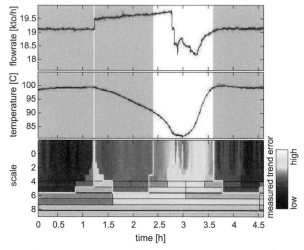

Fig. 8 Identification of linear trends in two time series through recursive calculation of the trend error using a multi-resolution

gle time series through integration [20] and summation of all measured time series (cf. (9)). The trend is of the given order in the given time interval if the resulting trend error is smaller than the preset bound. Obviously, it is rather difficult to find time intervals in which the measured variables follow the postulated trend if there are very many measurements (e.g. several hundred) and long time series (e.g. several hundred thousand data points). This problem was solved efficiently by developing a new multiscale method, which calculates recursively the trend error proceeding from the coarsest to the finest scale [24]. Starting and end times of the interval can be efficiently identified with this algorithm, as exact as the quantization of time in the data set if necessary. Results of the search for a linear trend in two time series of an industrial process are shown in Fig. 8 exemplarily. Time intervals showing a linear trend are colored in grey.

This trend detection technology can be applied in many ways. In real time application, for example, the start of a noteworthy drift of a critical process quantity can be detected in the context of fault diagnosis. The search for stationary trends in a data archive is required to determine a data driven model through correlation of dependent and independent process variables, for example, for the prediction of product quality indicators. In an industrial case study, phases of a certain minimum length were searched for in a data archive, which are supposed to be stationary. The values of the process quantities in these phases were used to determine such a model for the prediction of the melt index of a polymer. In the data set of about 700 000 data points 32 time intervals with the specified characteristics were found at low computational effort. A time interval which just about meets the requirements is shown in Fig. 9.

An evaluation of these wavelet-based methods for data analysis in various industrial case studies has shown that the algorithms scale very well with the size of the data set. Furthermore, results of distinguished quality are obtained which cannot nearly be reached by typically used filter [26] or statistical methods [14].

Fig. 9 Intervals with a stationary trend in time series of two chosen process quantities

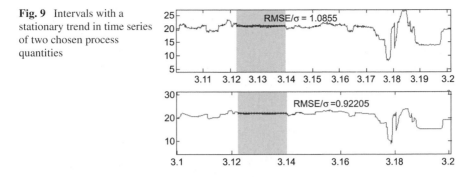

4.3 Solving Optimal Control Problems with Adaptive Wavelet Discretization

The numerical solution of optimal control problems constitutes another completely different application which benefits from exploiting multiscale discretization and localization of phenomena on multiple scales. We consider the optimal control problem (CDYNOPT)

$$\min_{\mathbf{u}(t), \mathbf{y}_0} \quad \Phi = \Phi(\mathbf{y}(t_f)) \tag{CDYNOPT}$$

$$\text{s.t.} \quad \dot{\mathbf{y}} = \mathbf{f}(\mathbf{y}(t), \mathbf{x}(t), \mathbf{u}(t)), \qquad t \in [t_0, t_f],$$

$$0 = \mathbf{g}(\mathbf{y}(t), \mathbf{x}(t), \mathbf{u}(t)), \qquad t \in [t_0, t_f],$$

$$0 = \mathbf{y}(t_0) - \mathbf{y}_0,$$

$$0 \geq \mathbf{h}(\mathbf{y}(t), \mathbf{x}(t), \mathbf{u}(t)), \qquad t \in [t_0, t_f],$$

$$0 \geq \mathbf{e}(\mathbf{y}(t_f), \mathbf{x}(t_f), \mathbf{u}(t_f)),$$

$$\mathbf{u}(t)^{\min} \leq \mathbf{u}(t) \leq \mathbf{u}(t)^{\max}, \qquad t \in [t_0, t_f].$$

which is transcribed into a nonlinear program by discretization of the decision variables $\mathbf{u}(t) \in [\mathbf{u}_{\min}, \mathbf{u}_{\max}]$ of the simulation model. $\mathbf{y}(t)$ is the vector of the differential states of the models, $\mathbf{x}(t)$ is the vector of the algebraic states, both of them are defined on the time horizon $t \in [t_0, t_f]$. The process models are described by the functions $\mathbf{f}(\cdot)$ and $\mathbf{g}(\cdot)$. The system's initial condition is given by $\mathbf{y}_0$. The states can additionally be restricted on the whole time horizon by $\mathbf{h}(\cdot)$ or at the end time by $\mathbf{e}(\cdot)$.

The simulation models often consist of thousands of state variables while the process is controlled only by a few decision variables. The n_u decision variables are discretized by

$$u_i(t) = \sum_{k \in \Sigma_i} p_{i,k} \phi_{i,k}(t), \quad i = 1 \ldots n_u, \tag{10}$$

All the other variables, which obviously depend on the decision variables $z := [p_{u_{i,k}}]$, are solved by numerical integration. An optimization algorithm has to find optimum values for all of the parameters which discretize the decision variables. This way the optimal control problem can be restated as the following non-linear programming problem:

$$\min_{\mathbf{z}} \quad \Phi = \Phi_M(\mathbf{y}(\mathbf{z}, t_f)) \tag{NLP}$$

$$\text{s.t.} \quad \mathbf{0} \geq \mathbf{h}^y(\mathbf{y}(\mathbf{z})),$$

$$\mathbf{0} \geq \mathbf{e}(\mathbf{y}(\mathbf{z}, \mathbf{t_f})),$$

$$\mathbf{z}^{\min} \leq \mathbf{z} \leq \mathbf{z}^{\max}$$

The solution of (NLP) requires a numerical integration of the model for the progressively improving trajectories of the decision variables $\mathbf{u}(t)$. In addition to the solution of the simulation model the sensitivity equations need to be solved to provide necessary gradients to the optimization algorithm [36]. These sensitivity equations result from differentiation of the model with respect to the parameters of the discretized problem:

$$\dot{\mathbf{s}}_i^y = \frac{\partial \mathbf{f}}{\partial \mathbf{y}} \mathbf{s}_i^y + \frac{\partial \mathbf{f}}{\partial \mathbf{x}} \mathbf{s}_i^x + \frac{\partial \mathbf{f}}{\partial z_i}, \tag{11a}$$

$$\mathbf{0} = \frac{\partial \mathbf{g}}{\partial \mathbf{y}} \mathbf{s}_i^y + \frac{\partial \mathbf{g}}{\partial \mathbf{x}} \mathbf{s}_i^x + \frac{\partial \mathbf{g}}{\partial z_i}. \tag{11b}$$

The solution quality depends strongly on the resolution of the chosen discretization. Our approach considers a whole sequence of optimization problems which results from different discretization resolutions [33–35]. At first, coarse scale contributions are collected on the coarsest discretization level, therefore only short calculation times are needed. When solving the optimization problem on the next finer discretization level the previous calculated coarse solutions can be used as initial guess. This is done repetitively until the solution only changes marginally. Only where high temporal resolution is required, local refinements of the discretization are carried out adaptively, such that a maximum solution quality can be achieved with a preferably low number of degrees of freedom. The foundation of this adaptive refinement strategy is a clever exploitation of the characteristics of bi-orthogonal spline wavelets. The algorithm has to provide two elements in order to provide a grid which is exactly adapted to the solution of the system considered, i.e.

- the local refinement of the solution representation, to achieve an improvement of the approximation quality and
- the elimination of detail, if no degradation of the solution occurs.

For adaptive discretization, the control variables are expanded in a wavelet series:

$$u_i(t) = c_{0,0}\phi_{0,0}(t) + \sum_k d_{i,k}\psi_{i,k}(t). \tag{12}$$

Elimination

The optimum solution $u_i^{\star,l}$ of the refinement step l can be transformed in an equivalent multiscale formulation, such that

$$u_i^{\star,l} = \mathbf{d}_{\Lambda_i}^{\star,l,^T} \Psi_{\Lambda_i} \tag{13}$$

where Λ_i is the set of wavelet indices. To find basis function candidates for elimination, an appropriate approximation $\bar{u}_i^{\star,l}$ is searched for, which only deviates from $u_i^{\star,l}$ subject to the tolerance ϵ':

$$\frac{\|\bar{u}_i^{\star,l} - u_i^{\star,l}\|_{L_2}}{\|u_i^{\star,l}\|_{L_2}} \leq \epsilon' \tag{14}$$

The norm equivalence allows transforming this requirement into conditions for the wavelet coefficients corresponding to basis functions, which can be eliminated, if $d_{\Lambda_i}^{*l} < \epsilon$ applies. The remaining coefficients are significant and are used for the next optimization sequence.

Refinement

Next to the elimination of basis functions, the systematic refinement is the second part of the adaptive strategy. The refinement approach is similar to elimination. While small wavelet coefficients are eliminated from $\mathbf{d}_{\Lambda_i}^{\star,l}$, large coefficients indicate the necessity for refinement to allow improvement in the solution of the next optimization sequence. For this purpose, the neighbors $\psi_{i,k-1}$ and $\psi_{i,k-1}$ of basis function with large coefficients or the basis functions on the next higher level $\psi_{i+1,2k}$ and $\psi_{i+1,2k+1}$ can be used. From these candidates only those wavelets are used for refinement, which contribute to a certain percentage κ to the norm of the wavelet coefficients of the last optimization sequence.

Example

An adequate refinement sequence is schematically shown in Fig. 10. The upper diagrams of the figure show the refinement sequence of the discretization of the glucose feed rate to a bioreactor, where a bacterial culture produces a specific protein from the glucose substrate.

For an optimum production of protein the dosage of glucose has to be optimized. The upper diagrams in Fig. 10 show the optimum dosage of glucose. Wavelets on higher levels are added to the existing control vector expansion to introduce more detail into the control profile in every iteration of the optimization sequence (lower diagrams). The algorithm detects in which time intervals the solution is imprecise (shown by the color of the small boxes), and improves the resolutions where needed.

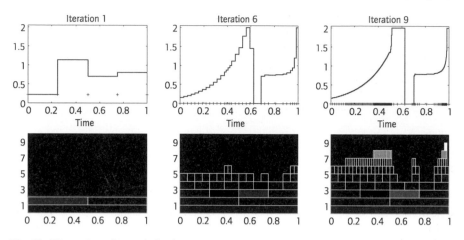

Fig. 10 Illustration of an optimization sequence

In the ninth and last iteration (diagrams on the right) a higher yield is obtained compared to the first optimization of the sequence. It can be observed that only a moderate resolution is necessary in the largest part of the time interval. The strong increase of the feed rate at time 0.5 and 1.0, as well as at the end of the "starvation period" at 0.7 have to be resolved with high precision. The lower diagrams of Fig. 10 show the results of the wavelet analysis of the feed profile, where light colors indicate the necessity for refinement. The problem size is significantly reduced through adaptive discretization resulting in much shorter calculation times. The calculation time is even reduced to an extent allowing the solution of the optimal control problem in real time, such that the fermentation process can be permanently operated at an optimum point.

4.4 Adaptive Methods for Partial Differential Equations in Fluid Mechanics

As mentioned before, numerous variants of fluid flow simulations play a central role in many application areas. Through the chain

$$\text{application} \to \text{mathematical model} \to \text{analysis} \to \text{discretization}$$
$$\to \text{algorithms} \to \text{mapping onto computer architectures}$$

one is often confronted with huge algebraic systems of equations that can usually not be treated by off-the-shelf black-box methods. A typical approach for compressible flows will be illustrated next by a simple example which nevertheless reflects the conceptual links to the above mentioned image or signal processing concepts.

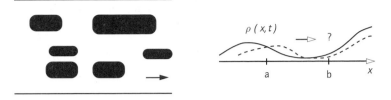

Fig. 11 Traffic flow

Model

In order to understand phenomena experienced in every day life like wave propagation in road traffic, consider individual cars as particles in a fluid, (see the left part of Fig. 11).

In the limit of denser and denser traffic, the collection of particles forms a continuum, i.e. the concentration of cars is represented by a nonnegative function of the spatial position. The road itself is thereby reduced to a line. Since the movement of the traffic changes, the local (car) *density* this function depends also on time. More precisely, this density $\rho(x,t)$ as a function of the spatial variable x and of time t can be thought of as the number of cars per unit car length, see the right part of Fig. 11. In particular, one has $0 \leq \rho(x,t) \leq 1$. One would like to understand now how this function evolves in time to see what might contribute to the formation of traffic jams. This is where the *flow model* comes into play. First, it is clear that the velocity $u(x,t)$ of the particles, as a function of space and time, is related to the density, that is one should have $u(x,t) = U(\rho(x,t))$. The simplest reasonable choice for this functional relation would be

$$U(\rho(x,t)) := u_{max}(1 - \rho(x,t)),$$

which just says that higher speed is permitted for lower densities. In general such a relation is called *constitutive equation*. In order to describe the dynamics of the process one employs a *conservation property*. To this end, note first that the quantity

$$f(\rho) := U(\rho)\rho$$

represents a flux rate (amount of mass per unit volume and time crossing an interface). In fact, for any stationary interval $[a,b]$ on the line (the road) (see Fig. 11), the temporal rate of change of the total mass in the interval just equals the difference of the flux rates at its end points, that is

$$\frac{d}{dt}\int_a^b \rho(x,t)dx = f(\rho(a,t)) - f(\rho(b,t)) = -\int_a^b \underbrace{\frac{d}{dx}\ f(\rho(x,t))}_{:=U(\rho(x,t))\rho(x,t)}\ dx. \quad (15)$$

From this one deduces formally that for every interval (a,b)

$$\int_a^b \partial_t \rho(x,t) + \partial_x f(\rho(x,t))dx = 0 \quad (16)$$

must be valid. Under the assumption that the integrand is continuous one concludes then that the function $\rho(x, t)$ must satisfy the *partial differential equation*

$$\partial_t \rho + \partial_x f(\rho) = 0 \tag{17}$$

typically subject to an initial condition $\rho(x, 0) = \rho_0(x)$ describing the traffic state at some initial time $t = 0$.

This is the simplest example of a scalar *conservation law* which in this case provides (a so far oversimplified) model for traffic flow. More generally, balance laws for mass, momentum and energy complemented by a constitutive equation reflecting for instance material properties lead even for much more complex situations, as encountered with compressible fluid flow in aerodynamics, to systems of partial differential equations of similar type, i.e.,

$$\partial_t u + \partial_x \big(f(u) + g(\partial_x u)\big) = S, \tag{18}$$

where now the solution u denotes a vector field of conserved quantities like density, momentum or energy. Equation (18) is then no longer a scalar equation but a system of partial differential equations for functions of several spatial variables collected in x and again of time t. The additional term $g(\partial_x u)$ contains for advanced models viscous effects and heat conduction involving typically spatial derivatives of order two. Finally, the term S models possible sources or sinks of the balanced quantity.

Whereas for scalar equations a fairly advanced theory concerning the existence, uniqueness and further properties of solutions is available such fundamental questions remain largely open for the general case of systems in three spatial variables. In general, it is impossible to obtain closed form solutions to such equations even when their existence is known. One has to approximate solutions with the aid of numerical techniques. Having a solid faith in nature, possible gaps in theoretical foundations for the practically relevant scenarios do not prevent us, of course, from attempts to solve such problems at least approximately by *numerical techniques*.

Discretization

Here the principle of *discretization* comes into play that allows us to turn equations for functions into systems of equations for possibly large arrays of unknown numbers, which in a certain sense approximate the unknown function. The better one wishes to approximate the function the more unknowns are usually needed, that is the larger the systems of equations are going to be. More accuracy therefore entails more computational effort because larger systems require more work. It would be already a great achievement to keep computational cost proportional to the number of unknowns which is usually not the case. Using off-the-shelf methods would typically result in a prohibitive increase of computational cost that grows like N^β where N is the number of unknowns and β is strictly larger than one.

In order to sketch now a simple discretization principle we return to traffic flow. Instead of considering a single interval (a, b) we decompose the whole computational domain into smaller subintervals (x_i, x_{i+1}) for $i = 0, \ldots, N$, say. For each

Fig. 12 Cell averages

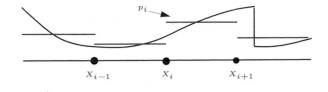

of those subintervals the balance law (15) must hold. Multiplying this balance by $1/\Delta x$, $\Delta x := x_{i+1} - x_i$, one obtains

$$\frac{d}{dt} \underbrace{\frac{1}{\Delta x} \int_{x_i}^{x_{i+1}} \rho(x,t)dx}_{:=\rho_i(t)} + \frac{1}{\Delta x}\left(f\big(\rho(x_{i+1},t)\big) - f\big(\rho(x_i,t)\big)\right),$$

$$i = 0, \dots, N. \tag{19}$$

Apparently, this yields a temporal evolution equation for the *cell averages* (see Fig. 12)

$$\rho_i(t) := \frac{1}{\Delta x} \int_{x_i}^{x_{i+1}} \rho(x,t)dx, \quad i = 0, \dots, N, \tag{20}$$

which is the starting point for so called *finite volume methods*. It should be emphasized that from the outset this does not aim at evolving in time the searched density itself but *functionals* of the density, namely its cell averages forming a piecewise constant function. The density itself will have to be recovered—possibly with higher order accuracy—from those cell averages.

Discretizing now the temporal variable t in the simplest case by $\frac{d}{dt}\rho_i(t) \approx (\rho_i(t+\Delta t) - \rho_i(t))/\Delta t$, setting $\rho_i(t_n) := \rho_i^n$, and choosing appropriate "numerical fluxes" $F(\rho_{i-1}^{\tilde{n}}, \rho_i^{\tilde{n}})$ as approximations to the true fluxes which depend though on point values of the unknown density and not just on its cell averages, one obtains the following discrete analog to (17):

$$\rho_i^{n+1} - \rho_i^n + \frac{\Delta t}{\Delta x}\left(F(\rho_i^{\tilde{n}}, \rho_{i+1}^{\tilde{n}}) - F(\rho_{i-1}^{\tilde{n}}, \rho_i^{\tilde{n}})\right) = 0, \quad i = 0, \dots, N. \tag{21}$$

Here $\tilde{n}$ either equals n or $n+1$. In the former case the method is called explicit because time propagation, i.e. the evaluation of cell averages on time level $n+1$, is realized by just substituting the known quantities from level n. However, this simple procedure is not always adequate. For instance, when dealing with fluid-structure-interaction, models of the form (18) arise in a coupled form that need to be discretized *implicitly* to properly treat stiff components. In this case $\tilde{n} = n+1$, such that the unknowns ρ_i^{n+1} appear on both sides of (21). Thus, on each time level one obtains a *nonlinear system of equations* in $N+1$ unknowns ρ_i^{n+1}, $i = 0, \dots, N$.

The larger the number of unknowns N is chosen on each time level, the smaller are the intervals and the more accurate one expects the agreement of the cell averages with the corresponding local density distribution. On the other hand, the density will exhibit a locally varying behavior. In some areas it may be nearly constant.

Here, the cell average for a relatively *large* interval will nearly coincide with the density on such an interval. In other areas it may change abruptly, for instance, when the traffic comes to a sudden halt as in traffic jams. There one needs *many small* intervals to ensure that the cell averages stay close to the true density, see Fig. 12.

The objective of an *adaptive* solution method is to choose in the course of the solution procedure the cell sizes *automatically* in such a way that a desired overall solution accuracy is achieved at the expense of possibly few cells and hence a possibly small number of unknowns. Such tasks require a very close intertwining of the analysis of the underlying differential equations and the design of adaptive algorithms and address therefore inherently mathematical issues [15, 16]. An efficient implementation of such techniques require novel data structures because storage needs vary dynamically and quick data access becomes highly nontrivial. Moreover, as a further example of the importance of advanced computer science concepts, large scale applications require parallel computing which again is harder to organize with adaptive schemes because load balancing becomes more tricky.

A New Concept for Adaptivity—How far Does the Analogy to Signal Compression Carry?

Adaptive schemes that have been used so far for compressible fluids are primarily of heuristic nature or confined to rather special cases. The approach developed in the SFB 401 is completely different and is based on concepts from *computational harmonic analysis*.

We list now a few basic aspects of related developments: As mentioned before, the new adaptive discretization scheme aims at realizing a desired accuracy tolerance for the numerical approximate solution of the fluid equations at the expense of a possibly small number of degrees of freedom, viz. unknowns. The development, theoretical foundation and implementation of adaptive solution concepts in concrete application scenarios has been a research focus at the Lehrstuhl für Mathematik of the Institute for Geometry and Practical Mathematics (IGPM), see [15, 16]. It turns out that in the present context the problem size may be reduced at times by a factor of up to three digits. The basic concept differs from common adaptive methods because of the underlying multiscale decomposition of the flow field that hinges on modern wavelet methods, quite similar to those employed for image compression, time series analysis or optimal control, see [21, 32]. Moreover, this approach has been complemented by a new grid generation concept that supports on one hand adaptivity, by permitting efficient grid refinement and coarsening, while on the other hand, it accounts for the necessity of efficiently handling varying domain geometries due to wing deformation. As in common block concepts the computational domain is first decomposed into hexagonal blocks which, however, in contrast to common approaches, are generated as images of *B-spline* based mappings of a parameter domain. The main point is that the (still large) set of mesh points which is constantly varying due to a laptivity is never stored, but only the relatively small set of *control points* determining the mappings needs to be stored whose size remains constant.

Returning to the example of traffic flow, at each time t the flow field is represented by an array of cell averages, i.e. by a piecewise constant functions, in analogy to the pixel array of a digital image. Just as in image compression where a wavelet decomposition reveals the importance of the various scale components, one can decompose the flow field into its scale components using again a suitable piecewise constant wavelet basis. "Suitable" means here that a sufficiently high order of vanishing moments is realized in order to effectively compress smooth flow components while still retaining high accuracy, see [21].

Of course, the analogy to image compression is limited by the fact that the complete information about a digital image is given explicitly while the flow field is unknown and only given implicitly through the partial differential equation. The detection of relevant scale components requires taking the differential equation into proper account [13, 21, 32]. A detailed discussion of the essential concepts and their analytical foundation can be found in [18] and [32].

The resulting method consist of the following main steps:

0. Wavelet decomposition of the initial values and generation of an initial grid; in principle this is similar to image compression;
1. prediction of the mesh for the next time level; this uses the dominantly hyperbolic character of the partial differential equations which ensures an essentially finite speed of information propagation and thus allows one to estimate the extent of fine scale modes that could be excited by the time step;
2. time integration: this is done using the locally refined representation in terms of cell averages;
3. wavelet decomposition of the propagated flow field, compression and return to (1).

This method applies in the same way to the balance laws of gas dynamics and is used by the new flow solver QUADFLOW [13].

The following examples are to illustrate how the method works in real applications. Figure 13 addresses a classical benchmark computation for a BAC-12 profile. The right picture shows the pressure distribution resulting from a simulation of an inviscid flow at Mach 0.95 and the angle of attack $0°$. The left picture displays the adaptively refined mesh generated during the computation. It indicates the span of relevant length scales in the flow.

The main objective of this simulation was to resolve well the complex shock interaction even far away from the profile. A common flow solver based on uniform grid refinements would have to work on approximately 77 million grid cells to realize the accuracy required here. By contrast the adaptive method accomplishes this target accuracy at the expense of only roughly 55 thousand cells. This indicates the massive complexity reduction that can be achieved by well founded adaptive strategies.

Figure 14 shows a high lift configuration in a much lower Mach regime. In this case viscous effects are taken into account by the underlying model based on the Navier-Stokes equations. The right side of the figure confirms the excellent agreement of experimental and computed drag and lift values. The left side displays again the corresponding adaptive mesh.

Fig. 13 Inviscid transonic
flow

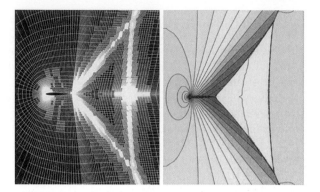

BAC 3-11, L1/T2 High Lift Configuration
$M_\infty = 0.197, Re_\infty = 3.52 \cdot 10^6, \alpha = 20.18°$

BAC 3-11, L1/T2 High Lift Configuration
$M_\infty = 0.197, Re_\infty = 3.52 \cdot 10^6, \alpha = 20.18°$

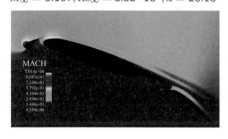

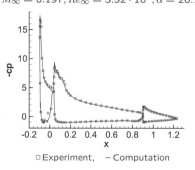

□ Experiment,   − Computation

Fig. 14 Viscous flow for a high lift configuration

A likewise reliable resolution of relevant flow scales is illustrated in Fig. 15 in a different physical regime, namely for hypersonic flows of much 8.5 in this case.

An Adaptive Grid Generator

An essential ingredient of the wavelet based adaptation concepts is a tailor-made grid generator that supports grid movement as well as efficient refinement and coarsening. In contrast to common strategies it hinges on B-spline based domain mappings onto the grid blocks [28]. Refining the grid means evaluating this mapping, while moving the grid amounts to just varying the mapping not the grid points which in turn requires changing only the control points in the B-spline representation.

The close linkage between grid generation and CAGD concepts was, in particular, crucial for designing the wing model for the wind tunnel experiments in a way that the shape and accuracy constraints for controlling the milling process were met. In the future course of the project, the wing cap will have to be replaced by winglets of variable geometries regarding the bending angle and profile slope.

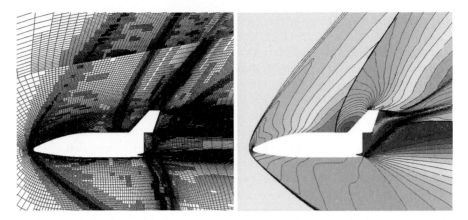

Fig. 15 Inviscid hypersonic flow

Fig. 16 Winglet-cap

In the meantime we have developed methods for generating such winglets automatically, see Fig. 16. On account of particular design constraints as well as demands posed by the flow solver, commercial CAD systems turned out not to provide sufficient quality. For a more detailed account of the modeling and fairing methods developed for that purpose we refer to [11, 12].

5 Analysis of Strengths and Weaknesses

Interdisciplinary research groups as they have evolved at RWTH Aachen University, in particular under the roof of the "Center for Computational Engineering Science", offer in principle an excellent framework for accomplishing long term methodological progress of lasting impact in the area of modeling and simulation sciences. This renders results of breakthrough character feasible that, quite in the spirit of catching several birds with one stone, apply simultaneously in several applications. An essential prerequisit is a specific science culture shared by all partners that appreciates and builds on the synergies emerging from complementary expertise instead of counting on unidirectional service cooperation. The projects outlined above indicate the potential and success of such an approach, in particular, when not being restrained by ever and ever shorter production cycles of industrial development. A currently observed tendency for outsourcing research tasks can certainly be explained by increasing economic pressure. The more so should universities establish complementary research tracks that can contribute to lasting innovation. In our experience one faces in this regard the following obstructions. In foundational research the conceptual potential is often far ahead of application practice. The chronically low level of staff coverage and a minimal infrastructure support in basic sciences at universities often boil down to administrating deficiencies. As a consequence, staff capacities are far from being able to raise innovative approaches to a level of routine practical applicability by non-experts. Aside from the question of user expertise, the software developed in such research groups, often comprising several hundred thousand lines of code, has usually only pilot character that is neither sufficiently optimized to exploit its true potential nor has the interfaces needed for a versatile usability in real-world applications. Since most of the software developers have only temporary employment the continuity of such developments is highly jeopardized. These structural deficits that will surely persist in the near future are in addition enhanced by accelerating industrial cycles which make it even harder to transfer complex innovations into industrial practice. This in turn diminishes a fruitful feedback to basic research.

6 Visions and Recommended Course of Action

In our opinion a substantial improvement of the status quo could be reached if better interfaces between research, in particular fundamental research, at universities and potential users in industry were established. The interaction between academia and industry could be very much improved if the decision makers in industry were more open towards the current developments in relevant research fields and if they truly accepted as well as accounted for the multiscale nature of methods-oriented academic research and development in their longer-term planning. The establishment of so-called transfer projects by DFG (funding joint research between academia and industry) can be seen as a recognition of these deficits. However, an even more pronounced symmetry between academic and industrial partners still needs to be implemented in these projects.

Acknowledgements   We gratefully acknowledge the continued financial support of Deutsche Forschungsgemeinschaft (DFG) especially in the Leibniz program. Furthermore we feel especially indebted to Dr.-Ing. F. Alsmeyer (AIXCAPE e. V.) and Dipl.-Ing. A. Hartwich (AVT-PT) as well as Dr. K.-H. Brakhage and Dr. S. Müller (IGPM) for their invaluable assistance in preparing this manuscript.

References

1. Abramovich, F., Silvermann, B.W.: Wavelet decomposition approaches to statistical inverse problems. Biometrika **85**, 115–129 (1998)
2. Alsmeyer, F.: Trend-based treatment of process data: application to practical problems. In: Proc. of the World Congress of Chemical Engineering 2005, Glasgow, UK, 10–14 July 2005

3. Alsmeyer, F.: Automatic adjustment of data compression in process information management systems. In: Proc. of the 16th European Symposium on Computer Aided Process Engineering and 9th International Symposium on Process Systems Engineering, Garmisch-Partenkirchen, pp. 1533–1538 (2006)
4. Biegler, L.T.: Solution of dynamic optimization problems by successive quadratic programming and orthogonal collocation. Comput. Chem. Eng. **8**, 243–248 (1984)
5. Binder, T., Cruse, A., Villar, C., Marquardt, W.: Dynamic optimization using a wavelet based adaptive control vector parameterization strategy. Comput. Chem. Eng. **24**, 1201–1207 (2000)
6. Binder, T., Blank, L., Bock, H.G., Burlisch, R., Dahmen, W., Diehl, M., Kronseder, T., Marquardt, W., Schlöder, J.P., v. Stryk, O.: Introduction to model based optimization of chemical processes on moving horizons. In: Online Optimization of Large Scale Systems, pp. 295–339. Springer, Berlin (2001)
7. Binder, T., Blank, L., Dahmen, W., Marquardt, W.: Towards multiscale dynamic data reconciliation. In: Kravaris, C., Berber, R. (eds.) NATO-ASI on "Nonlinear Model Based Process Control", Antalya, Turkey, 10–20.8.1997, pp. 623–665. Kluwer Academic, Dordrecht (1998)
8. Binder, T., Blank, L., Dahmen, W., Marquardt, W.: Iterative algorithms for multiscale state estimation, Part 1: Concepts. J. Optim. Theory Appl. **111**(3), 501–527 (2001)
9. Binder, T., Blank, L., Dahmen, W., Marquardt, W.: Iterative algorithms for multiscale state estimation, Part 2: Numerical investigations. J. Optim. Theory Appl. **111**(3), 529–551 (2001)
10. Binder, T., Blank, L., Dahmen, W., Marquardt, W.: On the regularization of dynamic data reconciliation problems. J. Process Control **12**(4), 557–567 (2002)
11. Brakhage, K.-H., Lamby, P.: Application of B-spline techniques to the modeling of airplane wings. Comput. Aided Geom. Des. **25**(9), 738–750 (2008)
12. Brakhage, K.-H., Lamby, P.: Modeling of airplane wings with winglets. In: 10th International Conference on Numerical Grid Generation in Computational Field Simulations, 16–20 September 2007, FORTH, Greece (2007)
13. Bramkamp, F., Gottschlich-Müller, B., Lamby, Ph., Hesse, M., Müller, S., Ballmann, J., Brakhage, K.-H., Dahmen, W.: H-adaptive multiscale schemes for the compressible Navier-Stokes equations—Polyhedral discretization, data compression and mesh generation. In: Ballmann, J. (ed.) Notes on Numerical Fluid Mechanics, Flow Modulation and Fluid-Structure-Interaction at Airplane Wings, vol. 84, pp. 125–204. Springer, Berlin (2003)
14. Cao, S., Rhinehart, R.R.: An efficient method for on-line identification of steady state. J. Process Control **5**, 363–374 (1995)
15. Cohen, A., Dahmen, W., DeVore, R.: Adaptive wavelet schemes for nonlinear variational problems. SIAM J. Numer. Anal. **41**, 1785–1823 (2003)
16. Cohen, A., Dahmen, W., DeVore, R.: Adaptive wavelet techniques in numerical simulation. In: De Borste, R., Hughes, T., Stein, E. (eds.) Encyclopedia of Computational Mechanics, pp. 157–197. Wiley-Interscience, New York (2004)
17. Cohen, A., Daubechies, I., Feauveau, J.-C.: Biorthogonal bases of compactly supported wavelets. Commun. Pure Appl. Math. **45**, 485–560 (1992)
18. Cohen, A., Mahmoud Kaber, S., Müller, S., Postel, M.: Fully adaptive multiresolution schemes for conservation laws. Math. Comput. **72**, 183–225 (2003)
19. Cohen, A., Dahmen, W., Daubechies, I., DeVore, R.: Tree approximation and optimal encoding. Appl. Comput. Harmon. Anal. **11**, 192–226 (2001)
20. Dahmen, W., Micchelli, C.A.: Using the refinement equation for evaluating integrals of wavelets. SIAM J. Numer. Anal. **30**, 507–537 (1993)
21. Dahmen, W., Gottschlich-Müller, B., Müller, S.: Multiresolution schemes for conservation laws. Numer. Math. **88**, 399–443 (2001)
22. Donoho, D.L., Johnstone, I.M.: Adapting to unknown smoothness with wavelet shrinkage. J. Am. Stat. Assoc. **90**, 1200–1224 (1995)
23. Dünnebier, G., van Hessem, D., Kadam, J., Klatt, K.-U., Schlegel, M.: Optimization and control of polymerization processes. Chem. Eng. Technol. **28**, 575–580 (2005)
24. Flehmig, F., Marquardt, W.: Detection of multivariable trends in measured process quantities. J. Process Control **16**, 947–957 (2006)

25. Foss, A.S.: Critique of chemical process control theory. AIChE J. **19**, 209–214 (1973)
26. Gertler, J., Chang, H.S.: An instability indicator for expert control. IEEE Control Syst. Mag. **6**, 14–17 (1986)
27. Hartwich, A., Marquardt, W.: Dynamic optimization of the load change of a large-scale chemical plant by adaptive single shooting. Comput. Chem. Eng. (2010). doi:10.1016/j.compchemeng.2010.02.036
28. Lamby, P.: Parametric multi-block grid generation and application to adaptive flow simulation. PhD Dissertation, RWTH Aachen (2007)
29. Massjung, R., Hurka, J., Ballmann, J., Dahmen, W.: On well-posedness and modeling for nonlinear aeroelasticity. In: Notes on Numerical Fluid Mechanics, vol. 84. Springer, Berlin (2003)
30. Massjung, R.: Numerical schemes and well-posedness in nonlinear aeroelasticity. PhD Thesis, RWTH Aachen (2002)
31. Massjung, R.: Discrete conservation and coupling strategies in nonlinear aeroelasticity. Comput. Methods Appl. Mech. Eng. **196**, 91–102 (2006)
32. Müller, S.: Adaptive Multiscale Schemes for Conservation Laws. Lecture Notes on Computational Science and Engineering, vol. 27. Springer, Berlin (2002)
33. Schlegel, M., Marquardt, W.: Adaptive switching structure detection for the solution of dynamic optimization problems. Ind. Eng. Chem. Res. **45**(24), 8083–8094 (2006)
34. Schlegel, M., Marquardt, W.: Detection and exploitation of the control switching structure in the solution of dynamic optimization problems. J. Process Control **16**(3), 1731–1751 (2006)
35. Schlegel, M., Stockmann, K., Binder, T., Marquardt, W.: Dynamic optimization using adaptive control vector parameterization. Comput. Chem. Eng. **29**, 1731–1751 (2005)
36. Schlegel, M., Marquardt, W., Ehrig, R., Nowak, U.: Sensitivity analysis of linearly-implicit differential-algebraic systems by one-step extrapolation. Appl. Numer. Math. **48**, 83–102 (2004)
37. Strang, G., Nguyen, T.: Wavelets and Filter Banks. Wellesley-Cambridge Press, Wellesley (1996)
38. Tsang, T.H., Himmelblau, D.M., Edgar, T.F.: Optimal control via collocation and nonlinear-programming. Int. J. Control **21**, 763–768 (1975)
39. Watson, M.J.A., Liakopoulos, A., Brzakovic, D., Georgakis, C.: A practical assessment of process data compression techniques. Ind. Eng. Chem. Res. **37**, 267–274 (1998)

Active Flow Control—A Mathematical Challenge

Rudibert King, Volker Mehrmann,
and Wolfgang Nitsche

1 Executive Summary

The sustainable development of ground and air vehicles requires a significant reduction in the use of energy and also lower pollution and noise emission. To meet these challenges, non-gradual improvements are necessary. A central role in achieving these future goals plays the active control of the occurring flow processes. While in the past, improvements were mainly achieved through passive methods, in the future the required drastic improvement can only be achieved via an active control. A characteristic feature in this development is the intensive interdisciplinary interaction of several scientific disciplines.

This article sketches the state of the art in active flow control, depicts the challenges and describes the necessary activities to meet the challenges. The contribution of mathematics in this process is multidimensional. It ranges from mathematical modelling of the flow processes, numerical simulation on modern high performance computers, model reduction, the development of methods for model-based control and optimization of processes, to the feature based visualization of the huge experimental and numerical data sets.

Supported by *German Research Foundation (DFG)*, Collaborative Research Center SFB557.

R. King (✉)
Institut für Anlagentechnik, Prozesstechnik und Technische Akustik, P 2-1, TU Berlin,
Hardenbergstr. 36a, 10623 Berlin, Germany
e-mail: Rudibert.King@TU-Berlin.de

V. Mehrmann
Institut für Mathematik, MA 4-5, TU Berlin, Str. des 17. Juni 136, 10623 Berlin, Germany
e-mail: mehrmann@math.tu-berlin.de

W. Nitsche
Institut für Luft- und Raumfahrt, F2, TU Berlin, Marchstraße 12, 10587 Berlin, Germany
e-mail: wolfgang.nitsche@tu-berlin.de

M. Grötschel et al. (eds.), *Production Factor Mathematics*,
DOI 10.1007/978-3-642-11248-5_4, © Springer-Verlag Berlin Heidelberg 2010

While the mathematical modelling of flow physics is an important research area for more than 100 years, the numerical simulation of flows has only received a major thrust in the last 30 years through the development of modern computers. However, despite all the developments in computational fluid dynamics, the computing times for realistic vehicle geometries are still tremendously high. For this reason the improvement of simulation methods is still of central importance.

Furthermore, to perform active flow control in real time, reaction times in milliseconds on micro-processors are necessary. To achieve these, it is on the one hand necessary to replace the models by simplified (reduced) models and on the other hand to develop algorithms that are able to incorporate sensor information in control and optimization.

For the economical and ecological development of the high-tech country Germany it is central to intensify the research support in the area of flow control.

2 Success Stories

The analysis of turbulent flows is a central question in the development and in the operation of modern aircraft as well as ground vehicles. Besides the experimental investigation in wind tunnels and the theoretical physical analysis of the models, the numerical simulation of flows has established itself as the third column for the analysis of non-stationary control processes. On the basis of the model equations, respectively strong simplifications of these equations that are achieved by physical or mathematical model reduction methods, it is now possible to develop and apply methods for control and optimization.

Using model based control, it is possible today to reduce the noise of turbines, to increase the lift of wings, to reduce the resistance of vehicles and to design the mixing of fuel and air in the engines so that less pollution arises [3, 7, 10].

If, for example, one blows periodically small portions of air out of the leading edge of the wing flap of an airplane into the surrounding flow, see Fig. 1, then it is possible to increase the lift significantly. Figure 2 gives an example of the achieved lift increase.

Via such an approach, it is possible in the future to have steeper take-off and landing trajectories, which can reduce the noise emission on the ground.

The biggest challenge of today is the transfer of these concepts to real, geometrically complex wings including engines nacelles. First steps in this direction are currently taken in the DFG funded collaborative research center SFB557 *Control of complex turbulent shear flows* at TU Berlin in cooperation with Airbus Germany [9].

Another example of success of active flow control is the reduction of damaging pressure variations in a burner, e.g. a turbine, see Fig. 3, that may arise in low emission operation [8]. Via secondary injections it is possible to reduce the pressure variations significantly see Fig. 4.

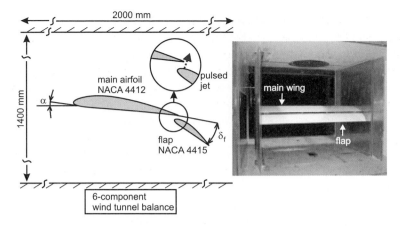

Fig. 1 Experimental setup

Fig. 2 Normalized lift of a
wing plotted via the angle of
attack

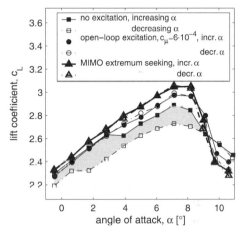

3 Active Flow Control, Status Quo

Active flow control plays a central role in the sustainable development of ground
and air vehicles. In the report "Vision 2020" of the European aviation initiative
ACARE [1] it is stated that the required reductions in emission, fuel consumption,
and noise paired with a tripling of the air traffic in the next 15 years cannot be
achieved via gradual improvements such as e.g. turbulators on the wings. It is necessary to obtain non-gradual improvements as they are possible by the manipulation
of the shear layers on airfoils. For this an intensive cooperation between different
scientific disciplines is necessary (experiment, control theory, sensor and actuator
technology, computational fluid dynamics, analysis, numerical mathematics, mathematical visualization, scientific computing).

The contributions of mathematics, e.g. in the development cycle of modern airplanes is depicted in the following list.

Fig. 3 Pressure variations in
a turbine

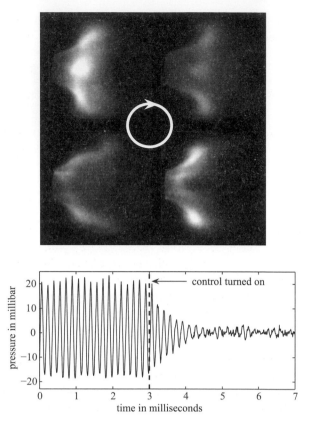

Fig. 4 Control of pressure
variations in a turbine

- Development and analysis of the mathematical model equations that describe turbulent flows in conjunction with experiments.
- Representation of an (ideally 3D) virtual mathematical model of the physical system in the computer.
- Development and analysis of numerical methods for the simulation and visualization of the flow around the virtual computer model.
- Simulation of real test configurations, comparison with experiments, improvement of the mathematical models and the numerical simulation methods.
- Systematic change and optimization of design parameters in the mathematical model, in the passive control via physical experiment and numerical simulation.
- Development of approximative reduced mathematical models for active flow control.
- Development of (real time) closed loop methods for active flow control.

Clearly this complex development process is not static, but via development loops and comparison with experiments, the mathematical models as well as the numerical methods for simulation and control are continuously modified. Such basic research requires an interdisciplinary cooperation as it is carried out in several DFG (German Research Foundation) funded collaborative research centers (SFB) (such as SFB 401 *Flow control and fluid-structure interaction at airfoils* at RWTH Aachen or in SFB 557 *Control of complex turbulent shear flows* at TU Berlin).

3.1 Modelling

A typical class of mathematical model equations for the description of turbulent flows around an airfoil are the non-stationary Navier-Stokes equations

$$\left.\begin{array}{c} \frac{\partial}{\partial t}v + \nabla \cdot (v \otimes v) + \frac{1}{\rho_f}\nabla p - \nu \Delta v = 0 \\ \nabla \cdot v = 0 \end{array}\right\} \quad \text{in } \Omega, \tag{1}$$

where v is the velocity of the flow and p the pressure in a computational domain Ω of the three-dimensional space. Here ρ_f is the density and ν the kinematic viscosity of the air. The quantity $1/\nu$ is called the Reynolds number Re. The system of equation is completed by boundary and initial conditions.

In order to manipulate the flow by blowing out or sucking in air at a point on the airfoil as shown in Fig. 1, this leads to time-varying boundary conditions. On the basis of this model, in the simulation one can use the intensity or the frequency of the blown air to increase the lift or to re-attach a detached flow, see Fig. 5 from [11].

For very fast flows with a lot of turbulence (this corresponds to Reynolds numbers of 100 000 or more), it is even on modern high performance computers not possible to simulate the flow with high-resolution methods like Direct Numerical Simulation (DNS) or Large Eddy Simulation (LES), not to mention an optimization of the flow on the basis of these methods. For a 3D simulation one would need a grid with 10 billion nodes (see, e.g. [6]) and would have to solve many nonlinear systems of equations of this size.

As an alternative to high-resolution methods, today one uses simplified methods like the Reynolds-averaged Navier-Stokes equations, typically coupled with statistic turbulence models. Due to the averaging, not every detail of the flow can be resolved, but the qualitative behavior of the flow can be determined. Such techniques are currently implemented in all academic or commercial CFD codes.

However, even these methods are, due to their long computing times, difficult to use in optimization or closed loop control. Realistic methods (see e.g. [11]) therefore use hybrid approaches, for which high resolution techniques such as direct numerical simulation are only used in small physically relevant parts of the domain (e.g. near the airfoil), while in the remaining domain averaged models or simplified methods are used. An example of such methods in flow control is shown in Fig. 6 from [11].

Another, mathematically motivated research approach that has high potential for control and optimization is the adaptive solution of the non-stationary flow equations. In this process the grid that is used for the flow simulation is continuously adapted to the behavior of the solution. By estimating the simulation error, a high resolution is obtained only where it is needed, see e.g. Fig. 7 from [5].

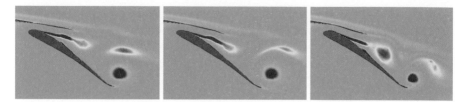

Fig. 5 Velocity field for different excitation frequencies

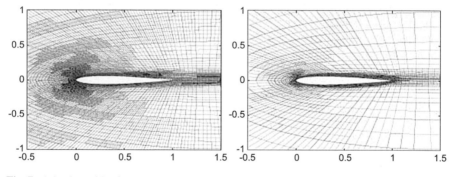

Fig. 6 Increase of lift via controlled blowing

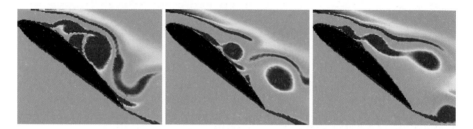

Fig. 7 Adaptive grid refinement

A third, currently very active research topic, is the approximation of the high dimensional mathematical model (e.g. the semi-discretized model with 10 billion equations) by a simplified low-dimensional model with only few, e.g. 10 equations, that capture the quantities that are relevant for control and optimization. These model reduction methods are not only used in flow control but are central research topics in almost all areas of science and technology [2, 4]. An important contribution of the mathematical research is here the development and analysis of new model reduction techniques, in particular the construction of provable error bounds.

With low dimensional mathematical models it is possible today to determine, in real time via pressure measurements on the surface, the occurrence of large eddies in the flow of two consecutive simple vehicle models, see Fig. 8. These eddies are responsible for irregular driving conditions and therefore for reduced driving comfort, see Fig. 9. In a second step it is then possible to actively manipulate the flow.

Fig. 8 Simplified vehicle model

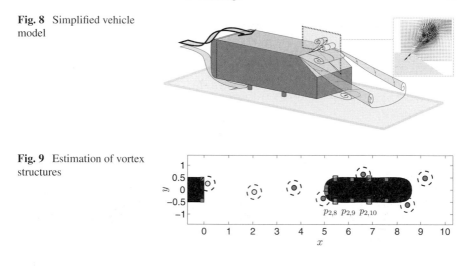

Fig. 9 Estimation of vortex structures

4 Analysis of Strengths and Weaknesses, Challenges

The modelling, simulation, optimization and control of complex flow processes is of central importance for the economical and ecological development. The necessary improvements can only be achieved with active procedures, that use model based control and optimization. The current models and the direct simulation methods as well as simplified hybrid methods cannot be applied in real time and can therefore not be used in active flow control.

Of central importance is, hence, the development of faster and more efficient methods for flow simulation that have the potential to be combined with model reduction methods. This is not possible in many of the current commercial CFD codes. Methods have to be developed that allow an adaptation of simulation algorithms with the goal of control and optimization, as they are discussed e.g. in [12].

Another approach is the further mathematical development of efficient model reduction methods that include error estimates, as well as the direct development of low-dimensional flow models that can be used in real time control and optimization.

At the same time it is necessary to develop new real-time control and optimization methods that work (on the basis of low-dimensional models) for real air-foils with engine nacelles and that are flexible to work with simulation and sensor data.

5 Visions and Recommendations

Active flow control is a prime example for the future scientific and technological development, in which the interdisciplinary cooperation for the solution of challenging problems has a direct impact on society as a whole. Mathematical research is a core component in this development.

An intensification of interdisciplinary basic research and the transfer of these results into industrial practice is essential. To meet these enormous challenges, it is necessary to educate young scientists that are able to work in interdisciplinary teams. This requires the establishment of interdisciplinary study programs (master/PhD) that fulfill these demands.

References

1. European Aeronautics: A Vision for 2020. Meeting Societies Needs and Winning Global Leadership. Report der europäischen Luftfahrtinitiative ACARE. Europäische Kommission (2001)
2. Antoulas, A.: Approximation of Large-Scale Dynamical Systems. SIAM, Philadelphia (2005)
3. Becker, R., King, R., Petz, R., Nitsche, W.: Adaptive closed loop separation control on a high-lift configuration using extremum seeking. AIAA J. **45**(6), 1382–1392 (2007)
4. Benner, P., Mehrmann, V., Sorensen, D.C. (eds.): Dimension Reduction of Large-Scale Systems. Lecture Notes in Computational Science and Engineering, vol. 45. Springer, Heidelberg (2005)
5. Bangerth, W., Hartmann, R., Kanschat, G.: deal.II—a general-purpose object-oriented finite element library. ACM Trans. Math. Softw. 33(4) (2007)
6. Griebel, M., Dornseifer, T., Neunhoeffer, T.: Numerical Simulation in Fluid Dynamics. SIAM, Philadelphia (1998)
7. Henning, L., Kuzmin, D., Mehrmann, V., Schmidt, M., Sokolov, A., Turek, S.: Flow control on the basis of FEATFLOW-MATLAB coupling. In: King, R. (ed.) Active Flow Control. Notes on Numerical Fluid Mechanics and Multidisciplinary Design, vol. 95, pp. 325–338. Springer, Berlin (2007)
8. Moeck, J.P., Bothien, M.R., Guyot, D., Paschereit, C.O.: Phase-shift control of combustion instability using (combined) secondary fuel injection and acoustic forcing. In: King, R. (ed.) Active Flow Control. Notes on Numerical Fluid Mechanics and Multidisciplinary Design, vol. 95, pp. 408–421. Springer, Berlin (2007)
9. King, R. (ed.): Active Flow Control. Notes on Numerical Fluid Mechanics and Multidisciplinary Design, vol. 95. Springer, Berlin (2007)
10. Petz, R., Nitsche, W.: Active separation control on the flap of a two-dimensional generic high-lift configuration. J. Aircraft **44**(3), 865–874 (2007)
11. Schatz, M., Günther, B., Thiele, F.: Computational investigation of separation for high lift airfoil flows. In: King, R. (ed.) Active Flow Control. Notes on Numerical Fluid Mechanics and Multidisciplinary Design, vol. 95, pp. 173–189. Springer, Berlin (2007)
12. Schmidt, M.: Systematic Discretization of Input/Output Maps And Other Contributions to the Control of Distributed Parameter Systems with a Focus on Fluid Control. Vdm Verlag Dr. Müller, Saarbrücken (2008)

Data Mining for the Category Management in the Retail Market

Jochen Garcke, Michael Griebel,
and Michael Thess

1 Executive Summary

Worldwide the retail market is under a severe competitive pressure. The retail trade in Germany in particular is internationally recognized as the most competitive market. To survive in this market most retailers use undirected mass marketing extensively. All prospective customers receive the same huge catalogues, countless advertising pamphlets, intrusive speaker announcements and flashy banner ads. In the end the customers are not only annoyed but the response rates of advertising campaigns are dropping for years. To avoid this, an individualization of mass marketing is recommended where customers receive individual offers specific to their needs. The objective is to offer the right customer at the right time for the right price the right product or content. This turns out to be primarily a mathematical problem concerning the areas of statistics, optimization, analysis and numerics. The arising problems of regression, clustering, and optimal control are typically of high dimensions and have huge amounts of data and therefore need new mathematical concepts and algorithms.

The underlying concept is the (semi-)automatic knowledge discovery via the analysis of huge databases, also known as data mining. The algorithmic core of the data mining process is called machine learning. This subject area originally belonged to computer science; the connection to statistics played a significant role

J. Garcke (✉)
TU Berlin, MA 3-3, Straße des 17. Juni 136, 10623 Berlin, Germany
e-mail: garcke@math.tu-berlin.de

M. Griebel
Universität Bonn, Institut für Numerische Simulation, Wegelerstr. 6, 53115 Bonn, Germany
e-mail: griebel@ins.uni-bonn.de

M. Thess
Prudsys AG, The Realtime Analytics Company, Zwickauer Straße 16, 09112 Chemnitz, Germany
e-mail: info@prudsys.de

M. Grötschel et al. (eds.), *Production Factor Mathematics*,
DOI 10.1007/978-3-642-11248-5_5, © Springer-Verlag Berlin Heidelberg 2010

from the beginnings. In recent years, further mathematical aspects were being considered especially in research, an example is the field of statistical learning theory. Algorithms with such a background are used successfully in many applications, not least due to their mathematical foundation.

The underlying assumption is that similar customer data signifies similar customer behaviour, this allows to assess new customers on the basis of the behaviour of former customers. Of fundamental importance is that many modern machine learning approaches use the representation of functions over high dimensional attribute spaces. This allows the coupled non-linear treatment of different attributes like income, debt, number of children, or type of car which results in an improved estimation of the likely customer behaviour. Approximation theory and numerics already play a substantial role for the development of new and the improvement of existing machine learning approaches. This will intensify in the future.

Nowadays the new numerical approaches for moderate high dimensional problems are advanced far enough that they can be used in first real live data mining applications for the category management in the retail market. Further research to extend the approaches to really high dimensions is necessary to allow the efficient treatment of even more attributes. Focused research programs should be setup to accelerate this development. With such additional support the development of new data mining methods could be pushed sufficiently far that even fully automatic interactive systems for the category management could be brought into maturity.

This article describes the role of mathematics for the category management in trade, in particular the role of approximation theory and numerics. Current state of art and success stories are outlined, and new developments and challenges are given.

2 Category Management in the Retail Market: Overview and Status Quo

The retail market is an especially dynamic one. This is traditionally due to the similarity in the offered products since all retailers have access to more or less the same range of products via their distributors. In the last years the internet allowed new business concepts and further intensified internationalization and increased competitive pressure. For the application of a typical data mining process much, mostly anonymous, data of the customer behaviour is available, which can be used for the optimization of the offers.

The problems arising in category management can be separated into four different areas:

- campaign optimization (i.e., selection of target groups and customers),
- cross- and up-selling (i.e., additional sales to customers),
- assortment optimization (i.e., product assortment and categories),
- price optimization (i.e., optimization of product prices and promotions).

A few years ago retailers started to apply mathematical approaches for the analysis of customer behaviour, but its use is sporadic and differs according to the line of business and the marketing activity. While the e-business shows a remarkable adoption of, arguably often too simple, algorithms, the mail order business uses mathematics to a large degree only for the optimization of mailing activities, that is the selection of customers with a high response probability to special offers. Last is the stationary retail market, but exactly here the current technological revolution of interactive digital shopping devices opens up new interesting possibilities for the development and application of new mathematical methods for the category management.

The high degree of customer interaction in retail is beneficial for the use of mathematical approaches since a large amount of customer data is available. At first the use of mathematics proved useful in some classical data mining fields. Here, it is very common to use classification algorithms for the optimization of mailings. Clustering methods for the segmentation of customers into thematic groups are increasingly successful. Other areas like real-time analysis and offers only apply the simplest methods. In the strategic field of management of commodity groups, that is the optimization of the range of products and their prices, the use of modern mathematical instruments is still the exception. But exactly here is, in combination with real-time approaches of optimal control, an important upcoming application area for the interdisciplinary cooperation of business, computer science and mathematics. Further information on data mining approaches in retail, marketing and customer relationship management can for example be found in [2].

In the following, we focus on the first two problems, i.e., campaign optimization and cross- and up-selling, and existing success stories of the use mathematical approaches in these fields.

2.1 Optimization of Campaigns

With regard to the use of mathematics the optimization of campaigns is the most advanced. The goal is to apply marketing campaigns with a clear focus on the target customers. This concerns both, the definition of the aims and procedure of the campaign, as well as the analysis of the results. One distinguishes here between target group (segmentation) and target customer (individualization). While target groups are strictly defined according to one or several attributes (for example female), target customers are selected based on an individual assessment in form of a numerical value, the score. An example for segmentation is the mailing of a catalogue of sporting goods only to customers interested in sports, i.e., those who bought sporting goods before. For the individualization on the other hand each customer is checked for affinity to this specific catalogue of sporting goods, independent of being part of the segment of sport affine customers. For segmentation mostly clustering algorithms are used, while for the individualization mainly classification and regression algorithms are applied. In the following we will discuss the case of mailing optimization in more detail.

Table 1 Profit calculation for a classical mailing campaign

Fixed costs	50 000 EUR	= 50 000 EUR
Costs for M	100 000 · 1.50 EUR	= 150 000 EUR
Costs for NF	98 500 · 1.50 EUR	= 147 750 EUR
Costs for IP	2 500 · 5 EUR	= 12 500 EUR
Total costs		= 360 250 EUR
Income	1000 · 500 EUR	= 500 000 EUR
Profit		= 139 750 EUR

Break-even-point: 721 responders

Success Story: Optimization of Mailings

For the mailing of catalogues mostly still all customers in the address list are selected, independent of their response probability. This is called unpersonalized mailing. The costs of such a mailing campaign consists of fixed costs (primarily creation and printing of the catalogue), the shipping costs (M), the follow up (NF), and the order processing (IP). Table 1 shows the calculation of the profit for an exemplary mailing campaign with 100 000 recipients, 1% response probability and 500 EUR income per responder. For simplification the revenue is considered profit.

To select only the customers with the highest response probability by means of data mining can increase the overall profit. The phases of such an optimized campaign are sketched in Fig. 1. The historical mailings for a catalogue are analysed based on the existing customer data. Models for the response probability are machine-learned based on the customer profile and then are evaluated on the 100 000 prospective recipients. This results in a list of scores out of which the 40 000 with the highest score (that is with the estimated largest response probability) are selected and sent the catalogue. If done correctly such a data mining approach can achieve more than twice the conventional response rate, here we assume 950 responders (ca. 2.4%).

The calculation of profit in Table 2 shows that, in comparison to the classical mailing campaign, in the end a higher profit of ca. 297 000 EUR instead of ca. 140 000 EUR is generated with less income. As an additional benefit customers overall receive less catalogues. Such personalized mailing campaigns are successfully used by several mail order companies.

Table 2 Profit calculation for an optimized mailing campaign

Fixed costs	50 000 EUR	= 50 000 EUR
Costs for M	40 000 · 1.50 EUR	= 60 000 EUR
Costs for NF	38 600 · 1.50 EUR	= 57 900 EUR
Costs for IP	2.100 · 5 EUR	= 10 500 EUR
Total costs		= 178 400 EUR
Income	950 · 500 EUR	= 475 000 EUR
Profit		= 296 600 EUR

Break-even-point: already after 357 responders!

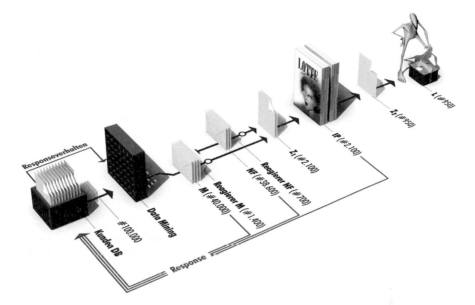

Fig. 1 The phases of an optimized mailing campaign

The optimization of a mailing campaign is based on the classification of customers using a model which is learned from existing customer data. An example for such a classification method is described in Fig. 2. Mathematically it is based on an approximate reconstruction and evaluation of functions over a high dimensional state space of customer attributes. Here, the method of sparse grids [5, 11] can be used for the approximation of such high dimensional functions. Alternatives are for example kernel based approaches with radial basis functions or neural networks. Many more classification algorithms exist in the literature, see [8] and the references cited therein.

2.2 Cross- and Up-Selling

As a second example for category management in the retail market we consider cross- and up-selling. Every salesperson knows that it is easier to sell additional products to an existing customer than to gain new customers. Cross- and up-selling addresses this core topic of increasing the customer value. The goal is to offer additional products (cross-selling) or higher valued products (up-selling) to existing customers based on their preferences which are indicated by their interests or former purchases. Besides the increase of revenue for the merchant, good cross- and up-selling also leads to higher satisfaction of the customer. Since the customer is receiving offers he is actually interested in he can save time and can avoid searching on his own.

Classification and regression with regulariza-
tion networks.

The optimization of a mailing campaign poses
a *classification problem* which mathematically
can be formulated as a *high dimensional ap-*
proximation problem.

The set $T = \{x_i \in \Re^d\}_{i=1}^M$ consists of M customers, who were recipients of a former mail-
ing campaign and which are characterized by
d attributes like sex, age, or profession. Furthermore, for each customer the target value
$y_i \in \{-1, +1\}$ is known, which indicates if the
customer responded to the campaign or not. The
goal is to reconstruct a function from the given
data which describes the likely relationship between attributes and the class label and therefore allows the prediction of the likelihood of
a response for new customers. An example of
such a function in two dimensions is shown in
the adjoining figure.

The solution process consists of two phases. In
the training phase the classifier is computed using the historical data, which describes the relationship between the customer attributes and
the response probability. In the evaluation phase
the learned model is evaluated for new customers.

To compute the classifier f a minimization
problem

$$\min_{f \in V} R(f)$$

is solved in a suitable function space V, here
R is an operator from V to $\Re$. For the most
advanced current methods this can be written
using the so-called regularization network approach [7]

$$R(f) = \frac{1}{M} \sum_{i=1}^M C\big(f(x_i, y_i)\big) + \lambda \phi(f)$$

where $C(x, y)$ is a cost functional measuring
the distance between the given data and the
classifier, e.g., $C(x, y) = (x - y)^2$, $\phi(f)$ is
the regularization operator, which describes a-
priori assumptions on f, e.g., $\phi(f) = \|\nabla f\|^2$,
and λ is a regularization parameter which balances these two terms.

Classification with a sparse grid function

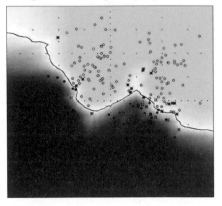

In the actual numerical computation a discrete
approximation of f is obtained. To cope with
the high number of dimensions for example the
sparse grid method can be employed [5, 11].
This approach scales only linearly with the
number of data, for details see [6]. The method
of sparse grids breaks the curse of dimension
of discetization approaches based on conventional grids due to its use of a truncated tensor
product multi-scale basis. Here, the number of
employed grid points still grows exponentially
with the number of dimensions but with a much
smaller basis.

In the evaluation phase the classifier f is applied on the data of new customers. The higher
the value of f, the so-called score, the higher
the likelihood of response. Therefore, the customers are sorted according to their estimated
score in descending order and the first n customers are chosen for the campaign. The choice
of a suitable n depends on the cost of the campaign and the expected revenue per customer.

Nowadays up to several million of customers
are analysed for mailing campaigns to achieve a
selection as good as possible. The use of classification approaches achieves up to 100% higher
returns. Note that response rates for traditional
mailings are typically below 1%.

Fig. 2 Classification with sparse grids for the optimization of mailing campaigns

Cross-selling starts with the disposition of the products into the market. This is traditionally the role of the category manager, although mathematical approaches are being used for several years as well. In particular, clustering approaches are used for basket analysis. These methods work transaction based and analyse for example cashier data with regard to cooperative sales. Products which are frequently bought together can therefore easily be placed near to each other in the store. Alternatively content based methods are used to analyse products and categories according to their attributes (colour, description, sound, . . .) and appropriate product clusters are formed.

In addition to the disposition of products the e-business brought new forms of interactive and automated cross-selling: recommendation engines and avatars lead the customer to related products and services. Well known and at the forefront is the online shop Amazon.com. While they are shopping, customers are presented with overviews of related products based on their current shopping basket and product searches ("customers who bought this product also bought . . . "). Although the early algorithms of Amazon.com were based on simple correlation analysis, they led the way for modern recommendation engines and adaptive analysis systems. Recommendation engines are nowadays established in e-business and are used in generalized forms for a wide range of applications like searches, matching, personalized pages, and dynamic navigation. At the same time it became the topic of academic research and meanwhile a large amount of publications exists.

Current methods range from clustering and text mining, Bayesian nets and neural nets up to complex hybrid solutions. Although the mathematical foundation of many approaches is still lacking, there is no doubt that currently an exciting research topic for applied mathematics is built here. In the following we discuss this example of dynamic programming for product recommendations in more detail.

Success Story: Product Recommendations

Recommendation engines nowadays play an important role for automated customer interaction. A recommendation engine offers, based on click and purchase behaviour of a customer, automatically related product recommendations. The recommendation engine learns online directly from the customer interaction. Recommendation engines increase the sales up to 20% and lead to enlarged customer satisfaction. But their application is not limited to this, modern recommendation engines vary design, product assortment and prices dependent on the user and allow totally new possibilities of personalization.

In stationary retail the use of automatic recommendation engines appeared until now technically infeasible, although interest exists since most buying decisions take place in the store. But change is on the horizon. In the first shopping malls electronic tools like the personal shopping assistant are available, a device which is placed on the shopping cart. Customers can access detailed information for a product from the shelf by using the scanner of the personal shopping assistant, the display then shows the corresponding information and additionally related product recommendations,

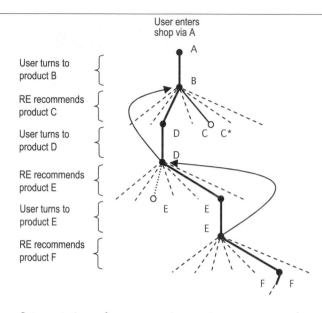

Interpretation of customer interaction as an interplay between recommendation engine and customer

Product recommendations as a reinforcement learning problem

The problems of adaptive product recommendations can be posed as a reinforcement learning problem which is mathematically based on the theory of optimal control. Besides theoretical studies already successfully commercial implementations exist. In reinforcement learning one considers a set of states s. For each state an action a from a set of actions can be chosen which leads to a new state s' with a scalar reward r. The sum of all rewards during a certain episode is to be maximized. In the simple case of a recommendation engine the states are the observed products, the actions are the recommended products and the reward corresponds to the price of a product in the event of a purchase. As long as the so-called Markov-property holds one can formulate a Markov decision process which is described by a discrete Bellman-equation:

$$V^\pi(s) = \sum_a \pi(s,a) \sum_{s'} P^a_{ss'}[R^a_{ss'} + \gamma V^\pi(s')]$$

with $P^a_{ss'}$ transaction probabilities
         $R^a_{ss'}$ transition reward
         γ discount rate.

Here $V^\pi(s)$ is the state value function, which assigns the expected cumulative discounted reward over the remaining episode to each state s. One now searches for the optimal policy $\pi(s,a)^*$, which describes the stochastic choice of the best action a at state s. There $\pi(s,a)^*$ gives the optimal recommendation.

Generalized methods from optimal control are used, in particular policy iteration and value iteration, to compute the optimal policy.

In many applications the state value function $V^\pi(s)$ cannot be described in tabular form due to the large number of states. In these cases it is approximated with regression approaches. Again, the efficient and approximate representation of high dimensional functions plays a significant role. Furthermore one has to distinguish between reinforcement learning approaches which know a model and use it, that is which know the transition probabilities $P^a_{ss'}$ (e.g. via regression). Here classical methods from optimal control can be used which allow the computation for several million of states in a few hours. On the other hand there are model-free approaches—in particular Monte Carlo and TD-methods—which can be used for online learning.

Fig. 3 Customer interaction using stochastic optimal control

Fig. 4 Product recommendations in a personal shopping assistant

see Fig. 4. Such systems allow for the first time fully automatic interaction with the customer in the store, for example in form of real time couponing on the receipt depending on the purchases or in form of dynamic price changes using electronic displays. This results in an interesting and broad new field for automatic analysis systems and recommendation engines for the generation of product recommendations and for price optimization.

A modern mathematical formulation of such adaptive product recommendations can be achieved with tools from stochastic optimal control and is discussed in Fig. 3. Here, a reinforcement learning formulation is used which leads to a discrete Hamilton-Jacobi-Bellman-equation. Again high dimensional state spaces arise for whose treatment often neural networks, kernel based approaches or decision trees are used [10]. The method of sparse grids can be applied here as well [9].

3 Outlook

An important aspect of successful data mining applications in retail is the approximate and efficient representation of high dimensional functions. Especially in the last decade significant improvements have been made to break the so-called curse of dimension [1], that is to develop and analyse methods whose complexity does not scale exponentially with the dimension of the underlying space. In addition to the sparse grid method [5] let us mention low-rank tensor product approaches [3, 4] as well as nonlinear approximation approaches like neural networks, kernel methods and LASSO [8]. Each technique has its own specific assumptions on the function and the data to be handled such that the curse of dimension can be avoided. Nevertheless, further research on numerical methods for high dimensional problems is absolutely necessary to improve the efficiency of the algorithms and to better understand their application possibilities.

A significant new perspective for the presented category management in the retail market is the management of the full customer life cycle. Until now, marketing activities like campaigns, cross- and up-selling as well as product assortment and price optimization are treated independently. Often such measures to increase the revenue are not sustainable. In an ideal situation an optimization over all five dimensions (customer, content, time, channel, price) of the whole problem should take place to maximize the customer value over the whole customer life cycle.

A mathematical contribution can contain a quantification and optimization of the value of each customer. A revenue calculation per customer would be possible which then would be maximized using stochastic optimal control or reinforcement learning. Such extended optimization problems require the representation of functions (the customer value) over a high-dimensional space (the state space). Here, the state space represents the customer, the product, the price as well as the time and form of marketing activities. A company can take actions and place suitable ads at given times for suitable products with attractive prices. Of special interest for the future are the development of new adaptive regression algorithms, for example to optimize campaigns in real time, and optimal control approaches, for example to optimize the revenue of the whole customer life cycle. Both these problems involve high dimensional approximation problems.

Finally, some critical remarks on the commercial use of personal data and sociodemographic information are in order. Banks, insurance companies, and, as shown, increasingly the retail industry use customer profiles and risk groups for decision making. Good conditions are only given to good customers. Growing databases, loyalty cards, and the trade of customer data increase the trend. Studies warn of an increasing across the board assessment of customer groups. There is the danger that some people have no access to certain services or products due to companies who exempt them according to a data mining model which is based only on the available data.

Recent examples are customers who have to pay higher credit interest rates based on their place of living, insurants who can not get an occupational disablement insurance only due to some unspecified former illnesses, or people who were not able to rent an apartment due to late payments of mobile phone bills and a resulting bad credit entry. To avoid such undesired developments a highly transparent data mining solution with explanations is necessary (why did this recommendation happen). Existing laws and practices on data privacy protection need to be critically revisited under these aspect and might need to be modified.

4 Visions and Suggested Actions

For the future it is quite possible that data mining approaches like analytic systems and recommendation engines for product and price optimization will be commonly used for a fully automatic interaction with customers in stores. The focus will be on interactivity and adaptivity, online learning will play an important role. Another trend is surely the management of the customer life cycle and the maximization of the customer life time value. To a large degree this would technically be possible nowadays, all data for the customer life cycle are in principle present somewhere in the company, but since they are not fully integrated in the IT-process they are typically not used this way. But the technical requirements for the use of data mining methods for the category management in the retail market are fulfilled. What needs to be done is a mathematical investigation of interactive and adaptive online learning methods to further their development. Until now mainly heuristic ad-hoc approaches exist. A mathematical formulation of adaptive product recommendations can be obtained with approaches from the field of stochastic optimal control via a reinforcement learning formulation which leads to a discrete Hamilton-Jacobi-Bellmann-equation. To solve this equation efficiently and fast new algorithms and methods especially for high dimensional problems need to be developed, and existing methods need to be substantially improved. This can only be achieved in an interdisciplinary cooperation of mathematicians, computer scientists and end users. For this task specific funding programs need to be set up which allow interested groups to be active in this field. In the priority programme 1324 "Mathematical methods for the extraction of quantifiable information from complex systems" of the Deutsche Forschungsgemeinschaft as well as in the programme "Mathematics for innovations in industry and services" of the German Bundesministerium für Bildung und Forschung first important steps are undertaken in practical as well as theoretical aspects. This needs to be continued and extended into the future. Besides institutionalized funding financial support from industry and retail is necessary, which until now happened infrequently and quite risk averse.

References

1. Bellman, R.: Dynamic Programming. Princeton University Press, Princeton (1957)
2. Berry, M.J.A., Linoff, G.S.: Data Mining Techniques: For Marketing, Sales, and Customer Relationship Management. Wiley, New York (2004)
3. Beylkin, G., Mohlenkamp, M.J.: Algorithms for numerical analysis in high dimensions. SIAM J. Sci. Comput. **26**, 2133–2159 (2005)
4. Börm, S., Grasedyck, L., Hackbusch, W.: Hierarchical Matrices. Lecture Note, vol. 21. Max Planck Institute for Mathematics in the Sciences, Leipzig (2003)
5. Bungartz, H.J., Griebel, M.: Sparse grids. Acta Numer. **13**, 147–269 (2004)
6. Garcke, J., Griebel, M., Thess, M.: Data mining with sparse grids. Computing **67**, 225–253 (2001)
7. Girosi, F., Jones, M., Poggio, T.: Regularization theory and neural network architectures. Neural Comput. **7**, 219–265 (1995)

8. Hastie, T., Tibshirani, R., Friedman, J.: The Elements of Statistical Learning. Springer, Berlin (2001)
9. Munos, R.: A study of reinforcement learning in the continuous case by the means of viscosity solutions. Mach. Learn. **40**(3), 265–299 (2000)
10. Sutton, R.S., Barto, A.G.: Reinforcement Learning: An Introduction. MIT Press, Cambridge (1998)
11. Zenger, C.: Sparse grids. In: Parallel Algorithms for Partial Differential Equations, Proceedings of the Sixth GAMM-Seminar, Kiel, 1990. Notes on Num. Fluid Mech., vol. 31, pp. 241–251. Vieweg, Weisbaden (1991)

Networks

Planning Problems in Public Transit

**Ralf Borndörfer, Martin Grötschel,
and Ulrich Jäger**

1 Executive Summary

Every day, millions of people are transported by buses, trains, and airplanes in Germany. *Public transit* (PT) is of major importance for the quality of life of individuals as well as the productivity of entire regions. Quality and efficiency of PT systems depend on the political *framework* (state-run, market oriented) and the suitability of the *infrastructure* (railway tracks, airport locations), the existing *level of service* (timetable, flight schedule), the use of adequate *technologies* (information, control, and booking systems), and the best possible deployment of *equipment and resources* (energy, vehicles, crews). The *decision, planning, and optimization problems* arising in this context are often gigantic and "scream" for mathematical support because of their complexity.

This article sketches the state and the relevance of *mathematics in planning and operating public transit*, describes today's challenges, and suggests a number of innovative actions.

The current contribution of mathematics to public transit is—depending on the transportation mode—of varying depth. *Air traffic* is already well supported by mathematics. *Bus traffic* made significant advances in recent years, while *rail traffic* still bears significant opportunities for improvements. In all areas of public transit, the existing *potentials* are far from being exhausted.

R. Borndörfer (✉)
Dres. Löbel, Borndörfer & Weider GbR, Churer Zeile 15, 12207 Berlin, Germany
e-mail: ralf.borndoerfer@lbn-berlin.de

M. Grötschel
Zuse Institute Berlin, Takustraße 7, 14195 Berlin, Germany
e-mail: groetschel@zib.de

U. Jäger
WSW mobil GmbH, Bromberger Straße 39-41, 42281 Wuppertal, Germany
e-mail: ulrich.jaeger@wsw-online.de

M. Grötschel et al. (eds.), *Production Factor Mathematics*,
DOI 10.1007/978-3-642-11248-5_6, © Springer-Verlag Berlin Heidelberg 2010

For some PT problems, such as vehicle and crew scheduling in bus and air traffic, excellent mathematical tools are not only available, but used in many places. In other areas, such as rolling stock rostering in rail traffic, the performance of the existing mathematical algorithms is not yet sufficient. Some topics are essentially untouched from a mathematical point of view; e.g., there are (except for air traffic) no network design or fare planning models of practical relevance. PT infrastructure construction is essentially devoid of mathematics, even though enormous capital investments are made in this area. These problems lead to questions that can only be tackled by engineers, economists, politicians, and mathematicians in a joint effort.

Among other things, the authors propose to investigate two specific topics, which can be addressed at short notice, are of fundamental importance not only for the area of traffic planning, should lead to a significant improvement in the collaboration of all involved parties, and, if successful, will be of real value for companies and customers:

- discrete optimal control: real-time re-planning of traffic systems in case of disruptions,
- model integration: service design in bus and rail traffic.

Work on these topics in interdisciplinary research projects could be funded by the German ministry of research and education (BMBF), the German ministry of economics (BMWi), or the German science foundation (DFG).

2 Success Stories

What good is mathematics in public transit? Three examples elucidate the benefits of mathematics for the customer, the planner, and the stakeholder.

Electronic Trip Planners

Thumbing through thick timetables and railway guides in order to determine the best connection in a bus, railway, or flight network is a matter of the past. Today, bus companies, railways, and airlines offer *electronic trip planners*, which provide this information via the Internet or via mobile phones in a comfortable and fast way, always up-to-date, and at no charge. To make this service work, correct and comprehensive data is needed first and foremost. The "intelligence" to utilize this data is provided by mathematics: good methods to compute *shortest paths in networks*.[1] Appropriate algorithms for this problem are know since the nineteen-fifties [17]. Their use in customer-friendly systems became a reality because of the rapid progress in information technology in recent years.

---

[1] Customers sometimes complain that they can find better or cheaper routes than a trip planner. This is, however, not a mathematical problem, but usually due to parameter settings such as "minimum transfer times".

Details	Date	Departure	Arrival	Duration	Changes	Products
		◄ sooner				
☐	16.06.10	15.:24	15:47	0:23	2	🚌 Ⓤ Ⓢ
☑	16.06.10	15:31	15:58	0:27	1	🚌 Ⓜ
☐	16.03.10	15:44	16:07	0:23	2	🚌 Ⓤ Ⓢ

Fig. 1 Berlin's "trip info" recommends a route

Examples for electronic trip planners are the "trip info" ("Fahrinfo") of Berlin's public transport company Berliner Verkehrsbetriebe (http://www.fahrinfo-berlin.de), see Fig. 1, the Hafas system, which is used by the German railway company Deutsche Bahn (http://reiseauskunft. bahn.de), and the flight search of Lufthansa (http://www.lufthansa.de). The basic method to compute shortest paths is *Dijkstra's algorithm*. This method has undergone many refinements and improvements over the years in order to deal with large networks and complex constraints, see [12].

Revenue Management

In the middle of 2010, Lufthansa offered flights to various destinations in Europe for 99 Euros, see Fig. 2, Air Berlin promotes flights to Paris as cheap as 29 Euros, and sometimes one can find tickets for less than 10 Euros. How do such prices come about? They are the result of sophisticated ticket sales strategies known as *revenue* or *yield management*. The idea is as follows. Once an airline has published a flight in its schedule, it is essentially clear what the costs will be. The goal is then to maximize the revenue. There were originally two strategies to do this: the classical carriers charged high prices (and had empty airplanes on certain flights), while the low cost carriers filled their airplanes solely with cheap tickets. Today, all airlines pursue (depending on the company) different mixtures of these strategies. The clou about

Lufthansa betterFly

Fliegen Sie mit Lufthansa günstig in die schönsten Städte Europas.

→ ab Berlin	ab 99 €*
→ ab Düsseldorf	ab 99 €*
→ ab Frankfurt	ab 99 €*
→ ab Hamburg	ab 99 €*
→ ab München	ab 99 €*
→ ab Stuttgart	ab 99 €*
→ Alle Ziele	ab 99 €*

*inkl. 10 € Lufthansa Ticket Service Charge

Fig. 2 Flight bargains in the middle of 2010 (sources: http://www.lufthansa.com/online/portal/ lh/de/specials, 16 June 2010; http://www.airberlin.com, 16 June 2010)

it is in the permanent adjustment of the booking classes and prices to the demand that has already materialized and the demand that is yet expected. Many airlines use mathematical methods of *stochastic optimization* to do these adjustments. On the basis of such forecasts, it can be reasonable to sell, at certain points in time, residual capacities at very low prices, such that at least some revenue is generated instead of flying empty seats.

The above described and at present commonly used form of revenue management was developed around 1990. In this context, the famous competition between American Airlines and the low cost carrier PeopleExpress is often mentioned, because AA finally won the fight by introducing "Super Saver" and "Ultimate Super Saver" tickets, which were sold using yield management methods. On the occasion of the bestowal of the INFORMS Edelman Award in 1991, AA provided evidence that revenue management created an additional revenue of 1.4 billion USD in the period from 1988 to 1991 [32]. After additional improvements, a benefit of even 1 billion USD per year was reported [9]. The most popular revenue management method is the *EMSR rule* (expected marginal seat revenue), which states that one should sell tickets of some booking class for a flight as long as the expected profit is positive [26]. Starting from this basic form, researchers and practitioners have developed a large variety of methods to control ticket sales, ranging from the consideration of individual flights ("leg control"), via the inclusion of simple network effects ("segment control"), to the treatment of entire itineraries ("origin destination control"), see [34] for a recent overview of the state of the art in this area of research in stochastic optimization.

Vehicle and Crew Scheduling

Figure 3 illustrates the development of Berlin's public transit company BVG over recent years by means of three statistics: the number of employees descended from 27 002 in 1991 to 10 982 in 2006, labor costs fell in the same period of time from 734 mil. Euros to 485 mil. Euros, and state subsidies sank from 762 mil. Euros in 1993 to 318 mil. Euros in 2006 [33]. These remarkable reductions could be achieved

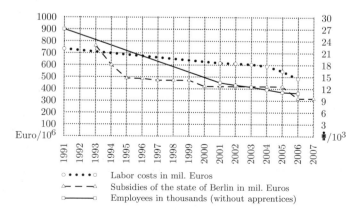

Fig. 3 Berlin's public transit company BVG in numbers (source: [33])

without changing the level of service. They are not only, but to a significant extent, results of mathematical optimization. As Andreas Sturmowski, CEO of the BVG, put it [33]: "The use if IT-based planning systems allows for significant improvements in the planning processes of the BVG! Resource allocation is optimized. Increases in the productivity of vehicle and crew utilization [are achieved] by minimizing deadhead trips, [...], better use of depot capacities, [..., and] reductions in staff requirements by optimized duty schedules [...]." (translation by the authors). All this is made possible by methods of *combinatorial* and *mixed integer optimization*, which can deal with the enormous problem sizes that are typical for this area. This "computing power" leads to both significant speed-ups of and to quality improvement in the planning process. Customers do not directly notice this progress; however, the indirect impact on ticket prices and state subsidies is significant.

Mathematical methods for *vehicle* and *crew scheduling* are today sold as standard modules in the market leading planning systems. In the area of public transit, such systems are, among others, ivu.plan (formerly MICROBUS 2) of IVU Traffic Technologies AG (http://www.ivu.de) from Berlin, HASTUS of the Canadian company Giro Inc. (http://www.giro.ca), and Turni by the Italian company Double-Click (http://www.turni.it/page001.htm), in air traffic the systems NetLine by Lufthansa Systems GmbH (http://www.lhsystems.com) and Carmen of the US Jeppesen group (http://www.carmen.se), in rail traffic the system railRMS by Jeppesen or the optimization modules of the US company Innovative Scheduling, Inc. (http://www.innovativescheduling.com). The state of the art of mathematical research is documented by the proceedings of the tri-annual international CASPT and the German Heureka conferences [10, 11, 14, 18, 35, 36].

3 PT Planning Problems: Survey and Status Quo

Public transit (PT) is of high relevance for every society. Table 1 illustrates at the example of some statistics on Berlin's public transit company (BVG), the German railway company Deutsche Bahn (DB), and the German airline Lufthansa (LH) for the business year 2006, that PT is also an important sector of the economy.

In the future, the importance of public transit will increase because of the surging costs of individual traffic. In Germany, in the summer of 2008, a single tankful of gasoline cost about 75 Euros, which is nearly the price of a monthly BVG public

Table 1 Public transit 2006 in Germany; sources: [4, 15, 16]. The respective figures must be interpreted, because some of them are derived in different ways. The BVG, e.g., reports for the number of employees the "average number of employees over the year", including apprentices. The figures of DB and LH include, among other things, cargo traffic

Company	pkm/year	Employees	Turnover/Euros	Profit/Euros
BVG	4.074 bil.	12 685	0.636 bil.	23 mil.
DB	74.788 bil.	229 200	30.053 bil.	1680 mil.
LH	110.330 bil.	93 541	19.849 bil.	845 mil.

transit ticket for the tariff zone AB, which contains the entire area of the state of Berlin (about 40 square kilometers).

Public transit can be subdivided into the following three areas:

- public mass transit, i.e., bus, tram, subway, and commuter trains (BT),
- long and short distance passenger rail traffic (RT),
- civilian passenger air traffic (AT).

We will denote these in the following shortly as *bus*, *rail*, and *air traffic*. The planning problems appearing in these areas can be classified from a business point of view as follows:

- scheduling (e.g., vehicle and crew scheduling)
- control (e.g., delay and disruption management)
- (network) design (e.g., infrastructure, timetable)
- regulation of competition (e.g., tenders).

Mathematical methods are already used in all these fields. However, the maturity of the technology and the penetration of practice is different. In *scheduling*, mathematical optimization methods are well established as an industry standard, the mathematical *control* and *design* of public transit systems is (with the notable exception of air traffic) in a research state, while the mathematics of PT *regulation* is still in its infancy. There are also differences depending on the transportation mode. One reason is that the planning problems in bus, rail, and air traffic may be similar, but they are not identical. Technical and organizational conditions as well as the market environment influence the structure of the planning process, as well as the definition and the characteristics of the individual tasks, which in turn has its effects on mathematical solution approaches. At present, the use of mathematics is most advanced in air traffic, and least common in rail traffic.

We survey in the following the state of the most important developments.

3.1 Scheduling

Scheduling determines—on the basis of a fixed timetable—the use of vehicles and crews. Operating a bus (including general overhead) costs approximately 250 000 Euros per year in Germany. About 70% of this amount are crew costs, and 25% vehicle costs. In rail and air traffic, the vehicle costs take a larger share of the total costs, in particular, because of higher fuel expenses.

The problems to plan vehicles and crews can, in their basic forms, be formulated in terms of standard models of combinatorial optimization. One can view them as *multi-commodity flow problems* or as *path covering problems*. These models are well studied and can be solved in dimensions which make a treatment of realistic and relevant scenarios possible at a high level of detail. However, technical constraints, legal prescriptions, and company agreements increase the complexity of the problems to such an extent that specialized methods have to be developed. These have

Fig. 4 Bus scheduling with
MICROBUS 2 (source [28])

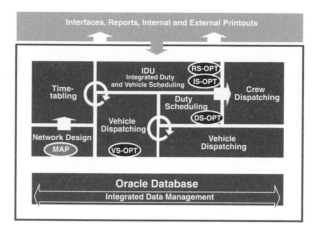

reached a mature state and are today (as already mentioned) offered by software
companies all over the world.

The particular structure of the planning process and its subdivision into individual problems depends on the perception of the operation and differs by countries,
transportation modes, and software companies. Figure 4 illustrates the view of the
software company IVU Traffic Technologies AG on public transit. The planning
system MICROBUS 2 can be used to schedule, on the basis of a given network
and timetable, vehicle rotations and crew duties for a day of operation, which are
subsequently concatenated to vehicle and duty rosters. These individual steps are
supported by optimization tools, which are marked in red. We give in the following
a synopsis and explain the problems both from an applied and from a mathematical
point of view.

Vehicle Scheduling

The first successful applications of mathematical methods in public transit dealt
with *bus scheduling*. This planning step is about the construction of sequences
of timetabled and deadhead trips, the so-called vehicle *rotations*, such that every
"timetabled trip" (a trip listed in the timetable) is operated by a suitable vehicle. The
planning horizon is usually a single day of operation. A bus departs from its depot
in the morning and returns in the evening. The objective is to service all timetabled
trips and to minimize the corresponding costs (of driving and waiting). This task can
be formulated mathematically as a *multi-commodity flow problem*, see Fig. 5. In the
case of Berlin's public transport company BVG, this leads to an optimization problem with 100 million variables, which can be solved by modern algorithms in less
than one hour on a standard desktop computer to proven fleet optimality [27]. Using
this method, BVG reports savings of 38 buses in a single depot (Spandau) in 2003,
i.e., about 20% of the vehicles, and of 377 hours of unproductive waiting time [30].

In air traffic, vehicle scheduling is usually subdivided into two steps, namely, *fleet
assignment* and *tail assignment*. Fleet assignment decides which type of aircraft is
used to operate a specific flight. In the subsequent tail assignment step, rotations for

Vehicle scheduling graph:

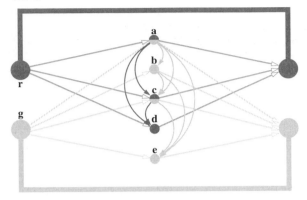

The vehicle scheduling problem in public transit can be mathematically modeled as a *multicommodity flow problem*. The model is based on a scheduling graph, which connects the timetabled trips by all possible deadhead trips. Vehicles of different types "flow" through this graph in such a way that every timetabled trip is covered by a rotation of a vehicle of a suitable type. Formally, the model can be described as follows. We denote by $\mathcal{D}$ a set of vehicle types, by $\mathcal{T}$ a set of timetabled trips, by A^d a set of arcs for all possible deadhead trips for every vehicle type d, and by κ^d the number of available vehicles of type d. We introduce a 0/1 variable x_{ij}^d for every arc (i, j) and every vehicle type d. x_{ij}^d encodes if a deadhead trip (i, j) is operated by a vehicle of d (in which case $x_{ij}^d = 1$) or not ($x_{ij}^d = 0$). The corresponding formulation as an integer linear program is:

$$\min \quad \sum_{d,(i,j)} c_{ij}^d x_{ij}^d$$

$$\sum_{d,(t,j)} x_{tj}^d = 1, \qquad \forall t \in \mathcal{T},$$

$$\sum_{d,(t,j)} x_{tj}^d - \sum_{d,(i,t)} x_{it}^d = 0, \quad \forall t \in \mathcal{T}, \ d \in \mathcal{D},$$

$$\sum_{(d,j)} x_{dj}^d \leq \kappa^d, \qquad \forall d \in \mathcal{D},$$

$$x \geq 0 \text{ and integer.}$$

Berlin's public transit company BVG distinguishes about 50 different types of vehicles, 28 000 timetabled trips, and approximately 100 million deadhead trips. This translates via the above model into a multi-commodity flow problem with about 100 million variables and several hundred thousands of equations and inequalities.

Fig. 5 Vehicle scheduling in public transit (see [27])

the individual airplanes of each fleet are constructed, including necessary maintenance activities. There are two reasons for this subdivision. First, operation costs for large and small airplanes differ much more than for buses, such that fleet assignment is much closer related to network design than bus scheduling to bus network design. Second, the planning horizon of the tasks is different (tail assignment is closer to the day of operation). Fleet assignment can be modeled as a multi-commodity flow problem, similar to bus scheduling, while tail assignment leads to path covering problems that are solved using *column generation methods*, see Fig. 6. Today, optimal vehicle rotations for an entire bus company or airline are computed routinely.

Vehicle scheduling in rail traffic features additional technical constraints, which depend, among other things, on the infrastructure at railway and shunting stations and on the driving dynamics of individual trains. Such constraints increase the complexity of the associated models substantially. At present, the mathematical methodology is not good enough to deal with problems of relevant sizes. Vehicle scheduling in rail traffic is currently to a wide extent still a matter of manual planning and heuristic methods.

Crew Scheduling

Crew scheduling is commonly subdivided into two consecutive steps: *crew/duty scheduling* and *crew/duty rostering*. Crew scheduling is about the construction of a number of crew rotations, the so-called *duties* (in bus traffic) or *pairings* (in air traffic), which are not yet assigned to specific crews; this step accounts for legal regulations and aims at cost minimization. *Crew rostering* subsequently concatenates anonymous duties to longer rosters over some planning horizon and assigns them to concrete persons; here, fairness considerations and employee preferences are taken into account.

Airline crew scheduling is the area where the first big successes in mathematical crew scheduling were achieved. Labor costs of large airlines range in the billions and are, next to fuel costs, the second largest individual cost factor [1]. At the beginning of the nineteen-nineties, it became possible for the first time to solve large crew scheduling problems with thousands of flights to proven optimality [21]. Since then, optimization technology has made rapid progress and is now established as an industry standard. Today, all major airlines use mathematical "crew optimizers" [37]. Methodologically, these systems are based on *set partitioning models*, which can deal with the complex rules for breaks, rest periods, etc. These models are solved by *column generation methods*, see Fig. 6.

These methods have been transfered with some delay to bus and in some places also to rail traffic. However, the problems in these areas have special combinatorics of their own, which, among other things, are caused by different degrees of freedom for crew reliefs (e.g., in bus traffic one changes crews "on the fly", i.e., during a trip, which is obviously impossible in airline crew scheduling); on the other hand, duty schedules in bus traffic are constructed for a single day of operation, while airline crew scheduling is typically done for a period of several weeks or a month.

Summarizing, crew scheduling is the area in which the use of mathematical methods in public transit is most advanced. Occasionally this is criticized, as documented, e.g., by the headline of the Atlanta Constitution of August 13, 1994: "Delta to furlough 101 more pilots in bid to cut costs". But such criticism has no bearing on the mathematical methodology. Indeed, it is no problem to optimize for social criteria. Managers, however, usually focus on cost reductions.

3.2 Control

Figure 7 illustrates that controlling the implementation of a schedule is as important as constructing it. Bad weather, accidents, technical damages, and strikes can cause

Duty scheduling graph in public transit with morning and afternoon peaks:

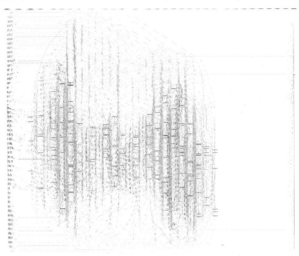

The duty scheduling problem in public transit can mathematically be understood as a *path covering problem*. Similar to vehicle scheduling, one constructs a scheduling graph, in which indivisible units of (driving) work, the so-called *duty elements*, are linked by all possible connections. Driver duties correspond to "paths" in this scheduling graph, and every duty element has to be covered by such a duty. The difference to vehicle scheduling is that complex rules for the feasibility of duties such as break rules have to be observed, which depend on the form of the duty in its entirety and which can not be decided locally. Consequently, a mathematical duty construction can no longer efficiently be done in terms of variables for individual concatenations of duty elements, but must introduce variables for entire duties, which makes these models larger and more difficult to solve. Following this line of thought leads to a formally very simple model, a so-called *set partitioning problem*. It considers for some set of duty elements I the set of all possible duties J. For every such duty j, a 0/1 variable x_j is introduced. It indicates whether duty j is included in the duty schedule (in which case $x_j = 1$) or not ($x_j = 0$). For every duty element, a constraint stipulates that exactly one duty covering this element must be chosen. For large, industrial problems it is not possible to handle this formulation explicitly, because there can easily be billions of duties even for small problems with only 200 trips (duty elements) [22]. However, such models can be solved implicitly using so-called *column generation methods*, which generate (in a mathematically precise sense) at any time the potentially interesting duties in a dynamic way [3, 13].

The corresponding formulation as an integer linear program is

$$\min \quad \sum_{j \in J} c_j x_j$$
$$\sum_{j \ni i} x_j = 1, \quad \forall i \in I$$
$$x \geq 0 \text{ and integer.}$$

Nowadays duty scheduling problems in bus and air traffic with several hundred or even thousands of crews are solved on a routine basis [2, 6].

Fig. 6 Duty scheduling in public transit [6]

disruptions. For example, the Chicago Tribune reports on December 27, 2007: "According to its pilots union ..., United has run low on crews to fly its planes. That's a result of lean staffing, scheduling practices and freakishly bad weather that have

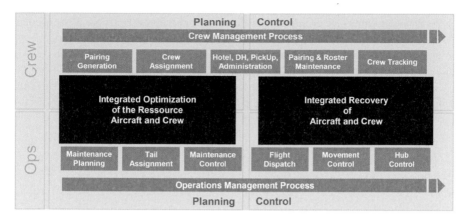

Fig. 7 Airline planning and control with NetLine (source [8])

caused large numbers of pilots to hit the maximum number of monthly duty hours allowed by federal regulators well before December's end." There are different possibilities to face such difficulties.

Stochastic and Robust Optimization

A theoretically satisfactory possibility is to use methods of stochastic optimization, that try to anticipate disruptions in a probabilistic way. A main obstacle to the application of stochastic methods, however, is the limited availability of expected values and probability distributions in practice. For this reason, a simplified variant of stochastic optimization, the so-called *robust optimization* approach, has been investigated in recent years. It only accounts for certain uncertainty intervals in the data by means of a worst case analysis. This method has not made it into practice yet. Real-world planning systems take some precaution by introducing buffers at critical points.

Online Optimization

Stochastic and robust optimization pave the way to overcome disruptions by instantaneous re-planning. Re-planning itself is a topic of *online* resp. *real-time optimization*, a method that searches for recovery operations that lead from a disrupted state back to a stable state as fast as possible and, in this way, reduce the damages caused by disruptions as much as possible. Online optimization methods are, e.g., currently successfully used for dispatching the "yellow angel" service vehicles of the German automobile club ADAC [20]. In principle, bus and rail traffic also have the necessary IT infrastructure at their disposal. However, in these areas disruption management is usually done by experienced dispatchers. In air traffic, first successes with mathematical methods are reported, e.g., at the American carrier Continental Airlines [38]. The software companies have started to take up the topic.

Inadequate control leads not only to customer dissatisfaction, but also to enormous costs. For example, airlines must pay passenger overnights and arrange ferry and extra flights, bus and railway companies do not get their full payment in case of quality impairments.

Robust and real-time optimization are new and topical branches of research, which can only have an impact if a close cooperation with practice can be established.

3.3 Service Design

Service design can be subdivided into three important areas: infrastructure construction, vehicle procurement, and line, timetable, and fare planning.

Infrastructure Construction

The arguably most important topic of planning in public transit are decisions about investments in infrastructure such as the construction of airports, railway tracks, railway stations, or bus depots. Infrastructure investment costs are huge. The new ICE (intercity express = high speed train) line from Cologne to Frankfurt, e.g., cost more than 5 bil. Euros, the construction of the new Berlin airport BBI is estimated at 2–5 bil. Euros. Once taken, decisions about infrastructure constructions can only be changed over a very long period of time. It is therefore necessary to conduct an extremely careful analysis of all possible alternatives.

The decision methodology that is currently used in infrastructure construction is based on the simulation of a few scenarios in the sense of case studies (usually accompanied by a super-charged political discussion). This method is not adequate to deal with projects that interact in a complex way with the entire transportation infrastructure. In rail traffic, to mention one example, there is currently a discussion in Germany whether the construction of new, expensive high speed tracks produces the biggest benefit or if a bundle of many small investments would be more useful. Infrastructure investments in bus traffic focus on the selection of sites for bus depots, a topic that is currently particularly important because of merger activities. Here, network optimization computations are necessary. To the best of our knowledge, however, this is not done to an extent that would be worth mentioning. From a mathematical point of view, problems of similar structure come up in telecommunications. The German research network association (DFN-Verein), an organization that operates the research Internet in Germany, e.g., has undertaken a simultaneous site and network optimization for the German science network (currently X-WIN) on a regular basis for many years [5]. The methods that have been developed in this context could be a starting point for mathematical approaches to infrastructure design in public transit.

Vehicle Procurement

The procurement of airplanes, trains, buses, etc. constitutes a major cost factor in public transit. An A380 passenger plane costs more than 200 mil. Euros, an ICE3 high speed train more than 20 mil. Euros, a 25 m articulated bus 500 000 Euros. The German railway company Deutsche Bahn operates 236 ICE-train units [15], Berlin's public transport company BVG more than 1300 buses [4], the German airline Lufthansa 411 passenger planes of various types [16]. The assessment of the fleet size that is necessary to operate a flight, rail, or bus network is sometimes still done using simple rules of thumb [24, 31], such as

$$\left\lceil \frac{\text{length of line} \times \text{frequency}}{\text{average speed}} \right\rceil + \text{reserve};$$

sometimes, instruments of operational scheduling are also employed. Usually, decisions are taken as a compromise on the political level, on which mathematical considerations do not always play an important role.

Line, Timetable, and Fare Planning

Line planning and timetabling deal with the specification of subway, tram, and bus line routes, travel frequencies, and the concrete departure and arrival times of the vehicles. These two planning steps belong arguably together, at present, however, they are carried out sequentially, mainly because the mathematical methods for a simultaneous treatment are not available. For both problems, research codes exist. In line planning, one can deal with medium sized instances, see, e.g., [7]. Somewhat more advanced is periodic timetable optimization, which was used, e.g., to compute the timetable of the Berlin subway [25]. All these approaches act on the assumption of a given, fixed demand.

Fare planning investigates how the demand for public transit changes in response to a change in ticket prices. To this purpose, one must model user demand patterns and, in this way, forecast the demand subject to varying prices. Very simple demand models are based on price elasticities. The problem with these models is that elasticities have only been empirically estimated in isolated places (at certain times and places) and with rather different results. Constant price elasticities can not be used to predict substitution effects. More modern models, based on discrete choice approaches, have been developed in the meantime and can now be solved to a moderate extent [29], but they have not been applied in practice. Air traffic is an exception. Here the already discussed revenue management methods are used with great success.

The long term goal must be to deal with line, timetable, and fare planning in an integrated way, because all these problems are intimately related to each other. However, as the individual problems are not completely mastered yet, we are still far away from an optimization at a system-wide level.

3.4 Regulation

Service design, as discussed in the previous section, is based on the current regulatory framework. All over the world, there is a discussion how far the public transit market can be deregulated and in what way competition can be introduced into this system. The deregulation in air traffic has led to a veritable boom. There was an enormous reduction in ticket prices and a surge in passenger volume. Similar effects are hoped for rail and bus traffic, however, it is not clear what the right deregulation recipes are. Hitherto experiences in different parts of the world produce a mixed picture. The deregulation of the railway sector in Great Britain was hailed as a major breakthrough in the beginning, then led to a disaster, and now, after changes in the general conditions, a success is beginning to show. Another example is the redesign of the bus transport network in Santiago de Chile, which took place in spring 2007. It resulted in a complete breakdown of the metropolitan traffic that nearly forced the government of Chile to resign. The mathematical analysis of regulatory measures can help to identify and prevent undesirable consequences of changes in the design of a traffic system.

An example of how this could look like in practice is given by the German "railway infrastructure usage regulation" ("Eisenbahn-Infrastruktur-Benutzungs-verordnung", EIBV), that is supposed to set the rules for a competitive access to the German railway network. The EIBV states in § 9 paragraph 5: "In the decision process ... the operator of the railway system must compare the fees for the controversial slots and ... in case of a conflict between more than two slots he must give precedence to those slots whose sum of fees is collectively maximal." (translation by the authors). This is an optimization decree. It obliges the German railway network operator DB Netz AG to solve a *track allocation problem*, and the German network regulation agency Bundesnetzagentur to make sure that an optimal solution is determined. Without mathematical optimization, a non-discriminatory access in the sense of this regulation is not conceivable. This is "overlooked" by all involved parties, usually with the justification that negotiated solutions are better suited. Basically very similar problems arise in airport slot allocation in air traffic and in the tendering of bus lines. These topics lead from a mathematical point of view to questions in combinatorial auction theory and in algorithmic game theory, which are at present debated in the economic research community and for which several Nobel prizes in economics, the last one in 2007, have been awarded.

There are by now successful applications of this methodology in the tendering of transportation contracts in logistics [23] and firms have emerged which market the corresponding auctioning platforms (combine.net). In public transport, however, such methods have not been applied yet.

4 Strengths, Weaknesses, and Challenges

Section 3 described planning problems in public transit from an operator's point of view. This section offers a synopsis from a mathematical perspective, elucidates strengths and weaknesses, and describes some of the current challenges.

4.1 General Conditions for the Use of Mathematics

Centralized Forms of Organization

The public transport modes bus, rail, and air traffic are characterized by relatively centralized forms of organization. Except for the area of operations control, there is sufficient lead time for all planning problems, such that it is possible to compute alternatives and variants. These are in principle excellent premises for the use of mathematical planning methods.

Availability of Data and Information Technology

In order to use optimization methods, detailed and precise data is required. Progress in information technology makes data available in ever larger quantity and in ever increasing quality. Radio data transmission, GPS, and electronic ticketing systems are abundant sources of information about user behavior and the state of operation of a public transit company. Far-sighted planning, however, needs considerable amounts of additional data. Vehicle scheduling, e.g., requires thousands of lengths of alternative deadhead trips, which can be used potentially. These so-called degrees of freedom are not automatically available, at least not in sufficient quality, and their procurement is often a major effort during the introduction of optimization systems. In some areas obtaining the necessary data is difficult in principle. In fare and service design in public transport it would actually be necessary to forecast user behavior in the presence of changes of prices or changes of the service level. Here, one usually uses statistical origin-destination matrices. The occasionally used price elasticities are based on fairly unreliable data. In these areas, a trustworthy data basis must be created in order to develop accurate methods of service design.

Public transit as well as the associated planning methods benefit hugely from the breathtaking developments in information technology. Cellular phones, radio data transmission, etc. made the use of quantitative, mathematical planning methods possible.

Complexity

Small public transit companies may very well produce good results using manual planning under constant general conditions. But even there, quality and efficiency requirements are increasing. For large public transit companies with thousands of buses, tens of thousands of employees, and countless connections, planning by "peering", by analyzing Excel sheets, or by arguing in terms of individual aspects is neither the state of the art nor best practice. The same holds for expensive investments in infrastructure constructions, which change the logistics of entire regions. It is again and again striking to see that airports are constructed without connections to the subway or commuter train system. Such mistakes are due to the lack

§8 Flight times of crew members

(1) The unrestricted flight time of every crew member between two rest periods is 10 hours. It is feasible to extend the flight time according to sentence 1 up to 4 times for up to 4 hours a time in 7 consecutive days; here, the sum of the extensions must not exceed 8 hours in any 7 consecutive days. The period of 7 consecutive days starts at any one time at 00.00 Greenwich Mean Time (GMT) of the first day and ends at 24.00 GMT of the seventh day. For a pilot, who acts in his flight time according to sentence 1 as an aviator all or in parts without support by another crew member, paragraphs 2 and 3 do not apply.

(2) The maximum flight time extension for crew members of 4 hours according to paragraph 1 is diminished

1. by 1 hour, if the flight time is longer than 2, but less than 4 hours,
2. by 2 hours, if the flight time has an intersection of 4 or more hours with the time window between 01.00 and 07.00 local time of the departure airport (winter time).

(3) A diminished flight time extension according to paragraph 2 is further diminished

1. by another hour, in case of more than 3, but less than 6 landings,
2. by another 2 hours, in case of more than 5 landings.

(4) If the aircraft crew is larger than the mandatory minimum, and if suitable sleeping accommodation is available on board in a compartment that is separated from the cockpit and the cabin, the regulation authority can accept a request in writing to extend the flight time according to paragraph 1, sentence 1, of up to 8 hours for two times within 7 consecutive days. Here, the time that a crew members spends in aviation and operation of the aircraft must not exceed 12 hours.

Fig. 8 From the 2nd executive order on work rules for aircraft crews (2. Durchführungsverordnung zur Betriebsordnung für Luftfahrtgerät (2. DVLuftBO), translation by the authors)

of a "global point of view" and the fact that network effects of individual decisions are not understood. Infrastructure policy decisions should actually be embedded in comprehensive network models, that describe the long term transportation demands of a region, and that, in particular, take inter-modal effects into account. Planning activities of this type are currently primarily done by engineering companies from the construction industry, that know all about the technical aspects, but that do not command the mathematical methods that are necessary to master the complex dependencies. Furthermore, planning is complicated by government regulations, company agreements, and other rules and standards, which are not written with an eye on the consequences that an increase in complexity incurs on the subsequent planning process, see Fig. 8 for a typical example.

Public transit planning problems are highly complex. Mathematics can not eliminate complexity, but it offers methods to cope with it. Indeed one can prove that the planning problems that come up in public transit are in a mathematically precise sense difficult ($\mathcal{NP}$-hard [19]). Even if mathematical methods may not necessarily be able to produce provably optimal solutions, they can often compute the deviation from optimality or similar quality guarantees, something that no other methodology provides. In general parlance: Nothing works without mathematics.

It is alarming that this insight is not yet realized by all public transit companies. In many cases manual planning is still the method of choice, even in large enterprises.

Really astounding, however, is that even some engineering consultancies and software developers in this area argue that optimization would only deal with special aspects, while manual planning would consider the problem in its entirety—a completely absurd claim from our point of view. Actually, the real goal must be to master the complexity of PT systems by combining traffic engineering know-how with the mathematical methodology. This is the only way to face the grand challenges in integrated planning and in the design of entire transportation systems. Investments in mathematical decision methods are marginal in comparison to investments in infrastructure construction, vehicles, and crews, the effects, however, that one can achieve in this way, are considerable.

Standardization

Many public transit companies tend to view themselves as unique and believe that the problems arising in their case are so special, that they need particular attention. In addition, the wish list that planners, managers, and work council members present to software companies is virtually unlimited. This type of atomization of problems is a major obstacle to the further development of the planning methodology. Only if planning problems are standardized and if data is available in standard formats, one can start a serious and sustainable investigation of the underlying mathematical problems over a longer period of time, analyze their mathematical structure, and develop special purpose methods, such that one can hope for practical solutions even for the problem dimensions in public transit. Areas that succeeded in standardization have enjoyed the development of powerful methods over time. The question about the "correct" definition of the individual planning problems is often controversial. However, the benefits of standardization are usually so large that it is better for a company to attune its processes, than to work with tailor-made individual solutions, which do not evolve at the same pace.

As usual, standardization is most advanced in air traffic. Here, the planning process is subdivided into generally accepted subproblems, for which data interfaces and powerful optimization methods are available. By now, bus traffic has established similar standards in the area of operational planning. In service design, standardization sometimes evolves as a result of the dominating market position of individual suppliers, e.g., by means of ptv's product VISUM. Such quasi-standards are in principle a possible basis for the development of optimization methods, but they would have to be extended, because they were primarily designed for simulation, and not for optimization, purposes. In rail traffic, on the other hand, nearly every company pursues an individual approach, such that results are very difficult to transfer from one railway company to another. First efforts on standardization are also being made, e.g., in the *OpenTrack* project of ETH Zurich.

Regulation and Deregulation

It is general consensus that public transport is a matter of public interest. However, it is controversial how far the public interest reaches and how the service is organized, i.e., how the state is supposed to get to the desired welfare maximum for its citizens. In the past it was believed that the citizens are best off if state-run, monopolistic operators run the public transportation system. This concept got under dispute. The European Union pursues a clear deregulation policy in public transport as well as in other areas. The EU proceeds on the assumption that a welfare maximum is more likely being achieved by market mechanisms than by monopolistic structures. Different countries implement this concept in different ways. It is often unclear, what kind of planning problems will evolve in the future and which individual market participant should plan what in the presence of deregulation measures.

For instance, the following scenarios are being discussed in bus traffic. Should a region specify the level of service in detail and only invite tenders for the operation of the trips, or should the regions evaluate offers of public transport companies, or should public transport companies act as providers of mobility in free competition? Air traffic has enjoyed an enormous boom because of the US-driven deregulation. Competition and the necessary, resulting optimization measures improved efficiency substantially. "Should wheel and track be split and what should be operated privately?" are controversial topics in the railway sector. The advocates of railway privatization refer to air traffic success stories, the opponents argue that an integrated system can be planned and controlled better, and is therefore more customer friendly.

A monopolistic approach offers a chance for an integrated system-wide optimization, while the creation of a market forces each participant to optimize his activities. No matter if and how regulation and deregulation take place, mathematics is always useful.

Regulation and deregulation, auctions and allocation mechanisms received a great deal of attention in the economic research community in recent years. One can, however, not by any means speak of a mathematical theory of regulation and deregulation yet. Up to now, quantitative implementations of theoretical concepts have only taken place in very special and individual areas such as auctioning truckload contracts, mining rights, or telecommunication frequencies—not always to the satisfaction of all parties.

4.2 Mathematical Models and Algorithms

Discrete optimization is in general perfectly suited to formulate many planning problems in public transit, because they are often about yes/no decisions that can be modeled as 0/1 optimization problems. Mathematical modelers have learned to express complex dependencies by means of linear equations and inequalities with integrality constraints and created, in this way, modeling tools that in principle allow to formulate problems in public transport at an arbitrary level of accuracy. This approach has several advantages. Traffic planning problems can be understood and described precisely, with all their constraints and prerequisites. In this way, discrete mathematics offers its services as a "language of traffic planning". It allows to communicate and analyze questions across companies and countries. Mathematical models help to structure problem areas, to recognize and point out special cases and generalizations. One should not conceal in this context that it is quite in-transparent for the layman whether a model is theoretically or practically easy or difficult to solve. Even experienced optimizers can often not tell a priori, whether and for which order of magnitude a concrete, difficult problem can be solved.

Standard Models and Techniques

Over time, a number of standard models of combinatorial optimization have proved to be useful in traffic applications (set partitioning, path covering, and network flows); these are complemented by general techniques such as the use of methods from linear and integer optimization. These methods and algorithms have been the subject of intensive work in the last decades. Today, problem sizes can be tackled that one would not have thought possible a few years ago. A study by Bob Bixby from 2004 showed that linear programs that took two months of computing time to solve in 1988, could be solved in a second in 2004. Faster computers contributed a speed-up factor of about 1600 to this, improvements in LP algorithms contributed a factor of about 3300. Simply writing down mathematical models and applying standard techniques, however, is only sufficient to solve the problems of small PT companies, and for special applications. Often, minor-looking modifications, which are motivated by practical requirements, increase the level of difficulty significantly and must be faced by well-directed research efforts. The consideration of time windows, precedence constraints, path length restrictions etc. are typical examples for this.

Data for Academic Research

Applied algorithmic research on $\mathcal{NP}$-hard problems can only succeed if practically relevant data is available, which can be used to test algorithmic alternatives experimentally. Availability of real-world data for academic research is of utmost value. One cannot avoid noticing that the situation in this area is very unsatisfactory. Even public transit companies that collect data with additional public funds, do (with very few exceptions) not even confidentially make this data available for academic research. Fear of leaking business secrets is a substantial obstacle to progress in the mathematical solution of planning problems in public transit.

Theory and Algorithms

Planning problems in the area of traffic and transport usually lead to combinatorial optimization problems with an extreme number of variables and/or constraints, which cannot be tackled by commercial linear or integer programming standard codes. In the last decade, techniques such as Lagrangian pricing or separation algorithms have been developed, which can deal with such problem sizes by means of dynamically generating the currently relevant variables and constraints on the fly, instead of dealing with every detail of the problem right from the beginning. These techniques allow to solve classical problems in traffic planning with sufficient quality.

These successes pave the way to the present research on a number of advanced topics:

- *Coupling of discrete models* (*integrated planning*). Up to now, success in the solution of coupled discrete models (e.g., vehicle and crew scheduling) has been limited to highly structured individual cases of limited size. The long-term goal is to overcome the artificial decomposition of the planning process into a hierarchical sequence of individual problems in order to produce an overall plan in one step, using a toolbox of models and methods that can be combined with each other. One approach to get the coupling of models to work uses methods of non-linear and non-differentiable optimization (sub-gradient and bundle methods). Substantial research and development efforts are still needed here.

- *Robust and stochastic planning.* The successes in classical optimization brought another aspect more and more into the focus of attention. It is that sophisticated plans are particularly vulnerable to disruptions. The overall goal must therefore be to compute plans that do not only look good on paper, but also turn out to be good in practice. Here, it is necessary to implement buffers in such a way that small disruptions can be intercepted without major operational changes. There are at present several, at least theoretically, competing approaches to deal with this question, in particular, methods of robust and stochastic optimization seem to be suitable. Practical applications and experiences, however, are virtually nonexistent.

- *Real-time planning.* Traffic systems sometimes get out of hand because one cannot take all potential disruptions into account. Exactly these disruptions annoy passengers the most, in particular, if no information on the further course of the journey is available. Real-time optimization tries to tackle this kind of problems. One must admit, however, that the theoretical tools are currently not particularly advanced and that, specifically, hardly any useful statements about the practical performance of online algorithms can be made. It is therefore an important matter to develop theoretically well-founded methods for the real-time optimization of transportation systems that work in practice.

- *Coupling of discrete, stochastic, and non-linear models* (*infrastructure planning*). An area that is, from our point of view, nearly devoid of mathematics is the construction of the infrastructure of transportation systems. If anywhere, traditional planning methods are dominating the scene here. Decisions are made in a political way, considering some technical details, past experiences, and, in good cases, the use of simulation software.

The development of mathematical methods of infrastructure design is undoubtedly difficult, because it requires a combination of models from different mathematical areas, that employ, in particular, different mathematical tools. User behavior can only be predicted and must therefore be modeled in a stochastic way, quantities such as revenue = price × demand lead to non-linearities, monthly tickets to staircase functions and hence to discontinuities, the consideration of the tradeoff between quality and price to multi-criteria optimization, and this list can be continued arbitrarily.

Considering the long-term impact of these decisions and the pure investment volume, however, it would seem appropriate to undertake serious attempts towards a model-based long-term planning. The research community is investigating some of the relevant basic problems such as non-linear network flows, integer programs with certain non-linearities, or the general acceleration of stochastic optimization approaches by means of, e.g., scenario reduction, algorithmic game theory etc. First successes were scored in the application of discrete choice models to fare planning in public transit or in airline revenue management.

In the end, decisions about infrastructure construction will clearly not be made on the basis of a single mathematical model. Alternative computations and complex scenario simulations must also be considered as decision criteria.

4.3 Transfer and Education

Communication and Education

An obstacle to the exploitation of synergies between practitioners, algorithms designers, and theoreticians is a common lack of willingness to approach each other. Mathematicians do not want to enter the "lower depths" of data acquisition, while practitioners find the considerations of the mathematicians too far from reality. This division starts already in the language. A practitioner who talks about optimization is often thinking of a different thing than a mathematician. This communication problem is reinforced by ignorance of the contributions that the respective other group could make. What hits the eye in the context of this article is, in particular, a striking deficit in the education of traffic engineers with respect to modern methods of optimization, graph theory, and discrete mathematics. Every engineer knows what a differential equation is, but many have not learned that decision problems can be formulated and solved as integer programs, albeit this is the modern approach to good traffic planning. This is (all over the world) a clear shortcoming in university education, that must be remedied.

Business Environment

With the advent of mathematical planning in the airline industry, many companies set up operations research (or similarly termed) departments, which investigated the planning problems mentioned in this article and developed proprietary solutions. Over the years, most of these departments have been sourced out. The best among them became software companies that offer standardized solutions which are used by many airlines. In bus and rail traffic, on the other hand, such OR departments have not evolved to the same extent. Software for bus and railway companies is developed by suppliers with a university background. All these companies are small in comparison to their customers and have problems to impose their ideas and internal standards against the manifold of special requests of their big customers. These special requests fragment the software market and produce high costs, because they require cost intensive individual developments and maintenance. Small software companies also have problems to make enough advertisements to promote their innovations successfully.

4.4 Conclusion

Problems of infrastructure design, crew scheduling, etc. are for more than 100 years topical subjects of traffic engineering. They were always solved using the available technical and mathematical tools of the time. The dean of the traffic engineering faculty of the university of Dresden wrote in his obituary to the eminent traffic engineer Rüger:

> "Using scientific insights to improve the work of public transit companies, their application and implementation in planning and operations, were always his maxim." (translation by the authors).

Rüger [31] in East Germany and Lehner [24] in West Germany wrote pioneering works on traffic planning, which describe, in principle, a mathematical model of a public transit company. At that time, the decision problems had to be reduced to (from today's point of view) simple, practically adequate, formulas, which constitute, even today, reasonable decision rules for practice. The progress in mathematics, however, now makes entirely new approaches possible, which derive decisions as solutions of complex network models. The goal must be to consequently develop these mathematics further, in order to continuously improve the planning process in the sense of Rüger and Lehner, such that resources are saved, the use of public funds is reduced, and the customers enjoy a user friendly and adequate level of transportation services.

We see the transition from formula- to model-based mathematical planning (as we have sketched it) as a leap similar to the transition from the Braun tube to the transistor, or from the drawing board to the CAD system, and we are convinced that mathematics will more and more become a substantial production factor in public transit.

Despite the mentioned shortcomings in inter-disciplinary communication, in education, and in the transfer to practice, Germany has, in our opinion, excellent premises to help mathematical planning methods achieve a breakthrough in public transit. German traffic systems are among the best of the world, German software companies have leading positions in the world market, and German universities feature a variety of departments where scientists work on the quantitative treatment of planning problems in public transit on the highest international level. Public transit offers substantial optimization potentials. The know-how to exploit them is available. If we succeed in getting the different parties to cooperate in a coordinated and well-directed way, Germany could become a showcase of efficient public transit.

5 Visions and Recommendations

Our vision is to establish discrete mathematics and optimization as an essential production factor in all areas of traffic planning. The mathematical models and methods have to be built up such that they can support the various planning processes in practice in an adequate and user friendly way. This requires mathematical progress (in some areas definitely significant progress), but also the interfaces to the involved engineers and practitioners must be improved in a joint endeavor.

As we have shown in the preceding sections, there is need for action in nearly all areas of public transportation planning. It reaches from mathematical modeling and the associated theory, via the implementation of this theory within algorithms, to the introduction of optimization systems in practice. This is an agenda for decades, which requires not only the overcoming of mathematical and company internal obstacles, but also the consideration of political conditions.

We want to propose here two concrete measures, which can be initiated on relatively short notice and which can be funded by the German ministries of education and research resp. industry and commerce (BMBF/BMWi) or the German science foundation (DFG) together with the industry. Both measures are characterized by the inter-disciplinarity that is typical for the traffic area. They combine mathematical research with engineering know-how, business management, practical experiences, and the innovative use of IT systems.

5.1 Discrete Optimal Control: Real-Time Re-Planning of Traffic Systems in Case of Disruptions

Operational disruptions of traffic systems are virtually unavoidable. Larger disruptions require adaptations of the schedule, which must be initiated immediately after the incident. From a mathematical point of view, this is a topic of online or real-time optimization. The application of online optimization methods needs many premises. All data about the planned and the actual state of the traffic system must be available. In addition, forecasts about possible evolutions of the system and existing alternatives of action are needed. The goal is to bring the traffic system back into a state in which plan and status quo coincide to the largest possible extent and in which the "damage" is minimal.

Online planning is done today in the control centers of public transport companies by experienced dispatchers, who are supported by control systems and their visualization tools. The dispatching itself, however, is mainly done "by hand". In most cases, decision tables and rule sets are used. These give clear instructions, but do not provide for optimization. Online optimization is supposed to advance the pursuit of goals instead of the execution of rules. A change to this type of mathematical planning requires, in particular, adaptations of the IT infrastructure and the linkage of different databases, a non-trivial endeavor in information technology.

A fundamental problem in online optimization is the recording of data in order to analyze and reconstruct the sequence of events. Ideally, such data should not only record the executed operation, but also the causes for disruptions and the countermeasures that have been taken. Such data is usually not available. A research program in online optimization should already start at this level and establish a high-quality database.

It must be possible to use this data in simulation systems, such that exemplary verifications of individual scenarios are possible. As long as there is no practically proven online theory, it is necessary to use simulation tools in order to reconsider solution approaches. Simulation has a long tradition in the railway industry, where a number of systems such as RailSys, OpenTrack, etc. allow to produce an accurate picture of the real operation. In bus and air traffic, such simulation systems are still missing or under development.

The real-time algorithms to be developed should take the dispatchers' years of experience and their heuristic scheduling rules into account. Cooperation between mathematicians, computer scientists, and experienced dispatchers is necessary. A possible additional progress that we see is that mathematical algorithms can monitor the state of a traffic system in the sense of a control circuit, in order to take preventative measures when changes in the state of the system are forecast or become apparent that can impair the operation. Ideas of this type could be summarized under the term "discrete optimal control". It refers to the continuous use of optimization algorithms to keep traffic systems on schedule. Similar questions come up, e.g., in mechanics and in production engineering. Their mathematical study is the subject of control theory, which typically uses methods from the theory of differential equations. The novelty in this case is that methods from discrete optimization play a prominent role as well.

Projects of this type could have been started 10 years ago. At that time, however, the IT base did not exist. Today we see a chance to bring know-how in information technology, mathematical methods, and operational experience together in a way that is beneficial for all participants. Such a project can range from the online optimization of subways, street cars, and commuter trains, via the real-time optimization of bus networks, to the real-time optimization of the entire German railway network or the entire European airspace; the last two of these applications are clearly still a long way off. From the customer's point of view it is important to combine online optimization and passenger information, such that there is an immediate benefit for the passengers.

5.2 Model Integration: Service Design in Bus and Rail Traffic

We have talked about various aspects of model integration and want to propose here the integration of mathematical models in the special area of service design in bus and rail transport. The goal is that the resulting models are algorithmically tractable and produce reasonable results from a practical point of view.

Similar as in the preceding case of discrete optimal control, the data that is actually needed is not available at a sufficient level of accuracy.

The special characteristic of service design is a tight correlation of several aspects. We have already pointed out that demand forecasting is a stochastic problem, while network design leads to discrete optimization problems, and the determination of fares to non-linear mixed-integer optimization models. The integration of models from different mathematical disciplines is therefore a prerequisite for the development of successful approaches to service design. Model integration is in general hopelessly difficult, however, the special structures of this concrete application seem to present a chance for the successful development of adequate solution approaches.

First steps in this direction have been taken in toll collection applications using so-called bi-level optimization techniques. Non-linear optimization models and stochastic demand models (discrete choice Logit models) have been combined in the area of fare planning.

The procedure would be based upon traditional planning structures. Beyond the computation of a few scenarios, however, integrated models could take feedback and substitution effects into account, e.g., by explicitly including individual traffic (i.e., car traffic) in the models.

References

1. Anbil, R.E., Gelman, B.P., Tanga, R.: Recent advances in crew-pairing optimization at American Airlines. Interfaces **21**, 62–74 (1991)
2. Barnhart, C., Cohn, A.M., Johnson, E.L., Klabjan, D., Nemhauser, G.L., Vance, P.H.: Airline crew scheduling. In: Hall, R.W. (ed.) Handbook of Transportation Science, pp. 517–560. Kluwer, Boston (1999)
3. Barnhart, C., Johnson, E.L., Nemhauser, G.L., Savelsbergh, M.W.P., Vance, P.H.: Branch-and-price: column generation for solving huge integer programs. Oper. Res. **46**(3), 316–329 (1998)
4. Berliner Verkehrsbetriebe AöR (BVG): Business report 2006. http://www.bvg.de/index.php/de/Common/Document/field/file/id/1409 (2006)
5. Bley, A., Pattloch, M.: Modellierung und Optimierung der X-Win Plattform. DFN-Mitteilungen **67**, 4–7 (2005)
6. Borndörfer, R., Grötschel, M., Löbel, A.: Duty scheduling in public transit. In: Jäger, W., Krebs, H.-J. (eds.) MATHEMATICS—Key Technology for the Future, pp. 653–674. Springer, Berlin (2003)
7. Borndörfer, R., Grötschel, M., Pfetsch, M.E.: A column-generation approach to line planning in public transport. Transport. Sci. **41**(1), 123–132 (2007)
8. Brieger, S.: Workflow oriented and integrated optimization. Talk at the AGIFORS Conference, May 2005

 9. Cook, T.M.: Sabre soars. ORMS Today **25**(3), 27–31 (1998)
10. Daduna, J.R., Wren, A. (eds.): Computer-Aided Transit Scheduling. Lecture Notes in Economics and Mathematical Systems. Springer, Berlin (1988)
11. Daduna, J.R., Branco, I., Pinto Paixão, J.M. (eds.): Proc. of the Sixth International Workshop on Computer-Aided Scheduling of Public Transport (CASPT) (Lisbon, Portugal, 1993). Lecture Notes in Economics and Mathematical Systems, vol. 430. Springer, Berlin (1995)
12. Demetrescu, C., Goldberg, A., Johnson, D. (eds.): 9th DIMACS Implementation Challenge— Shortest Path. DIMACS, Piscataway (2006)
13. Desaulniers, G., Solomon, J.D.M.M. (eds.): Column Generation (Gerad 25th Anniversary). Springer, Berlin (2005)
14. Desrochers, M., Rousseau, J.-M. (eds.): Computer-Aided Transit Scheduling. Lecture Notes in Economics and Mathematical Systems. Springer, Berlin (1992)
15. Deutsche Bahn AG: Business report 2006. http://www.db.de/site/bahn/de/unternehmen/ investor_relations/finanzberichte/geschaeftsbericht/geschaeftsbericht_2006.html (2006)
16. Deutsche Lufthansa AG: Business report 2006. http://konzern.lufthansa.com/de/downloads/ presse/downloads/publikationen/lh_gb_2006.pdf (2006)
17. Dijkstra, E.W.: A note on two problems in connection with graphs. Numer. Math. **1**, 269–271 (1959)
18. Forschungsgesellschaft für Strassen- und Verkehrswesen: Heureka'05: Optimierung in Verkehr und Transport, Cologne (2005)
19. Garey, M.R., Johnson, D.S.: Computers and Intractability. Freeman, New York (1979)
20. Grötschel, M., Krumke, S.O., Rambau, J., Torres, L.M.: Making the yellow angels fly: online dispatching of service vehicles in real time. SIAM News **35**(4), 10–11 (2002)
21. Hoffman, K.L., Padberg, M.: Solving airline crew scheduling problems by branch-and-cut. Manag. Sci. **39**(6), 657–682 (1993)
22. Klabjan, D., Johnson, E.L., Nemhauser, G.L., Gelman, E., Ramaswamy, S.: Solving large airline crew scheduling problems: random pairing generation and strong branching. Comput. Optim. Appl. **20**(1), 73–91 (2001)
23. Ledyard, J.O., Olson, M., Porter, D., Swanson, J.A., Torma, D.P.: The first use of a combined-value auction for transportation services. Interfaces **32**(5), 4–12 (2002)
24. Lehner, F.: Der maximale Wirkungsgrad des Personaleinsatzes. Alba Verlag, Düsseldorf (1978)
25. Liebchen, C.: Periodic Timetable Optimization in Public Transport. PhD Thesis, TU Berlin (2006)
26. Littlewood, K.: Forecasting and control of passenger bookings. In: AGIFORS Symposium Proceedings, vol. 12. Nathanya, Israel (1972)
27. Löbel, A.: Optimal vehicle scheduling in public transit. Shaker Verlag, Aachen. PhD Thesis, TU Berlin (1997)
28. Müller-Elscher, M.: Die MICROBUS-Optimierungskomponenten im Überblick. Talk at 1st PSYRUS User Forum, 30.6.2005, Fulda
29. Neumann, M.: Mathematische Preisplanung im ÖPNV. Master's thesis, TU Berlin (2005)
30. Partner für Berlin Gesellschaft für Hauptstadt-Marketing GmbH. Berlinbrief (2003). ISSN 1611-3284
31. Rüger, S.: Transporttechnologie städtischer öffentlicher Personenverkehr. Transpress, Berlin (1986)
32. Smith, B.C., Leimkuhler, J.F., Darrow, R.M.: Yield management at American Airlines. Interfaces **22**(1), 8–31 (1992)
33. Sturmowski, A.: Notwendigkeit der Effizienzsteigerung unter Einsatz von IT-Tools. In: Proceedings der IVU-Konferenz IT im ÖPNV am 05.10.2007. IVU Traffic Technologies AG, Berlin (2007)
34. Talluri, K.T., van Ryzin, G.J.: The Theory and Practice of Revenue Management. International Series in Operations Research and Management Science, vol. 68. Kluwer Academic, Boston (2005)
35. Voss, S., Daduna, J. (eds.): Computer-Aided Scheduling of Public Transport. Lecture Notes in Economics and Mathematical Systems. Springer, Berlin (2001)

36. Wilson, N.H.M. (ed.): Computer-Aided Transit Scheduling. Lecture Notes in Economics and Mathematical Systems. Springer, Berlin (1999)
37. Yu, G. (ed.): Operations Research in the Airline Industry. Kluwer Academic, Dordrecht (1998)
38. Yu, G., Argüello, M., Song, G., McCowan, S.M., White, A.: A new era for crew recovery at Continental Airlines. Interfaces **33**(1), 5–22 (2003)

Towards More Intelligence in Logistics with Mathematics

Rolf H. Möhring and Michael Schenk

1 Executive Summary

Solving logistical problems has been an important aspect of human activity since we first began working towards goals together. The foundations of what we call logistics today stem from the military sphere. The Roman Empire, for example, was to a large extent founded on its mastery of military logistics. It is not known today whether mathematical considerations played a role back then. However, Napoleon, for example—who befriended the most important mathematicians of his time— sought to optimize the transport of his troops and the dissemination of information and put this to strategic use.[1,2]

The function of logistics has undergone profound and far-reaching changes since the Napoleonic era, and the last forty years have seen it evolve from its heavy concentration on a company's physical processes to a holistic process- and customer-oriented management instrument. This has also meant a continual change of the mathematical challenges in logistics. An overview of the development of logistics over time and the associated areas of optimization is found in Fig. 1.

Today logistics is the third-largest industry in Germany after retail and the automotive industry, employing over 2.6 million with a gross annual sales volume of more than 166 billion Euros [6]. It shows a yearly growth of 3–10%. This growth is qualitative, not just quantitative, since logistics sees itself continually confronted

---

[1] http://www.geophys.tu-bs.de/geschichte/laplace.htm, online: 19 June 2008.

[2] http://www.napoleon-online.de, online: 19 June 2008.

R.H. Möhring (✉)
TU Berlin, MA 6-1, Straße des 17. Juni 136, 10623 Berlin, Germany
e-mail: Rolf.Moehring@TU-Berlin.de

M. Schenk
Fraunhofer-Institut für Fabrikbetrieb und -automatisierung IFF, Postfach 14 53,
39004 Magdeburg, Germany
e-mail: michael.schenk@iff.fraunhofer.de

M. Grötschel et al. (eds.), *Production Factor Mathematics*,
DOI 10.1007/978-3-642-11248-5_7, © Springer-Verlag Berlin Heidelberg 2010

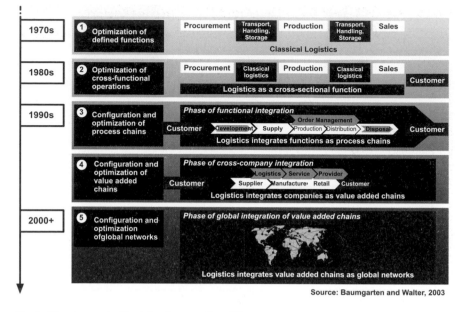

Source: Baumgarten and Walter, 2003

Fig. 1 Development of logistics (source: [2, p. 2])

by new challenges, which are to be presented in greater detail at the end of this paper.

To understand the contributions of mathematics to logistics, it is helpful to look at two of the many definitions of logistics that have been offered throughout its continual development. Plowman's Seven Rights definition states that logistics ensures the availability of the right goods, in the right amount, in the right condition, at the right place, at the right time, for the right client, at the right cost.[3] And another definition taken from Baumgarten states that "logistics includes the holistic planning, regulation, coordination, execution and monitoring of all flows of goods and information within a company internally and externally. Logistics provides process- and customer-oriented solutions for total systems and sub-systems of a company."[4]

The following section provides several examples of how the use of new mathematical methods has improved logistics and led to the success of the particular company in question. An initial process chain model of a value-added process should make it easier for the reader to make sense of the examples in context and hence of the mathematical challenges involved. Section 3 will then take a more detailed and somewhat more abstract look at the status quo of logistics and its relation to mathematics. The strengths and weaknesses at present and several current visions and challenges are discussed at the end.

---

[3] http://www.fml.mw.tum.de/fml, online: 6 June 2008.

[4] http://www.bvl.de, online: 6 June 2008.

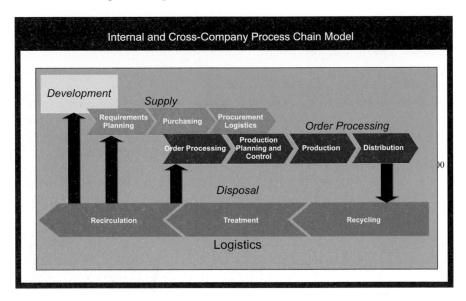

Fig. 2 A company-internal and inter-company process chain model (source: [2, p. 7])

2 Examples of Success

In almost every company a value-added process consists of the four ideal types of process chain: development, supply, order processing, and disposal. These build a cycle where the processes intermesh sequentially and in parallel. The set-up of a process chain model and the classification of the process chains is depicted in Fig. 2.

2.1 Logistics Chains

It is not just the supply department that has a supply network—every other process chain is based on a supply network consisting of delivery sites and receiving sites. These sites are connected to one another through supply chains and depending on the process in question are called supply chains, acquisition chains, transport chains, cargo chains, shipping chains, or disposal chains or just logistical chains in general [11]. An optimal supply network requires these supply chains to link the delivery and receiving sites to each other as well as possible, which calls for mathematical optimization. The following examples come from various areas that contribute to an optimal supply network.

The planning of very large and regularly recurring shipments, such as affects second-party service providers to the Deutsche Post or the transport of cargo or people in train and air traffic, presents a great challenge—these service providers must find an optimal way to integrate decisions about locations and fleets with decisions about transport (routing and scheduling).

   The conception of suitable allocation structures and the efficient organization of
the process flow in the chains is based on the combination of new techniques of data
acquisition and communication (integration of GPS, RFID, EDP and barcodes) and
sophisticated mathematical optimization procedures such as network algorithms and
(integer) linear optimization.

   The consistent and complete utilization of these results is an important economic
factor. The use of such methods (and the liberalization of the market) has transformed Deutsche Post from a national mail service to a global business in less than
15 years, for example. Its acquisition of the British logistics business Exel in December of 2005 made the Deutsche Post AG the global market leader in the fields of
air and ocean logistics (both DHL Global Forwarding) and contract logistics (DHL
Exel Supply Chain) with a yearly revenue of 55 billion Euros.[5] Other examples of
success include the optimization of shipping in the airline industry,[6] in container
logistics[7] and in rail transport. Hence, in 2008 the renowned Edelman Award of
INFORMS, the American operations research institute, was awarded for the optimization of the Dutch railway (Nederlandse Spoorwegen) timetable.[8]

   Large companies have also recognized the importance of optimizing logistic networks and have endowed their own chairs at universities, such as the Deutsche Post
chair at the RWTH Aachen or the DB Logistics Lab at the Technical University of
Berlin.

2.2 Shipping

Logistics originated in the field of shipping, where we can see the application of
mathematical methods from very early on. These methods are often allocated to
operations research due to their great practical relevance.

   A classical example is the vehicle routing problem (VHP)—the challenge of coordinating various shipping processes within a brief span of time with the goal of
optimizing the routes [12].

   The solution to this problem involves reducing it to a mathematical model than
can then be optimized. Two aspects need to be considered: the clustering, which
describes which commissions can be put together into one route, and the routing,
which describes the series of points within one route. The objective of VHP is to
minimize the number of vehicles used, for example, or the distance traveled, the time
of deployment or a more complex cost function. In the standard VHP problem all
starting points and destinations are in one depot and a limited or unlimited number of
identical vehicles with limited capacity are available there. Other variations consider

---

[5]http://www.dpwn.de/dpwn?skin=hi&check=yes&lang=de_DE&xmlFile=2004705, online: 2 July
2008.

[6]http://web.mit.edu/airlines/www/index2.htm, online: 2 July 2008.

[7]http://www.hhla.de/Altenwerder-CTA.64.0.html, online: 2 July 2008.

[8]http://www.informs.org/index.php?c=401&kat=Franz+Edelman+Award, online: 2 July 2008.

several depots or arbitrary starting points and destinations (so-called pickup-and-delivery problems).

The development of mathematical techniques for VHP has made great progress in the past decades, since the rapid development of information technology made new and faster possibilities of implementation available. Thus mathematical models are the basis of many software packets; see http://www.wior.uni-karlsruhe.de/bibliothek/Vehicle/com/ (18 June 2008). The availability of digital road systems and the use of GPS together with mathematical procedures have been the basis of the success of such systems.

2.3 Production Logistics

Production logistics deals with the entirety of all logistic activities, measures and issues related to the production of goods and services. The challenges of production logistics are the planning, regulation, shipping and storage of raw materials, operating and process materials, purchased or replacement parts, and half-finished or finished products.[9] Thus they form the core of every industrial business. The optimization of the entire production system is at the forefront of production logistics.

Thus as in the logistics of shipping, mathematical methods of operations research have long been in use in this field—such as scheduling, warehouse storage, linear (integer) optimization and flow models. Often these models are very specifically adapted to a certain type of production, such as in the case of controlling tool machines, minimizing set-up time or controlling high rack warehouses. With the increasing availability of data, in particular with ERP systems, the trend is tending towards the use of general techniques collected under the heading of APS systems (Advanced Planning and Scheduling systems) throughout the entire production chain. Figure 3 shows this chain using the example of Porsche AG.

Usually this involves scheduling under complex side constraints and with uncertain data. APS systems seek solutions on the basis of the planning data or estimated intermediate data and adapt these solutions as soon as the real data are available. The complex side constraints require a very general procedure for solutions; typically integer linear optimization, constraint programming and heuristics are used. The uncertainty in the original data requires prognostic techniques and a validation of the solution procedure, which is typically accomplished through simulation models that test the suitability of the optimization models for random variations in the data. This combination of prognostication, optimization and simulation often produces significant success.

Such is what the ILOG company has reported about the use of their ILOG Plant PowerOps tool for Danone in a complex production environment. A typical Danone site produces ten intermediate products which 120 finished products are made from in 15–25 production lines. The individual production steps use more than 100 tanks

---

[9]http://www.bvl.de/taschenlexikon/, online: 2 July 2008.

Fig. 3 SAP VMS—Vehicle Management System (importer) (source: [15, p. 13])

for preparation, pasteurization, fermentation and storage. All of the steps in this production chain are very time-sensitive, for reasons of hygiene and shelf-life, and require meticulous scheduling that simultaneously accounts for change-over processes (including cleaning), tank capacity, delivery deadlines and changes in demand [13]. Often such complex situations produce contrary criteria (e.g. saving change-over time vs. adhering to delivery deadlines). The tool is based on integer linear optimization, constraint programming and heuristics and calculates good production plans in less than 15 min. It is supported by complete integration to SAP, possibilities of sensitivity analysis (what-if analysis) and a good graphic presentation of the results (see Fig. 4, [13, p. 16]).

This is how the pharmaceutical company Merck KGaA planned the materials flow for a new solids factory under construction. The factory comprises 10 000 m^2 of production space on six levels and is designed for a maximum of 1000 tons of pills per year. With the help of a simulation examining various model variations matched with various production programs, an appropriate production logistics concept was generated. The core of the production facility is the highly automated internal logistics. This logistical concept enabled 80% savings on storage space and a more than 70% reduction of manual transport activities [16].

The Beumer Maschinenfabrik GmbH and Co KG implemented new control center software for their sheet and saw processing that created an optimized interface between the PPS system, storage and the machine tools. This resulted in an improvement in the machine capacity of up to 30% in saws and up to 60% in laser-beam cutting, a 20% reduction of the processing time and more than 2000 hours of unmanned operation for laser-beam cutting tools [19, p. 52].

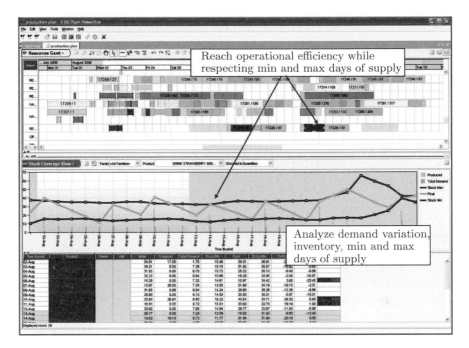

Fig. 4 Schedule in ILOG Plant PowerOps (source: [13, p. 16])

2.4 Controlling Logistical Networks

The applications so far have still been largely isolated and global, guided by a single company or planner. Accounting for various intercompany participants acting independently on one another produces a significantly higher level of complexity and leads to the concept of logistical networks.[10] These networks include very many stacked and interwoven layers with many interdependencies, including flows of materials and products, information flows concerning triggering new events, control, monitoring and documentation, and other activity chains that all place a demand on the resources provided.

A typical example of this is a supply chain over several stages with autonomous planners at each stage. This gave rise to a problem first recognized by Proctor & Gamble. The wholesaler ordered widely varying amounts of Pampers diapers from Proctor & Gamble, although the end customers' demand was almost constant. Thus Proctor & Gamble's demand for preliminary products fluctuated heavily. These fluctuations snowballed further and further, prompting Procter & Gamble to begin to look for the causes of this effect. It turned out that the orders at the individual levels no longer had any connection with the original demand of end customers, which was constant. A comparison of demand and inventory at the individual steps in the

---

[10]http://www.logistik.wiso.uni-erlangen.de/german/profil/selbst/index.htm, online: 2 July 2008.

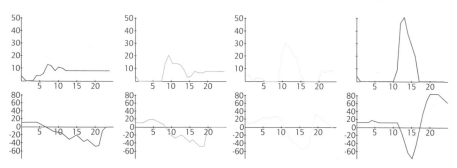

Fig. 5 Bullwhip effect. *Above*: demand from left to right. *Below*: warehouse stock from right to left (source: [17])

process showed that the further along one looked in the chain, the greater the distance to the end-customers, the greater were these fluctuations. This effect in the supply chain is now known as the Bullwhip Effect. The reasons for it are found in the complex and dynamic dependencies of the supply chain. The "agents" within the chain made seemingly rational decisions, locally, about ordering and products, due to false perceptions and distorted information about the system[11] which ultimately had nothing to do with the actual demand of the end customers (cf. Fig. 5, [17]).

This effect has been examined with the tools of probability theory [7] and stems from the fact that the participants along the chain observe different randomly fluctuating amounts of demand and orders (information asymmetries) and thus make incompatible decisions that can snowball further down the line. Corrective measures such as found in the logistic concepts JIT (Just in Time, the manufacture and delivery of parts/products occurs exactly by the appointed time without recourse to warehouse stock), ECR (Efficient Consumer Response, makes wares available to the customers on the basis of their information) and APS (Advanced Planning and Scheduling, uses optimization and prognostication and takes scant capacities into account). Barcodes and RFIG tags play an important role for the ability to make accurate predictions and intervene in real time.

ITWM has developed methods of providing decision-making support in complex logistical networks at the strategic level as well. LibStrat-SCM was used in the APO software at SAP AG and comprises the Module Supply Chain Design. It uses efficient optimization algorithms and interfaces with CPLEX, the Mixed Integer Programming Solver from ILOG.[12]

The following examples should demonstrate quite impressively the magnitude of the savings effects due to the improved control of logistical networks.

Festo AG & Co KG received the German logistics award in 2003 for its market supply concept in the field of pneumatics. Festo delivers components and systems for pneumatic automation and currently finds itself on a growth course with ca.

---

[11] http://beergame.t ni-klu.ac.at/bullwhip.htm, online: 2 July 2008.

[12] http://www.itwm.fhg.de/opt/Dokumente/LibStrat-SCM_de.pdf, online: 2 July 2008.

1.5 billion Euro revenue and around 12 000 employees world-wide. With its winning logistics concept Festo succeeded in increasing its delivery reliability within Europe, where it guarantees delivery within 19 hours, from 86.3% in 1996 to 97.7% in 2002. Moreover, while sales climbed 55% they were able to reduce decentralized inventory by 83% and total inventory by 19% from 1996 to 2002. Alongside a standardization of products and processes, a classification of the products according to logistical types and a learning organization, the concept also included the introduction of hierarchically structured mathematical planning and prognostic procedures that led to optimal capacity load [8].

The Würth Logistics AG coordinates a large share of the procurement logistics between the suppliers and the subsidiaries of the world-wide mother company Würth, which, with more than 51 000 employees, delivers an array of more than 100 000 products to around 2 800 000 tradespeople and workshops in 81 countries. The Würth Logistics AG does not have any of its own trucks, airplanes or ships. Generally the customer enters the shipping orders online in the Würth Logistics software system, and from there the orders are automatically calculated following a long series of optimization steps—route optimization, among others—that takes all data on the articles, customers, destination and price into consideration and only has to be approved by the employees with the push of a button. The ordering system and the shipping system are fully integrated. The total system has led to savings on the order of millions. The Würth subsidiary Melpa-Alfit reports around 20% savings on logistics costs [10].[13]

The first SCM activities at Daimler Chrysler AG in their Sindelfingen factory began with supply bottlenecks for door interior lining of their C-class during their 1997 model upgrade. Due to the unexpectedly high demand, system suppliers were unable to provide enough door interior lining. An extra facility had to be installed, which took several weeks to start operating, so that around 2000 C-class vehicles were unable to be produced in the Bremen and Sindelfingen factories and another 3000 vehicles had to be retrofitted. The lesson of this experience was that reactive methods in one-step procurement process chains are no longer sufficient, and so the SCM project began, with the goal of developing proactive logistics methods for production in procurement networks. Data acquisition revealed that behind the system supplier were ca. 100 additional sub-suppliers at a total of seven different levels. A simulation was created to represent and analyze the dynamic behavior of the network for the door interior lining of the E-class. The repercussions of program changes at the individual stations of the network could be investigated and critical procurement chains identified with a sensitivity analysis. Since 1999 the entire process chain from customers to the 6th-tier suppliers is mapped out using an Internet-based software tool, the IC Tool (Information Control Tool). Each supplier has access to the gross requirements, stock and capacities in the entire procurement chain. Any anticipated supply difficulties are immediately indicated online and thus the system supplier has a chance to introduce the appropriate measures together

---

[13] http://www.wurth-logistics.com, online: 18 May 2008.

with the sub-supplier in sufficient time. The analysis of the IC Tool leads to an overarching inter-chain cost and benefit analysis and a participation model, so that all suppliers have a share in the savings. Savings of over 20% have been achieved across the entire process chain. It is estimated that this SCM tool can guide ca. 5–10% of all procurement process chains more efficiently [9].

3 Logistics and Mathematics: the Status Quo

3.1 On the Development of Logistics

Since the 1970s logistics has developed into one of the important core competences in Germany. It was forced to continually reorient itself in the face of the new challenges and problems that kept arising.

As seen in Fig. 1, so-called classical logistics in the 1970s primarily involved problems of the flow of materials and wares. It was focused on securing the availability of materials and wares within the production process. Since at this point logistics was embedded in the prevalent corporate structures of the time, the areas of ware flow and material flow were split off from one another, which led to work structures that tended to dampen efficiency and to uneconomical sub-processes. In the 1980s logistics began to shift its focus to the points of interface between the individual areas. Now not just isolated functions were optimized, as in the 1970s, but overarching, interfunctional processes as well. In the third stage of development (the 1990s) logistics moved from a function-oriented approach to a flow-oriented point of view. It developed into a company-wide and inter-company coordinating function and concentrated primarily on optimizing the flow of materials and wares. New developments in information technology now made it possible to improve information flows and reduce information deficits. The planning and coordination of the flow of wares, materials and information no longer restricted itself to the areas of procurement, production and distribution and came to encompass development and waste removal. Since the year 2000 companies have begun cooperating with all the partners on their value-added chain to optimize the entire chain from suppliers to end customers (supply chain management). Integrated value-added chains were expanded to global networks and optimized [2, pp. 3, 4].

Today approximately 60 000 businesses are active in the field of logistics services, including global firms like the Deutsche Post (15.6 billion Euro yearly revenue) or the Deutsche Bahn (5.5 billion Euro) but also predominantly middle-scale businesses with yearly revenue of under 10 million Euros. 55% of these companies are from industry and trade and 45% are logistics service providers [6, p. 22].

With this development, logistics has evolved from a strategic instrument of company leadership to a management philosophy and an important competitive factor. It shows an above-average growth of 3–10% per year, depending on the particular branch, and invests ca. 15 billion Euros annually (see footnote 13).

Yet the development of logistics is not yet over, and the next years will bring many new challenges and responsibilities for logistics—whereby mathematical theories will continue to take on an important function. The current status quo of mathematics in logistics is described in the following.

3.2 Mathematics in Logistics

Mathematics found its way into logistics early on with Operations Research. Figure 6 represents the fields of mathematics that are relevant to the areas and activities of logistics.

Figure 6 shows that mathematics is already in use in many logistics tasks. The effectiveness of today's algorithms—particularly in integer linear optimization—and their availability in commercial solvers such as CPLEX and XPRESS and not least of all the enormous improvement of the data situation have all contributed decisively to this.

Nonetheless we also see clear limits, such as with control processes, early warning systems and reactions to disruptions (disruption management).

One cause of this is found in the uncertain or incomplete information available at the time of planning. For mathematics this means moving from deterministic to stochastic models—for which theories and methods are already available, but require a significantly higher computational effort. Moreover, the results are tainted with probability (probable scenarios, average values) with which decision-makers are often not adequately familiar and comfortable.

Another cause is found in the necessity of making decisions in real-time, for many processes—i.e. in a few seconds. This imposes a severe constraint on solution procedures, since typically they cannot make optimal decisions in such a brief time. However, this is not a deficit for mathematics: it can even prove (in complexity theory) that there is a trade-off between the quality of the solution and the calculation time. Hence very limited calculation time has a detrimental effect on the quality of the results.

But even here we can see progress being made. Firstly, in general, quality guarantees can be made for certain algorithms (empirically or by way of proof) through mathematical analysis; secondly, new areas are being developed—such as approximation algorithms and online algorithms and robust optimization—that examine improved ways of dealing with incomplete information in real time or account for minor data interferences in optimization and provide recovery methods for such cases. These sorts of methods have already been put to use in planning train and air traffic [1, 14].

4 Future Challenges

In many areas of operative logistics we can currently see the growth of the so-called real-time reaction ability. This tendency is facilitated above all by the rapid develop-

| *Areas and activities of logistics* | → | *Logistical problems* | → | *Areas of mathematics* |

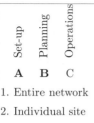

A B C

1. Entire network
2. Individual site
3. Transport system
4. Supply chain
5. Supplier
6. Production site
7. OEM
8. Warehouse or transition point
9. Customer

a. Product line analysis and classification of articles – 1A, 1B

b. Demand prognosis – 9A, 9B

c. Location problem – 1A, 1B

d. Selection of transport routes – 1A, 1B

e. Set-up of supply chains – 1A, 1B

f. Route planning – 3B, 3C, 4B, 4C

g. Layout planning – 6A, 7A

h. Storage and replenishment strategies – 5C, 6C, 7C

i. Supply strategies – 5C, 6C, 7C

j. Storage organization and operation strategies – 8B, 8C

k. Set-up and dimensioning of vehicle systems – 6A, 7A, 8A

l. Set-up of consignment processes – 8A, 8B

m. Formation of logistic units – 8B, 8C

n. Marginal efficiency and congestion effects – 6A, 6B, 8A, 8B

o. Order scheduling and production planning – 5B, 5C, 6B, 6C, 7B, 7C

p. Sequence problems – 6B, 6C, 7B, 7C

- Linear (integer) optimization – c, d, e, f, g, k, l, m, o, p
- Heuristic optimization methods – c, d, e, f, g, k, l, m, o, p
- Scheduling algorithms – l, p
- Special analytic models – a, b, h, i, l, m
- Graph theory – c, d, e, f, k
- Automata theory and Petri nets – o
- Mathematical statistics – a, b, i, k, n
- Queueing models – n
- Continuous simulation models – e, h, i
- Discrete event simulation models – g, h, i, k, l, n, o, p

1

Fig. 6 Mathematics in logistical problems

ment of automated identification and localization systems. A system is in real-time, in logistics, if it can make decisions synchronized with the time that processes take in the "real world". Ideally, any changes of state in the system (events) are registered instantaneously, saved in the form of protocols and interpreted and presented with the help of the appropriate software. The real reaction time only depends on the person or machine responsible for the selection and implementation of the necessary measures.

The so-called autonomous objects of logistics—products, packaging, carriers— are given an electronic identity and linked to their environment wirelessly with RFID technology. With the help of RFID sensors and tags, data is documented during the entire production cycle—from production to transport to delivery and storage at the site of sale or use—and conveyed to further IT processing systems for evaluation at regular intervals.

The lack of clear and generally accepted theoretical concepts and mathematical models to record and interpret real-time data often results in a situation where the scope and content of the registered data is not sufficiently oriented to the goals and techniques of the process analysis in logistical networks and material flow systems. Not infrequently a modern (e.g. RFID-based) data logging system generates so much data that both, information scientists and logisticians, are unsatisfied. The information scientists don't know where the data should go, and the logisticians ask themselves, "what use is this data to us, the planners and managers?". An underlying mathematical model is still needed that would allow for a successful dialogue between logisticians and information scientists in conceiving and implementing real-time systems. This basis can be established if, for example, both sides consider the temporal aspects of the processes to be observed and analyzed from an event-oriented view and the spatial aspects from an object-oriented view [18, pp. 222–226]. The development of foundational concepts for logistical real-time systems—monitoring, event management and early warning—urgently needs the cooperation of mathematicians.

Supply Chain Event Management will be another challenge in the coming years. SCEM deals not just with the efficiency and effectiveness of a supply chain but also, primarily, with optimization and guidance for the sake of a sustainable and stabile supply chain. An important element of this is the event-oriented measurement, which new DV systems should make possible. Rather than checking the performance at irregular intervals, the system is to inform those responsible for the process about exceptional circumstances and about any need for action [3]. However, the suitable methods and models to distinguish relevant from irrelevant data and to recognize and evaluate them are still lacking. Mathematical theories are needed to be able to develop new models for the development of such data processing programs, such as, e.g., pattern recognition or data mining.

5 Visions and Recommendations for Action

The repercussions of the modern RFIP technology for logistics processes in the near future are comprehensively described in the reader "Internet der Dinge" [5].

Autonomous objects, agent-based control, independently controlled systems, and intelligent loading equipment will set the course of logistics in the coming years. These promising technologies will ensure that wares and repositories get progressively more independent and "intelligent". They know where they have to go and control the systems within which they move.

An example of an intelligent carrier is the Smart Box developed in the RFID and telematic technology lab LogMotionLab at the Fraunhofer Institut für Fabrikbetrieb und -automatisierung IFF. This is a reusable box with RFID antenna structures and an autonomous electrical supply. It recognizes the entry or removal of objects so as to enable continual inventory. It can also control access to the wares with access systems. The intelligent carrier's position as well as all access operations are sent to the central office using GSM modules. However, here there is a need for further research in order to expand this technology to other receptacles such as Europalette [18, p. 77 ff].

Additional ideas and examples of the use of RFID in various branches of industry include improving security against counterfeit medications through the use of RFID labels on the packaging, as well as the use of RFID in air transport, for example in locating lost luggage or shortening check-in times by equipping plane tickets with RFID [4, p. 157 ff].

The continually improving data situation, the superior computer performance and the use of new mathematical insights and procedures ensure the potential for continued successes in logistics. However, this will require the common and concerted efforts of logisticians, mathematicians and information scientists, since the complex requirements of logistical systems can only be mastered through interdisciplinary work. Projects of the BMBF/BMWi or DFG in collaboration with industry would offer a suitable framework for this.

References

1. Algorithms for Robust and online Railway optimization: improving the Validity and reliability of Large scale systems, EU Projekt ARRIVAL, http://arrival.cti.gr/, online: 2 July 2008
2. Baumgarten, H., Walter, S.: (2000): Trends und Strategien in der Logistik 2000+
3. Becker, T.: Ereignisorientiertes Prozessleistungsmanagement. In: Ijioui, R., Emmerich, H., Ceyp, M. (eds.) Supply Chain Event Management, pp. 57–70. Physica-Verlag, Berlin (2007)
4. von Bögel, G.: Technologische Trends bei RFID-Systemen für den Einsatz im Internet der Dinge. In: Bullinger, H.-J., ten Hompel, M. (eds.) Internet der Dinge, pp. 157–177. Springer, Berlin (2007)
5. Bullinger, H.-J., ten Hompel, M. (eds.): Internet der Dinge. Springer, Berlin (2007)
6. Bundesvereinigung Logistik e.V. (ed.): Wachstum schaffen – Zukunft gestalten. Logistik als Motor für Wachstum und Innovationen in Deutschland. Thesen und Handlungsempfehlungen (2005)
7. Chen, F., Drezner, Z., Ryan, J.K., Simchi-Levi, D.: Quantifying the bullwhip effect in a simple supply chain: the impact of forecasting, lead times, and information. Manag. Sci. **46**(3), 436–443 (2000)
8. Gericke, E.: Klassenbester in der Marktversorgung im Bereich Pneumatik. In: Schenk, M. (ed.) Tagungsband Gastvortragsreihe Logistik 2004 – Logistik als Arbeitsfeld der Zukunft – Potenziale, Umsetzungsstrategien und Visionen. Fraunhofer IFF (2004)

9. Graf, H., Putzlocher, S.: DaimlerChrysler: Integrierte Beschaffungsnetzwerke. In: Corsten, D., Gabriel, C. (eds.) Supply Chain Management erfolgreich umsetzen – Grundlagen, Realisierung und Fallstudien. Springer, Berlin (2004)
10. Großer, T.: Die Strippenzieher. In: Antrecht, R. (ed.) McK Wissen 16 – Logistik. McKinsey (2006)
11. Gudehus, T.: Logistik 2. Netzwerke, Systeme, Lieferketten. Springer, Berlin (2007)
12. Günther, H.-O., Tempelmeier, H.: Produktion und Logistik, 6th edn. Springer, Berlin (2005)
13. ILOG: Improving plant performance and flexibility in process manufacturing with an example from the food and beverage industry. White Paper (2007)
14. Kohl, N., Larsen, A., Larsen, J., Ross, A., Tiourine, S.: Airline disruption management—perspectives, experiences and outlook. Technical Report, Carmen Systems (2004). http://www.carmen.se/research_development/articles/crtr0407.pdf, online: 2 July 2007
15. Kühlwein, R.: Einsatz von SAP-Lösungen entlang der gesamten Automobil-Prozesskette. Präsentation Branchenforum Automobillogistik, Stuttgart (2006)
16. Münch, G., Seltmann, B.: Produktionslogistik als Maßanzug für die Pharmaproduktion. In: LOGISTIK für Unternehmen, Juni 2001 (2001)
17. Nienhaus, J.: What is the bullwhip effect caused by? Supply chain World Europe, Amsterdam (2002). http://www.beergame.lim.ethz.ch/Bullwhip_Effect.pdf
18. Schenk, M., Tolujew, J., Barfus, K., Reggelin, T.: Grundkonzepte zu logistischen Echtzeitsystemen: Monitoring, Event Management und Frühwarnung. In: Jahrbuch Logistik (2007)
19. Ältere Lager- und Fertigungsanlagen mit Software zur Produktionslogistik aufrüsten. In: LOGISTIK für Unternehmen, October 2003

Optimization of Communication Networks

Jörg Eberspächer, Moritz Kiese,
and Roland Wessäly

1 Executive Summary

Global communication networks (e.g. telephone networks, cellular radio networks, or the Internet) have become the nervous system of today's economy and society. Due to ongoing technological progress in electronics, computer and communication technologies, not only the number of users in the networks keeps growing, but new applications like radio and TV in the Internet, e-business, and interactive games are upcoming.

Immense amounts of information are already traveling over today's copper, fiber, and radio links—but traffic keeps on growing with an exponential rate. Private service providers build and maintain these networks in a highly competitive commercial environment, forcing them to continuously grow their networks as efficiently as possible. Systematic planning is indispensable, because predictions for future traffic growth, applications or even technologies considerably suffer from uncertainty. Locations for network nodes, trunks and antennas have to be chosen, links have to be dimensioned and on top of that, resilience against failures has to be taken into account. Traffic flows with highly diverse characteristics (voice, data, multimedia) have to be routed, while failures should not have any noticeable effect on customers. The dominating factor, however, is cost-efficiency. Design and planning methods supporting these processes have to excel in terms of robustness, flexibility, and scalability while producing cost-effective solutions. Albeit this constitutes a

J. Eberspächer (✉) · M. Kiese
TU München, Institute of Communication Networks, Arcisstr. 21, 80290 Munich, Germany
e-mail: joerg.eberspaecher@tum.de

M. Kiese
e-mail: moritz.kiese@tum.de

R. Wessäly
atesio GmbH, Sophie-Taeuber-Arp-Weg 27, 12205 Berlin, Germany
e-mail: wessaely@atesio.de

M. Grötschel et al. (eds.), *Production Factor Mathematics*,
DOI 10.1007/978-3-642-11248-5_8, © Springer-Verlag Berlin Heidelberg 2010

perfect scenario for advanced mathematical methods, a closer look reveals the need for action for all parties involved—scientists, developers, and users.

This chapter provides an overview of mathematical optimization in communication networks and identifies future challenges. Building on the technologies and architectures of today's networks, we elaborate on the tasks of strategical, tactical and operational planning considering their individual time horizons. We focus on discrete optimization methods, where the sheer size of the problems (number of network nodes, links, ...) is a challenge on its own. Necessary constraints for resilience and QoS further increase the complexity. We expect that growing networks, new technologies, and adequate handling of the inherent uncertainties of the input parameters will attract most attention. Hence, mathematicians and network engineers should intensify their collaborations. Similarly, the communication industry (providers, manufacturers, tool vendors) should adopt more recent methodologies. Therefore, we emphasize the importance of integrating mathematical planning methods in engineering curricula, and we propose a research project, "Modeling, Planning and Evaluation of Communication Networks", bringing together mathematicians, communication engineers, economists and industrial users, jointly attacking the grand challenges of network planning.

2 Mathematics as the Foundation of Every Network

Mathematical methods in communications enjoy a long and successful track record, with Danish mathematician A.K. Erlang establishing the theoretical foundation for planning telephone networks with his groundbreaking statistical analysis of trunk groups. His findings are still in use to plan communication networks. The Internet profits from methods of graph theory and stochastics. Mathematical optimization has grown to become a key method in network planning.

2.1 Routing in the Internet

The global Internet is a gigantic network consisting of more than 27 000 individual networks, the so-called *autonomous systems*. How do innumerable data packets find their way through this maze day after day? How is their way determined and by whom? In most cases, a shortest path from the source to the destination is chosen. This is not necessarily equivalent to the fastest or let alone the cheapest way. There is no unique, centralized instance determining this path, but every node along the way individually decides what the next step will be. For every kind of data exchange, may it be e-mails, video conferencing or music downloads a vast number of possible routes exists. In contrast, network nodes will have to decide within milliseconds, if and how they forward a data packet. In simple cases, the path with minimal "virtual costs" is chosen, taking parameters into account like traveling time (see Fig. 1), number of hops, or even "real" monetary costs into account. Every node individually

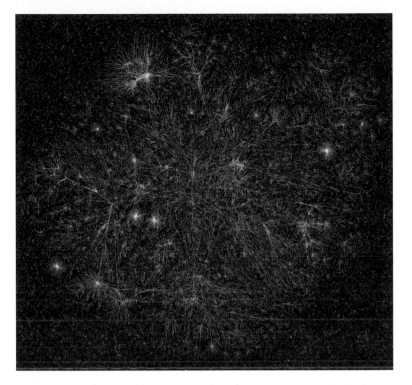

Fig. 1 Routes in the Internet: the color indicates the top level domain (e.g. pink for .de, .uk, .it, .pl and .fr), the length of the edges represents the delay time [9]

calculates these routes to all reachable targets by using algorithms originating from graph theory. This happens millions of times during every second in the Internet. The most prominent candidates for this task are derivates of Dijkstra's Algorithm [8] (originally invented in the 1950s), which is one of the most frequently used algorithms.

2.2 Quality Control in Phone Networks and the Internet

Making a phone call means: connecting customers from anywhere in the world at anytime. This scenario poses a great challenge for service providers because connection requests appear more or less randomly, call activities depend on the time of the day, the day of the week, and last but not least suffer from predictable but exceptional peaks such as on New Year's Eve or during interactive TV shows, and completely random events like natural disasters. These connections use resources in the nodes and links of the network, for example link capacities which have to be planned in advance. This so-called *network dimensioning* has been one of the key tasks of network providers since the very beginning. If the resources in the network are not sufficient, *blocking* will occur too frequently (signaling "busy line" to the user) and will not be tolerated by the customers. Overprovisioning however increases the cost of the network, which is not affordable for the providers due to fierce competition on the market. Thus, in the early ages of the communication industry, a mathematical theory was developed (pioneered by said A.K. Erlang) allowing providers to predict the blocking probability in their networks under certain assumptions on the statistical nature of the connection requests. The only two input parameters required for this calculation, namely the average call arrival rate and the average call duration time, can be determined by measurements with reasonably high accuracy.

In contrast to this, another parameter is gaining importance in packet networks: the time a packet needs to reach its destination. Along its path, a packet is mainly delayed in routers, because these routers contain buffers to suppress momentary traffic peaks. These buffers or *queues* cause delays just as their real-world counterparts and, in addition to that, due to their limited size, even losses. The mathematical *Queuing Theory* is essential for any quantitative information about service quality in modern networks. This field was pioneered by the US scientist L. Kleinrock [13]—one of the fathers of the Internet. The application of this theoretical foundation stone has been constantly extended to accommodate the rapid technological progress in communication systems and is still an important requisite for modern planning methods.

2.3 Cost Optimization of Network Investments

Innumerable interrelated decisions comprise the infrastructure planning of a communication network. In this first phase, strategic planning, site locations have to be found, the basic structure of access, aggregation, and core networks have to be designed, suitable hardware for a multitude of different technologies has to be selected for the individual sites, connections have to be dimensioned, and finally services have to be routed over the planned network with high fault-tolerance. Billions of Euros can be necessary to create and maintain such an infrastructure. *Integer Linear Optimization* has proven to be a valuable mathematical methodology for infrastructure planning. In 2005, planning tasks of the German Research Network X-WiN (cf. Fig. 4 in Sect. 3.2) consisting of quintillions of variables were solved to reach a provable cost minimum. Due to this optimization German researchers are connected via a highly available network offering high-speed access at comparatively moderate prices. Commercial providers rarely offer insight to achieved cost-savings for obvious reasons. A notable exception to this rule was an article in 2000, where AT&T stated that a planning procedure based on integer linear programming saved "hundreds of millions of dollars" [12].

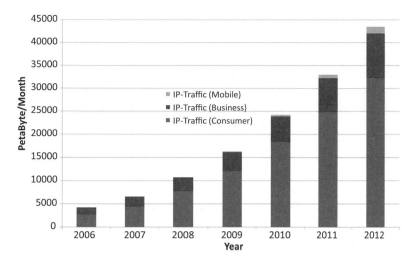

Fig. 2 Traffic growth in the Internet (source: Cisco [11])

3 Networks, Planning and Methods

Private and commercial use of the global communication networks is continuously increasing. The amount of traffic in the Internet doubles every 18 months—see Fig. 2. The economic importance of communication technologies can roughly be estimated by business volumes, where a single provider like Deutsche Telekom achieved a turn-over of € 60 billion in 2007. Due to the high investment costs and global competition, cost-effective realizations of services are of vital importance for providers. Accurate planning with well-founded methods can make a valuable contribution to this process.

3.1 Network Architecture and Planning Aspects

Today's communication networks employ a plethora of different network technologies structuring a network both horizontally and vertically as depicted in Fig. 3.

On the horizontal plane, networks are divided according to their physical size and functionality in access, aggregation, and core networks. Aggregation networks collect traffic from private or business customers and forward it into the core network. Hence, the aggregation networks serve as "feeders" for the national and international data highways connecting the global networks.

Layers, realized via a variety of technologies and protocols, divide a network vertically. An analogy in everyday life is transport of goods. A given item may be packed into a packet, many of these packets can be stored on a container. Trucks carry a container and transport the goods between distribution centers using available infrastructure, namely highways and interstates. Similarly, a modern network

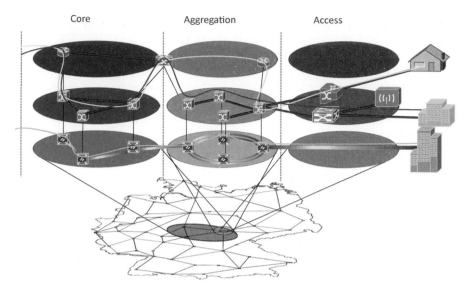

Fig. 3 Horizontal and vertical network layers

is able to transport different services like voice, data or video using a multitude of technologies such as IP, Ethernet, ATM, SONET, and WDM which cannot be explained in greater detail here. These technologies use fibers, copper lines, and radio links as their physical infrastructure.

Thus, network planning is comprised of tasks like the selection of the network architecture, locations to place hardware, technologies, vendor and hardware selection, topology planning, dimensioning of network elements and connections, and, last but not least, traffic forecasts and routing. The main objective is to create and operate a high-quality network, which is robust, scalable and cost-effective. Planning goals and time horizons determine the priorities of different tasks and goals. We differentiate between three different planning cases:

- Strategic Planning (infrastructure and network architecture),
- Tactical Planning (network development),
- Operational Planning (network configuration and operation).

We will elaborate on these cases in the following sections and present both state of the art of the relevant mathematical methods and their respective application in practical contexts.

3.2 Strategic Planning

The selected technologies for the horizontal and vertical structure determine the network architecture. In order to prepare architectural decisions, *strategic planning* has to assess different options and compare them considering the complex interplay between technical and economic constraints. Within this context, planning for the individual network layers as well as the different networks has to be performed.

Determining the locations of the core equipment and the respective aggregation structure is one of the most important and longest-ranging decisions for a network provider. Based on traffic forecasts, technological capabilities of suitable network hardware, and associated cost structures, a plethora of closely interrelated questions has to be answered:

- What are the core network sites?
- Which of those sites are peering points to other providers?
- How are access and aggregation network areas segmented?
- What hardware in which configuration is used?
- What are the link capacities?
- Which resilience methods are used?

In order to answer these questions, scientists have developed a number of mathematical models. Their basis is quite often a *Facility Location Problem* formulated as an *Integer Linear Program* (ILP). Figure 4 shows a model which is similar to those used for topological planning of the German Research Networks G-WiN and X-WiN. Both, Lagrangian Relaxation as well as branch-and-cut methods have been used successfully to solve these problems. Substantial progress in the polyhedral characterization of the solution space allows planning of networks with up to 1000 locations.

Today, an expansion of fiber-based access networks is the next logical step for providers in Germany. A nationwide network would contain roughly 40 million traffic sources, which is close to the number of households in Germany. Considering all these sources individually is impossible even with cutting-edge mathematics and the most powerful supercomputers. Mastering these problem sizes is still an enormous challenge for the mathematical research community.

3.3 Tactical Planning

In contrast to strategic planning, *tactical planning* deals with less far-reaching decisions. Based on a given network architecture and an at least partially existing infrastructure, tactical planning organizes mid-term (roughly one budgetary year) development. Classical tasks for tactical planning are

- development of a physical topology,
- introduction of new network elements, and
- expansion planning for link capacities.

The overall aim is to adapt the network to mid-term traffic increases with minimal investments. Forecasts for existing services are usually quite reliable. The same is true for the costs of expansions options, although the costs can vastly differ. For

The German Research Network (DFN) connects more than 700 universities and scientific institutions nationwide. Variations of the classical Facility Location Problem were used as a mathematical basis for the strategic planning of the expansion stages G-WiN (1998) [6] and X-WiN (2004) [7]. The following formulation closely resembles the one used for X-WiN.

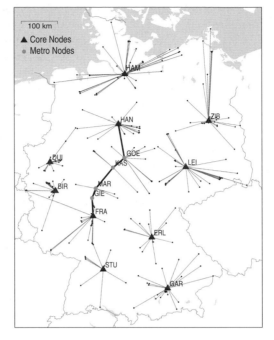

Example of a network hierarchy in X-WiN.

Problem: A set of possible aggregation sites A has to be chosen from the set of all possible locations N with a given subset of core network sites V. Every location has to be connected via a connection from set L with sufficient bandwidth to an aggregation site. The aggregation site itself has to be connected via a resilient path p to core sites (again with sufficient bandwidth).

Variables: $y_a \in \{0, 1\}$ with $y_a = 1$ iff a is selected as an aggregation site. $x_{ij} \in \{0, 1\}$ with $x_{ij} = 1$ iff i is a customer location and i is connected to the aggregation or core site j. $z_p \in \{0, 1\}$ with $z_p = 1$ iff $a_1, \ldots, a_m$ are selected as aggregation sites and are connected via the chain $p = (v_1, a_1, \ldots, a_m, v_2)$ to the core sites v_1, v_2. In a network with more than 700 locations and 100 possible aggregation and core sites, such a formulation contained quintillions of variables.

Mathematical Model: The corresponding formulation as an ILP is as follows:

$$\min \quad \sum_{a \in A} k_a^A y_a + \sum_{p \in P} k_p^P z_p + \sum_{ij \in L} k_{ij}^L x_{ij}$$

$$\sum_{ij \in L} x_{ij} = 1 \qquad \forall i \in N$$

$$\sum_{aj \in L} x_{aj} = 1 - y_a \quad \forall a \in A$$

$$x_{ia} \leq y_a \qquad \forall a \in A, ia \in L$$

$$\sum_{p \in P, a \in p} z_p = y_a \quad \forall a \in A$$

$+$ sufficient bandwidth

$+$ resilience

The application of complex reduction and aggregation techniques condensed the problems to *Mixed Integer Programs* with *only* 50 000 to 100 000 variables and inequalities. These MIPs were solved with cutting-plane and branch-and-cut methods.

Result: According to the DFN, "only mathematically well-founded solution approaches and highly automated planning methods could guarantee high-quality solutions with small remaining uncertainties and fully exploit the optimization potential."

Fig. 4 X-WiN Optimization Model

example, new trenches in urban environments for new fibers will be more costly than just upgrading the node hardware, in case the fiber has already been provisioned.

The network extension has to accommodate future demands with existing resources. The routing of the traffic streams has to meet the requirements of the service provider: In optical networks, transparent paths cannot be longer than certain limits dictated by physical properties of the transmission systems, packets in IP networks should not traverse too many nodes due to delay constraints, and so forth. Likewise, failures should not lead to traffic loss, or even worse, result in penalty fees. Due to the enormous link capacities in modern wide area networks, resilience is of vital interest to providers [10].

From a mathematical point of view, formulations based on multi-commodity flow problems with discrete link- and node-capacities model tactical planning tasks rather well [14]. Millions of alternative paths can be effectively handled using *column generation*. Constraints for resilience can be rather easily integrated as well, which is also true for most constraints imposed by hardware requirements.

In accordance with the vertical network structure, networks have to be planned across several layers. Thus, connections in higher layers are realized using (resilient) paths in lower layers. Considering several layers at the same time, leads to so-called *multi-layer problems*. From a mathematical point of view, these problems are extremely complex. While single-layer planning problems with up to 50 sites can be solved rather well, multi-layer problems can only be solved with acceptable solution gaps for small networks with about 10 nodes, even when sacrificing resilience constraints. Multi-layer planning and the integration of other planning aspects are future challenges, as we will accentuate in Sect. 4.

3.4 Operational Planning

Short-term aspects of network configuration and extension are part of the *operational planning*. In addition to mere hardware configuration, planning and optimization of the traffic routing in all layers is the fundamental task. Today's networks mainly use two approaches, namely *connection-oriented* and *connectionless* routing.

In *connection-oriented* routing schemes, network elements do not have their own intelligence for finding paths through the network, i.e. every route has to be explicitly planned and configured. Many lower layer technologies such as SONET use this approach, where an automated discovery of new routes is (yet) not possible. In order to establish a so-called *lightpath*, i.e. a ray of light transparently traversing several nodes, numerous settings closely related to the physical layer have to be manually configured. However, the latest hardware generation provides remote control of such features in *Optical Cross-Connects* (OXCs) and *Reconfigurable Add-Drop Multiplexers* (ROADMs), although automatic routing is still far away.

In contrast to this, in *connectionless* routing, network elements provide an autonomous process to discover paths in the network. IP packets for example are

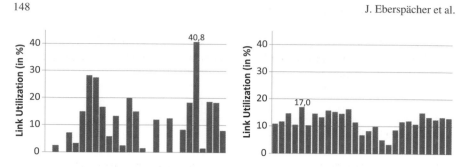

Fig. 5   Load distribution with factory defaults (*left*) and optimized weights (*right*)

routed according to "virtual distances". Core network technologies can also combine both approaches. For example, MPLS routes packets along the shortest paths, but service providers can manually configure diverging routes as well.

Modern mathematical methods are capable of solving realistic scenarios for both approaches. Variants of multi-commodity flow problems are a well-established standard formulation for connection oriented routing problems. In combination with column generation, additional constraints accommodating resilience or operational requirements can be included even in networks with several hundred elements.

Today's methods are furthermore capable of considerable improvements in connectionless scenarios compared to default settings which are routinely used in IP routers. Although the factory defaults may lead to acceptable results, an optimization of the link utilization is able to save costly resources, in particular, when taking failures into account. Consequently, expensive hardware upgrades can be postponed—and this without the use of "special" protocols, but merely by adjusting the virtual distances.

Even without going into detail of Fig. 5, the advantages of the optimized solution are quite obvious. One of the critical quality parameters in IP/OSPF networks, namely the maximum link utilization, can be reduced from 40.8% to 17%. Since this configurations do not require different hardware than before, they will be essential for future business models, where customers will get "bandwidth on demand".

Future network structures, realizing all services via a single *All-IP Network* will require a better understanding of stochastic properties of services and to integrate this information—which are currently based on measurements—into methods for operational planning.

3.5 Status Quo: Summary

In the past two decades, mathematical methods to solve planning and optimization problems have evolved considerably. This is particularly true for methods based on ILP. Powerful models and problem specific approaches have been developed for topology, routing and dimensioning planning. Combinations of delayed column generation, polyhedral methods, separation algorithms and Lagrangian methods are capable of solving problems with millions of routes or variables. Despite this impressive progress, a plethora of challenges remains open, on which we will elaborate in the following section.

4 Influences and Challenges

4.1 Market Influence, Competitive Conditions, Regulations

In the 1990s telecommunications monopolies fell in most countries—for example the monopoly of Deutsche Telekom for telephony services fell in 1998. Since then, a dynamic telecommunication market has emerged not only in Germany. In parallel, pressure regarding costs and margins has increased and will continue to do so in the future. Cell phone, cable, telephone, and data network carriers compete with service and content providers in a single common market with an increasingly broadening product range (telephony, data services, TV).

Hence, we can predict that integrated multi-service networks will gain further attraction and the conditions for mathematical optimization will improve likewise. Techno-economic analysis in preparation of regulatory decisions will lead to more accurate market assessments and technology evaluations.

4.2 Data Basis

Trustworthy input data is the foundation of any meaningful planning. Network information, data about hardware in network nodes or links, costs and traffic forecasts comprise the main input data needed for network planning. Availability and level of detail are highly varying. The data basis for node and link hardware is in good shape. Equipment manufacturers provide important information such as the data rate of a connection or the switching capacity of a node. On top, network operators maintain extensive information in their databases: Where are the network sites? What hardware is installed at which site? How are the network elements connected and with which media (copper, fiber, radio)? How are services routed over the network? The situation for cost and traffic data is notably worse. Price lists are usually not comparable due to vendor depending or intransparent rebate schemes. Furthermore, capital expenditures (CAPEX) are just a part of the total cost of ownership—operational expenditures (OPEX) have to be taken into account as well. First approaches to model these OPEX have appeared in recent publications, but the interdependencies are far from being sufficiently understood to be integrated into planning processes. Traffic forecasts are uncertain by definition. Standardized technology with integrated measurement and tracing capabilities thus can be a crucial advantage for operational (i.e. short-term) planning, as traffic demands can be quite reasonably deduced from measurements. However, long-term forecasts remain challenging—scenario-based analysis is the only practicable approach at the moment. From a provider's point of view, we can summarize that not only reasonable data for planning is already existing, but this data is also sufficient for modern mathematical optimization methods, capable of significant improvements. On the other hand, scientific progress requires a broad availability of realistic reference data for researchers. SNDlib [1], the BMBF projects EIBONE [2] and 100GET [3], and the EU projects NOBEL [4] took a first shot at providing such data sets. However, further efforts are necessary to provide adequate data for future planning challenges. The history of integer linear programming shows, how significant such contributions can prove to be: Typical optimization problems, published in the MIPLIB [15] can be solved a *million* times faster today, than at their original publication date [16]. Increases in computing power are responsible for a factor of 1000 in this figure. Significantly more powerful mathematical methods can be accounted for the remaining factor of 1000.

4.3 Convergence and System Complexity

In past networks, for almost every single service, a separate infrastructure was built—at least on layers above the physical transmission layer. This situation will completely change within the so-called *Next Generation Networks* (NGN). Future networks will converge towards a common basis of Ethernet and Internet technologies and will use one single WDM-based optical infrastructure. The result of this so-called *network convergence* will be large national and international multi-service networks: One network for all services. Whether this technical simplification will also simplify the necessary planning tasks is questionable, at least when considering the multitude of different services which have to be transported at the same time. While the clearly separated networks of the past could be planned independently, we now have to appropriately decompose one single, extremely complex problem first. This change cannot be mastered without innovative mathematical methods.

4.4 Planning Processes and Planning Tools

Current planning tools are powerful software packages supporting substantial portions of the network planning processes. Based on databases, which are also used for documentation purposes, these tools allow computer-aided, partly automated, but ultimately still manual planning. A good visual representation of the technological layers supports this process. Considering available mathematical methods for planning and optimizing networks, the used algorithms in said planning tools can only be characterized as outdated. Due to proprietary or nonexistent software interfaces however, the integration of new planning methods is extremely costly. Naturally, tool vendors have only very little own interest to integrate innovative methods, while the initial effort remains sufficiently high to scare network operators away from evaluating new methods in their planning environments. The looming change to *Service Oriented Architectures* (SOA) offers a new perspective on a mid-term timeline. This could lead to better standardized software platforms, which would allow easier combination of modular building blocks to form complex planning tools. Open interfaces would enable mathematicians and engineers to establish new and innovative methods, supporting providers to create cost efficient networks with higher resource utilization.

4.5 Collaboration Between Industry and Science, Engineers and Mathematicians

The telecommunication industry has traditionally been in close collaboration with universities. Sadly from a mathematical point of view however, these collaborations were focused on engineering sciences. In addition, large system manufacturers like Alcatel-Lucent, Ericsson and Nokia Siemens Networks as well as network operators like Deutsche Telekom have their own research departments. In national and international precompetitive projects, these departments collaborate to a considerable extent, as for example in the national BMBF programs EIBONE [2], 100GET [3], as well as in the European projects NOBEL [4] and EURO NGI [5]. Besides this, many bilateral projects exist between industry partners and engineering departments of universities. The transfer of both, know-how as well as employees, works quite well. Mathematics have only very rarely participated in these partnerships. However, this cannot be accounted to the engineering side; Mathematics have to open up regarding real life applications. Stray individual projects, such as the DFG research center MATHEON, participations of the Zuse Institute Berlin (ZIB) and a few mathematical spin-offs, are just too little, because engineers use mathematical methods with enthusiasm. The use of well-known optimization methods to solve complex planning tasks is getting more and more common and these well-established algorithms are readily available in software libraries—this situation is a perfect match! Mathematicians focus on theoretical and practical aspects of solution methods, thus enabling engineers to achieve higher-quality results.

4.6 Selected Challenges

If mathematicians want to increase their contribution to the development of communication networks, they will have to face the key challenges, which result from the migration to future networks.

Challenges for Network Operators

The challenges for future development are:

- mastering continuous change in the networks,
- techno-economic considerations,
- efficient utilization of existing resources.

Mastering Continuous Change

Next generation multi-service networks will transport enormous amounts of traffic. Within the next 5–10 years, the access data rates will raise up to 1 Gbit/s and for commercial customers even to 10 Gbit/s. Considerable investments in new technologies will be necessary to cope with these transmission capacities, however a huge base of installations is already existing and cannot easily be torn down. Hundreds of thousands of nodes comprise today's networks with complex connectivity via various different technologies. Despite these difficulties, current networks work rather well! Migration to new network architectures and new services raises new complicated questions: When should which services be provisioned using the new network? How do optimal structures of these networks look like? How dynamic will future traffic be? How can migration costs be minimized?

Techno-Economical Considerations

Network convergence, especially with regard to convergence of fixed and wireless networks, raises the importance of understanding the complex interactions determining the *Total Cost of Ownership* (TCO), which in return has a huge impact on turn-over and profits. Among the most significant factors are various technology and hardware options, traffic forecasts, expected price developments for the respective technologies, financial developments, expected demand for the various services and operational costs for the network. Consequently, network models have to cope with all layers.

From a provider's point of view, these models have to consider the most important technical aspects (description of IP, Ethernet, WDM nodes and physical connections) and the most important economic aspects (capital expenditures for the hardware, fibers, digging costs, operational expenditures like leased lines, energy costs, and human resources, financing and amortization options) as well. If alternative scenarios are compared using such techno-economic models in conjunction with turn-over and development expectations, management decisions will find a solid and transparent basis.

Efficient Utilization of Network Resources

Operation of today's communication networks is laborious, inflexible and quite resource consuming. Connection-oriented links are planned and established manually across all transport layers with high personal expenditures. A significant reduction of the setup time would set the field for a completely new business model: "Bandwidth on Demand" would allow customers to rent network capacities on-demand for comparably short time periods. In the end, customers could add more bandwidth just as simple as making a phone call. Furthermore, using transmission media as efficiently as possible is of great interest, no matter whether it's the classical copper line, cell phone channels or fibers for wide area networks. In order to provide the required transmission capacities, which go into the Tbit/s range, the number of wavelengths per fiber is critical—hundreds of wavelengths can already be reached under lab conditions, but there is significant interest in even larger increase. However, when pushing the fiber to its physical limitations, nonlinearities determine its characteristics, which in return have to be integrated in the transmission models.

Mathematical Challenges

Which mathematical challenges are created due to the described technical evolutions?

Fast & Scalable Optimization Methods

Network convergence towards a unified multi-service-platform has one decisive drawback for the planning processes. Borders between access, aggregation and core networks become more and more diffuse and the formerly inherent decomposition in tasks for individual network segments disappears. Thus, the size of the networks which have to be analyzed for planning grows significantly. Today's methods however have not been designed to accommodate these larger networks. Hence, improvements of the theoretical foundations and suitable integration with heuristic methods have to be the foremost goal in order to be able to optimize large networks in a fast and reliable manner. Consequently, innovative and computing friendly models are necessary as optimization methods, capable of handling non-linear problem classes which are caused by physical effects.

Integration of Planning Tasks

Up to now, the above described methods can support planning problems, dealing with a single layer, a single point of time, a single demand matrix or a fixed network hierarchy. This limitation is caused by insufficient performance of the mathematical methods. In particular for future Next Generation Networks more and more Multi-X problems have to be solved, where "X" can stand for layer, period, hour, level, etc. Loose models for these planning tasks exist even today and promising solution methods for small networks have appeared. A set of building blocks made up of modular models and methods, providing flexible, quick and scalable responses to telecommunication planning requests, has to be the ultimate goal.

Robust and Stochastic Optimization and Simulation

Forecasts of future network traffic and costs suffer from inherent uncertainties. A well tuned network planning process aiming to efficiently use resources, has to provide guarantees against uncertainties in the input parameters. Robust and stochastic optimization have the means to provide these guarantees, however significant research effort has to be put forth to increase the performance allowing to plan future networks. Simulation and scenario analysis as well-established methods in this context can certainly not be replaced. In fact, further research regarding these methods is necessary: The more accurately we can analytically formulate the stochastics of data traffic, the better we can build robust planning methods on top. Furthermore, *self organizing networks* will play an important role in the future, which leads to challenging questions: How can infrastructure planning provide sufficient and cost-effective resources, so that the self-organized operation can run flawlessly?

Recommendations

The possibilities of applying innovative mathematical methods are endless. Mathematical research should focus on strategic investments in the beginning. In this area, the attainable leverage will be the largest, as telecommunication networks will fundamentally change within the next ten years. Both, the expansion of fiber-based access networks with end user transfer rates up to 10 Gbit/s as well as the migration towards multi-service networks require enormous investments. In the long term, operational and strategic planning will be attractive for mathematics as well. However, tools and methods have to change to accommodate innovative methods to be able to exercise significant influence.

5 What Has to be Done? Proposals and Initiatives

Efficient creation and operation of telecommunication infrastructure is of utmost importance. Mistakes in the design or dimensioning will cause high costs or weaken complete geographical regions or commercial sectors. Accurate and methodical planning is the base for the use of the various continuously evolving network technologies. The relevant tasks can be solved significantly better by closely collaborating application-oriented mathematicians and methodically well-educated engineers. Thus, interdisciplinarity has to be facilitated within education, joint research projects and collaborations between science and industry.

5.1 Education

Mathematical education is partly lacking in practical relevance. In addition to the necessary theoretical foundations, the curriculum in the main studies should be en-

hanced with material about modeling, simulation and optimization of real-life processes.

The engineering education provides a solid mathematical foundation. The main
focus lies on analytical and numerical methods. In regard to these challenges of
planning, configuration and optimization of communication networks, there is a
significant need for education in graph theory, combinatorics, linear and integer
programming and operations research.

5.2 Economy

Increasing the usage of mathematical methods in commercial contexts is a difficult
task. The reason for that can be found in the different problem solving approaches
and a missing common language, but more practical reasons exist as well. The current IT infrastructure for planning tools is a huge obstacle for the integration of innovative methods. The joint research project proposed in the next section, is conceived
as a first step towards a more open process for planning and operating communication networks. Depending on the level of industry collaboration, a reference system
for an innovative planning process could be created. Obviously, this has to be considered as "a long-term vision" despite the immanent advantages. A more formal
standardization of planning processes or goals (in analogy to ISO 9001) might help
providers to advertise accurately planned and "mathematically certified" networks.

5.3 Research Proposal

We consider a joint research project with collaborations between engineering and
mathematics institutes to be a very promising option. Such a project—which could
be appropriately funded by DFG or BMBF—could be called "Modeling, Planning
and Evaluation of Communication Networks". The main task would consist of a
comprehensive modeling of the network, planning and optimization tasks in a collaborative effort by engineers and mathematicians. Adequate methods to simulate
and analyze networks would further augment this work. Methods for the challenges
described in Sect. 3 and Sect. 4, particularly for large networks (scalability), QoS
planning, and multi-layer planning would comprise further tasks. Based on these
models, processes capable of flexibly answering strategical, tactical and operational
planning questions, should be developed. Without creating a true SOA infrastructure, these models should be loosely coupled and data exchange via open interfaces
should be possible. In order to ensure practical relevance, strategic departments of
operators and manufacturers should play an advisory role, which will support the
creation of realistic reference data sets.

References

1. http://sndlib.zib.de
2. http://www.bmbf.de/de/6103.php
3. http://www.celtic-initiative.org/Projects/100GET
4. http://www.ist-nobel.org/Nobel/servlet/Nobel.Main
5. http://eurongi.enst.fr/en_accueil.html
6. Bley, A., Koch, T.: Optimierung in der Planung und beim Aufbau des G-WiN. DFN-Mitteilungen **54**, 13–15 (2000)
7. Bley, A., Pattloch, M.: Modellierung und Optimierung der X-WiN Plattform. DFN-Mitteilungen **67**, 4–7 (2005)
8. Dijkstra, E.W.: A note on two problems in connexion with graphs. Numer. Math. **1**, 269–271 (1959)
9. Lyon, B., et al.: January 2005. http://www.opte.org/maps
10. Grover, W.D.: Mesh-based Survivable Networks. Prentice-Hall, New York (2004)
11. Cisco Systems Inc.: Cisco visual networking index—forecast and methodology, 2007–2012 (2008)
12. Ambs, K., Cwilich, S., Deing, M., Houck, D.J., Lynch, D.F., Yan, D.: Optimizing restoration capacity in the AT&T network. INTERFACES **30**(1), 26–44 (2000)
13. Kleinrock, L.: Queueing Systems. Wiley, New York (1975)
14. Pioro, M., Mehdi, D.: Routing, Flow, and Capacity Design in Communication and Computer Networks. Elsevier, Amsterdam (2004)
15. Bixby, R.E., Boyd, E.A., Indovina, R.R.: MIPLIB: a test set of mixed integer programming problems. SIAM News 25(20) (1992)
16. Bixby, R.E.: Solving real-world linear programs: a decade and more of progress. Oper. Res. **50**(1), 3–15 (2002)

Mathematics in Wireless Communications

Holger Boche and Andreas Eisenblätter

1 Executive Summary

Mobile communication has a great economic importance in today's world. Since the introduction of GSM in the early 1990s—which is still the dominant mobile communications standard—it has had an enormous impact on our social life. Its importance will continue to climb with new fields of application, e.g. in the construction of machines and factories, the automobile industry and in domestic living. New systems of mobile communication are continually being introduced to live up to the great variety of applications and the desires for new services.

This development would have been impossible without the digitalization of mobile communication—whereby mathematics played an indispensable supporting role. Mathematics is the foundation of information and communication theory, the pathbreaker in the development of new transmission procedures and an essential instrument in the planning and optimization of mobile networks. Probability theory, for example, and discrete mathematics are used in information theory; linear algebra, convex optimization and game theory are used in the development of new methods of transmission; and linear, combinatorial and stochastic optimization are used in planning radio networks.

Despite this immense progress, essential foundational problems in wireless communication are still to be resolved. Examples include a network information theory for more complex multi-user systems (such as mobile ad hoc networks and wireless sensor networks), a quantum information theory for miniaturized systems, a theory of self-organizing networks, and the development of new paradigms for the use of

H. Boche (✉)
TU Berlin, HFT 6, Einsteinufer 25, 10587 Berlin, Germany
e-mail: holger.boche@mk.tu-berlin.de

A. Eisenblätter
Bundesallee 89, 12161 Berlin, Germany
e-mail: eisenblaetter@atesio.de

M. Grötschel et al. (eds.), *Production Factor Mathematics*,
DOI 10.1007/978-3-642-11248-5_9, © Springer-Verlag Berlin Heidelberg 2010

the scant frequency spectrum. In all these cases it is the collaboration of telecommunications engineering and mathematics that has to deliver the essential contributions to a solution.

2 Success Stories

The economic and social importance of mobile communication and thus of radio is extraordinary. Today radio networks are an important part of the telecommunications infrastructure. In many parts of the earth radio telecommunications offers the first economical basis for large-scale communications infrastructure.

Mobile telephony and communication through brief text messages has spread since the early 1990s with a practically inconceivable dynamism. This growth is based on the introduction of the second wireless telecommunications generation (2G) that uses digital data transmission, allowing a considerable leap in power compared to the previously prevalent analog technology. The Global System for Mobile Communication, GSM for short, was one of the first 2G systems and is today still the most successful. While only around a million wireless customers were counted in Germany in 1992 (the first GSM users, classic car telephones, etc.), by the end of 2007 they amounted to more than 97 million [29]. Thus, on average each resident holds more than one mobile subscription. Worldwide there are more than 3 billion GSM customers and more than a million new customers every day [6].

With UMTS many industrial nations have seen the introduction of the third generation (3G) system. This is supposed to replace GSM in the long run and already has more than 200 million users in 73 countries [16]. UMTS stands for Universal Mobile Telecommunications System and is especially designed for mobile data service with a maximal data rate of 384 kbit/s. Compared to GSM it increases the maximum data rate per connection by a factor of around 30. Moreover, in Europe enhancements to UMTS have already been installed that allow a further increase in the data rate as well as a reduction of reaction time. Reaction time is, roughly speaking, the time that elapses between a data request (for example a mouse-click on a website) and the receipt of the data (in this example, the next website). The UMTS extension High-Speed Downlink Packet Access (HSDPA) allows data rates of 7.2 Mbit/s from the base station to the user (downlink) and the High-Speed Uplink Packet Access (HSUPA) allows rates of 1.4 Mbit/s in the other direction (uplink). As an alternative, or in conjunction with UMTS, WiMAX (World-wide Interoperability for Microwave Access) has been introduced in several countries, with even better power properties for wireless communications than UMTS and its enhancements.

The UMTS Long Term Evolution, LTE for short, is introducing another international standard for mobile telecommunications systems. LTE will perform at the same level as WiMAX at least, should be seamlessly integrable with GSM and UMTS and in contrast to the existing systems should be considerably easier to operate.

2.1 Mobile Communications Needs Mathematics

The intensive application of mathematics is a cornerstone of this development. This article will describe the important contributions of mathematics for three areas:

- Foundations of mobile communications
- Development of mobile communication systems
- Planning and optimization of mobile networks

The development of new paradigms for the fair use of the scant radio frequency spectrum is difficult to fit under one of these areas. This will be discussed in Sect. 5.4.

The following three sections each present one exemplary contribution of mathematics to the development and operation of modern mobile communications systems for each of these areas. This should illuminate mathematics' share in the increased efficiency of modern telecommunications. Representing fundamental research, Sect. 2.2 will describe how multi-antenna systems can multiply transmission capacity. Section 2.3 will present multi-user receivers as an example of the contribution to the development of mobile communications systems. The mathematical contributions to frequency planning for GSM networks will be presented in Sect. 2.4 as an example of the planning and optimization of radio networks.

2.2 The Foundations of Communications Theory

The Shannon information theory forms the mathematical foundation for modern communications theory. It describes how to enable communication over a noisy channel and how users are to share communication resources, for example frequency, time, space and power. Essentially, information theory seeks to answer two fundamental questions: what is the maximum amount of information that can be transmitted in a given communication scenario, and which practical communication techniques have to be used to achieve the maximum rates of transmission. The Shannon noisy-channel coding theorem [31] is a central result of information theory.

The noisy-channel coding theorem states that for each transmission rate R smaller than the channel capacity C the probability of error-free decoding at the receiver can be made arbitrarily small. Conversely, for rates $R > C$ an error-free transmission is impossible. Thus the channel capacity sets the upper limit for the maximum possible error-free transmission rate.

Figure 1 describes this capacity limit and shows what methodical approaches can be used to overcome this limit of classical transmission systems. The focus is on the simultaneous low-noise transmission between two users, a transmitter and a receiver, over several antenna each. The theory of the noisy-channel coding theorem for communication between two users had in large part already been developed in Shannon's 1948 article.

The concept of channel capacity comes from C.E. Shannon, who created the foundations of today's information theory with his pioneering 1948 work *A Mathematical Theory of Communication* [31].

The channel capacity C is defined as the supremum of all rates that can be achieved error-free. Shannon showed that for a discrete, memoryless channel the capacity is given by the maximum of the mutual information over all possible input probability distributions $p(x)$:

$$C = \max_{p(x)} I(X; Y). \qquad (1)$$

The mutual information $I(X; Y)$ measures the strength of the static dependence between the input X and the output Y. The larger the mutual information, the better the inference from the received signal to the transmitted signal.

For the AWGN channel, the received signal y is given by

$$y = hx + n$$

where x represents the signal to be transmitted, h the channel influence and n the additive white Gaussian noise (AWGN) at the receiver. If the transmitter power is constrained by P then

$$C = \max_{p(x):\mathbb{E}\{|X|^2\}\leq P} I(X; Y)$$

$$= \log\left(1 + \frac{|h|^2}{\sigma^2} P\right) \qquad (2)$$

indicates the channel capacity, where $\frac{|h|^2}{\sigma^2} P$ describes the signal-to-noise ratio (SNR), see e.g. [11].

Equation (2) describes the case where transmitter and receiver are both equipped with one antenna each. This configuration is called a SISO-channel (single input single output). The more general case with N_T transmitter antennas and N_R receiver antennas is called a MIMO channel (Multiple Input Multiple Output). Here the reception vector Y is given by

$$Y = HX + N$$

The transmission vector X contains the transmission signals for the corresponding transmission antennas. The channel matrix H describes the channel influence. The (k, j)-th input of the matrix describes the influence of the channel on a signal between the j-th transmitting antenna and the k-th receiving antenna.

MIMO-channel:

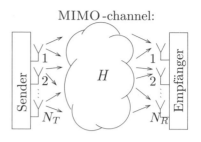

The capacity (1) of this channel with transmission power constraint P is

$$C = \max_{\mathrm{tr}(Q)\leq P} \log\det\left(I + \frac{1}{\sigma^2} H Q H^H\right), \qquad (3)$$

whereby $Q = XX^H$ describes the transmitting covariable matrix [33, 34].

The MIMO channel can be described as parallel SISO channels by means of the singular value decomposition. If the decomposition is known, the so-called water-filling solution provides the optimal covariance matrix

$$Q^* = V \Lambda_Q V^H$$

of (3), whereby the characteristic vectors of Q^* correspond to the singular vectors of the channel on the right side. With this choice of Q^* the resulting system can be interpreted as parallel SISO channels coupled via their transmitting outputs. The available total transmission output is distributed among the individual channels. This finally produces the capacity of the MIMO channel

$$C = \sum_{i=1}^{N} \log\left(1 + \frac{\lambda_i}{\sigma^2} P_i\right),$$

whereby λ_i describes the influence of the i-th SISO channel and P_i the corresponding transmission output. The capacity of the MIMO channel corresponds to the sum of the individual SISO channel capacities. Hence multiple transmission and reception antennas can significantly increase the capacity of the system.

Fig. 1 Channel capacity of single and multi-antenna systems

It has proven much more difficult to develop an information theory for several participants. Two important multi-user communication systems are the multiple-access channel and the broadcast channel. In a multiple-access channel several users want to send messages to one user simultaneously. This communication scenario models the uplink to a cellular network, i.e. several users send signals to a base station simultaneously. In a broadcast channel one user sends different messages to several users. This models the downlink of a cellular network, i.e. the base station send diverse information to several users. In the past 30 years it has become possible to characterize the optimal transmission strategies for the multiple-access channel and in important special cases for the broadcast channel as well. The understanding we have gained of the capacity regions of this multi-user systems has had a great influence on the development of UMTS, HSDPA, HSUPA and LTE.

New mathematical techniques will have to be developed to solve the many questions of information theory in practice. One example of this is the theta-function introduced by Lovász in [23]—an interesting discussion of which is found in [3]. With this function Lovász succeeded in calculating the capacity for error-free communication for certain classes of communication graphs. The Lovász theta-function has been generalized by Grötschel, Lovász, Schrijver in [20] and it is used today in many branches of mathematics in this general form.

2.3 Key technologies

For a new communication system to reach the specified throughput goals it has to achieve a certain spectrum efficiency. Spectrum efficiency indicates how many bits per second per Hertz can be transmitted. How efficient the system can be made depends on the physical signal dispersion, although this in turn can also be influenced by the system designer, for example through the number of antennas used.

Figure 2 describes this case for a multi-antenna system. The system designer can analyze, for example, how many transmitter and receive antennas the individual users should use. The spectrum efficiency that can be achieved depending on the number of antennas used is given by relation (3) in Fig. 1. Rayleigh Fading was used as the channel model [34]. This model is derived from the physical conditions

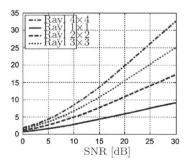

Fig. 2 The attainable efficiency of a multi-antenna system in dependence on the signal-to-noise ratio at the receiver

of wave propagation with the help of the limit theorems of probability theory. The entries in the channel matrix are independent and distributed identically according to a Gaussian distribution with the mid-point of zero. The model describes very well how a narrow-band signal disperses when there is no line of sight between transmitter and receiver. In general the analytical assessment of spectrum efficiency presents a difficult problem, since a suitable mathematical model of the channel has to first be developed.

The results discussed here so far essentially concern the optimization of communication between two users. However, signals sent by other users can interfere with this communication. One important approach within modern communication theory is not to see these interferences as noise, but rather to make use of their special structure to minimize the effective influence of interferences. This is just what multi-user receivers do. Multi-user receivers are able to suppress interferences with varying effectiveness depending on the complexity of the implementation. The optimal multi-user receiver for the multi-access channel (see Sect. 2.2) even allows the complete suppression of the interferences generated by selected users.

The new concepts for multi-antenna systems and multi-user receivers have made an essential contribution to the development of UMTS, HSDPA, HSUPA, WiMax and LTE. Without these technologies the rapid increase in maximum transmission rates would not have been possible. The most diverse mathematical disciplines were needed for the analysis and design of the corresponding communication systems— including, for example, combinatorics, discrete mathematics, mathematical optimization, functional analysis and the theory of random matrices.

2.4 Optimization of GSM Networks

The planning and optimization of GSM mobile networks is quite challenging. Setting up a network is not a short-term process of two or three years—it is a continual disassembly and reassembly. In Germany many hundreds of millions of Euros will be invested in infrastructure for the year 2008, i.e. 16 years after the start [19, 29]. In addition to the one-time costs of building the network, regular operation costs also continue to accrue, and thus network operators want to run an efficient and cost-effective infrastructure.

Methods of combinatorial optimization, a sub-field of mathematics, can be used at all levels of planning: the long-term planning of network architecture for many years, the medium-term planning the network structure, and the short-term operative planning of the configuration of individual network elements intended for a few months.

The medium-term planning of the coverage surface and the short-term adaptation of network capacity can be considered in isolation from one another in the case of GSM. Coverage is determined primarily by the selection of stations and the alignment of the typically three sector antennas, each of which defines a cell of the network. The capacity is determined for each cell by the number of frequency channels reserved.

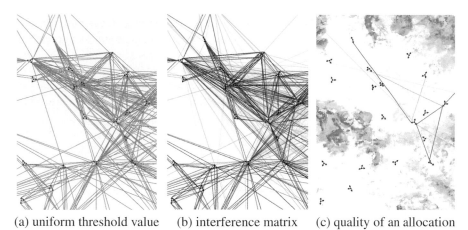

(a) uniform threshold value   (b) interference matrix   (c) quality of an allocation

Fig. 3 A GSM network with 19 base stations is depicted. *The small arrows* show the position and alignment of the sector antennas. The graph theoretical model (**a**) only accounts for interferences above a certain threshold. A differentiated consideration of the interference relations between cells is possible with the interference matrix in (**b**). *Red lines* represent stronger interference relations than *yellow*. In (**c**) *the lines* for a specific allocation indicate which of the interference relations are not triggered by different frequencies. *The color* of the background represents the interference strengths at different locations. *Red* indicates poor quality

Frequency planning involves choosing which frequency channels a cell should operate. For example, many tens of thousands of cells demand a channel or several channels from a small number of around one-hundred different channels. The channels have to be reused. Many preconditions of suitable channel allocation have to be considered. Since the early 1990s automatic frequency planning has thus been in development.

The first models were made purely in terms of graph theory and did not distinguish interferences according to their strength [17, 21]—see Fig. 3a. With the growing need for capacity, this initial modeling proved to be too simple. Starting in the mid-1990s a close collaboration between telecommunications engineers and mathematics introduced more detailed models that accounted for interference [13, 28]—see Fig. 3b. New allocation procedures were developed on the basis of these models [1, 10].

Since the end of the 1990s most network operators use correspondingly high-performance frequency-planning procedures, either of their own development or as commercially acquired products. In this way network operators achieve the intensive reuse of GSM frequencies that is necessary today.

3 Knowledge Brings Progress

Section 2 showed the considerable impact that mathematics exercises on telecommunications technology with three examples. This section will show how this impact is achieved for concrete applications. The section ends with a discussion of the connection between the technological progress achieved in the semi-conductor industry and the applicability of complex transmission technologies. The great importance of mathematics can be seen at various levels in these examples. It occupies a central role in the exact formulation of the telecommunications engineering problem; it then provides the technologies to analyze the structure of the problems; and it offers methods to efficiently calculate the corresponding solutions and provides the basis for a universal theory concerning the axiomatic approach to interference limited wireless systems.

3.1 Optimized Resource Allocation in Cellular Radio Networks

The technology of channel aware transmission enabled a considerable increase in spectrum efficiency in the development from UMTS to HSDPA (High Speed Downlink Packet Access) and HSUPA (High Speed Uplink Packet Access). This began with the observation that due to the physical nature of radio waves, the quality of a radio connection depends very much on where the user is located. When a user moves, the quality of the channel changes.

In a radio cell with many users, there is a high probability of finding a user with a good channel quality. The idea of channel aware transmission in HSDPA is to preferentially send messages to the user when the channel is in a good state, thus increasing the total transmission rate. However, this gives rise to a new problem for the telecommunications engineer—the problem of the fair allocation of resources. If the resources are distributed such that all users receive the same data throughput, this generally leads to a poor total throughput for the network.

A new approach to resource allocation aims at maximizing the total throughput. A utility function assesses the data throughput that is possible at that moment for each user from the point of view of the network. The recourses are then distributed so as to maximize the utility function. Yet this network operator-centered approach does not generally guarantee fairness for the users.

One challenge in developing optimal schedulers for HSDPA and HSUPA consists in bringing together the disciplines of information theory and queueing theory. Figure 4 describes the channel aware transmission for the downlink and emphasizes the importance of data buffering for this concept. Heuristics were developed for practically applicable schedulers on the basis of the theoretical results achieved thus far (see Fig. 5). Yet in general a satisfactory combination of information theory and queueing theory still presents an unresolved problem.

Channel aware transmission is an important example of the optimization of wireless systems through the use of cross-layer information. This approach is described in Fig. 6.

The next section will show that the approach to optimizing a utility function can be put to use just as successfully in optimizing meshed networks and ad hoc systems.

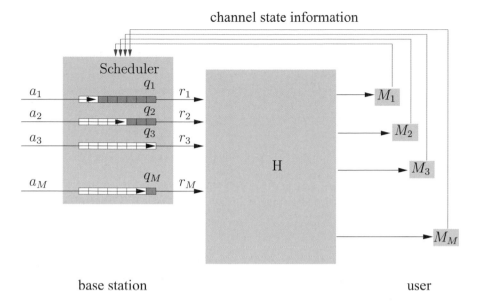

Fig. 4 Multi-user downlink system and the buffering structure. **H** describes the state of the time-dependent broadcast channel. The vectors $(a_1, \ldots, a_M)$, $(r_1, \ldots, r_M)$ and $(q_1, \ldots, q_M)$ are the momentary arrival rates, data transmission rates and buffering states

3.2 Controlling Transmission Power

The control of transmission power is a mechanism of resource allocation and interference management in wireless networks. Utility-based strategies to power control aim at coordinating transmitter power outputs from base stations or mobile communication devices such that a network utility function reaches its maximum for a given set of permissible transmitted power outputs [8, 32, 38].

Figure 7 shows a meshed communication network. The network consists of several base stations. Some of the base stations are not directly connected to the fixed line network and use other base stations as relay stations. Every base station has a multi-antenna system whose directional pattern is controlled by the so-called transmitting beamformer or receiving beamformer (the electronic control of the antenna in one or several preferential direction). The transmitted power outputs of the base stations are to be ascertained through the use of a distributed algorithm. The optimal allocation of outputs depends on the state of the channels. If the channels change, the transmitting power outputs have to be adapted. This also holds for the transmitting and receiving beamformers, which in addition can also be influenced by the transmitting power outputs. Figure 8 shows how power control can be achieved through the use of utility optimization.

The maximization of the throughput represents an important special case of this where the network utility function is a weighted sum of the data rates of the individual users. However, optimal throughput strategies can be very unfair, since depending on the channel states there is a possibility that they could deny some users

The scheduling problem can be formulated as follows: let $t \in \mathbb{N}$ be a discrete time parameter of the observed random process for M users. To be able to describe very general and diverse physical circumstances (multi-user antenna systems, coding SW) the physical level is described with a rate vector $r_t = (r_t^{(1)}, \ldots, r_t^{(M)}) \in \mathbb{R}_+^M$. This vector has to be within the capacity region $\mathcal{C}_t(p) \subset \mathbb{R}_+^M$ with a general power budget $p = (p^{(1)}, \ldots, p^{(M)}) \in \mathcal{P}_t$, for each time interval, whereby $\mathcal{P}_t$ is a permissible power region. It should be emphasized that the region $\mathcal{C}_t(p)$ nodes not necessarily need to present any special structure (for example convexity). To ensure the stability of the system, the buffering state $q_t = (q_t^{(1)}, \ldots, q_t^{(M)}) \in \mathbb{R}_+^M$ of the users is introduced and accounted for in scheduling. The rate vector r_t is the solution to the following optimization problem:

$$\begin{aligned} \max \quad & \mu(q_t)^T r_t \\ \text{s.t.} \quad & r_t \in \mathcal{C}_t(p), \\ & p \in \mathcal{P}_t \end{aligned} \tag{4}$$

Here $\mu = (\mu^{(1)}, \ldots, \mu^{(M)}) : \mathbb{R}_+^M \to \mathbb{R}_+^M$ is a representation of vector value containing the criterion aimed at in each situation (e.g. minimal delay). It can be shown that the description (4) is exhaustive in a certain sense. An example of this sort of problem is the maximum weight matching scheduling with $\mu_t(q_t) = q_t$. However, it is not in general clear which mapping $\mu_t(q_t)$ leads to the stability of the entire system. An interesting sub-problem is found in the context of information theory where $\mathcal{C}_t(p)$ is a so-called polymatroid. Even in this case the solution of (4) is a non-trivial combinatorical problem.

In practical systems the capacity region $\mathcal{C}_t(p)$ can be described as the combination of the realizable rates $r_{t,i}, i = 1, 2, \ldots$. This produces the following discrete optimization problem:

$$\begin{aligned} \max \quad & \mu(q_t)^T r_t \\ \text{s.t.} \quad & r_t \in \bigcup_i \{r_{t,i}\}, \\ & r_t^{(m)} \geq r_{min}^{(m)}, \quad \forall m \end{aligned} \tag{5}$$

The required minimal rates $r_{min} \in \mathbb{R}_+^M$ ensure minimal quality for the services. This problem is also essentially unresolved for many cases of practical application.

Fig. 5 Scheduling according to maximum utility

To support broadband services with high data rates and individual quality-of-service (QoS) requirements, radio resources have to be dynamically allocated in wireless communication systems. If this allocation occurs using system information from several functionally separate layers [7], it is called a cross-layer problem.

Cross-layer problems are a diverse component of current telecommunications research—particularly in connection with orthogonal frequency division multiplexing (OFDM) systems which allow a particularly "fine" resolution of the resources time, frequency, and space. Special resource allocation procedures in the media access control (MAC) layer, also called scheduling procedures, are particularly important—they take into account power budget, channel state, service requirements and minimal data rates in a common problem context. It should be noted that the data packets of each user arrive not continuously but rather randomly distributed over time and are saved for the meantime in buffers. This can negatively impact the stability of the entire system. Different criteria can be used to define stability (e.g. recursion in a Markov model) and influence the system behavior accordingly [36].

In general the scheduling problem can be described as a utility optimization problem. Frequently used utility concepts include system throughput and reaction time [37, 39]. The system stability has to then be implicitly accounted for.

Fig. 6 Optimization of mobile systems through the use of cross-layer information

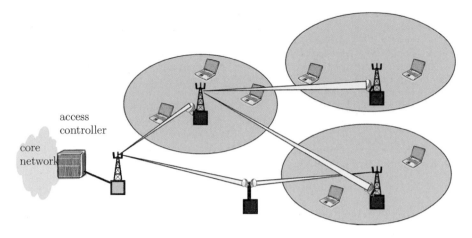

Fig. 7 Wireless communication with relay stations

access to the radio channels for a relatively long time. Since this in turn can leads to long reaction times for certain end-to-end connections, other network utility functions are under consideration that can ensure a better compromise between fairness and throughput performance [32].

3.3 Axiomatics for Interference Limited Radio Systems

Many approaches to the optimization of wireless communication systems are concerned with controlling the interference between the users caused by the dispersion of radio waves. The kind of interference experienced by an individual user depends on the user's receiver as well as the transmission strategies of other users. This variety gives rise to a seemingly unmanageable number of potential interference conditions.

Axiomatic theory allows for an elegant solution to this problem and helps to understand the essential properties of interference (see Fig. 9). The theory of interference functions has proven itself to be a high-powered method to help solve resource allocation problems. Technologies have been developed to control output and optimize reception in complicated communication scenarios, for example ad hoc systems.

The class of interference functions has an interesting algebraic structure. For example, the minimum of a set of interference functions is itself an interference function. The services that a multi-user system can support can be completely described with interference functions with the mathematical properties of general interference functions.

Moreover, certain utility optimization challenges of resource allocation can be reduced to the optimization of interference functions. One example of this is the Nash bargaining solution [22], where the product of the utilities of the individual users

Let us assume there are $K > 1$ users and $\mathbf{p} = (p_1, \ldots, p_K) \in P \subset \mathbb{R}_+^K$ is the power vector the coordinates of which describe the transmitted power outputs of individual users. The set P contains all permissible power vectors. Utility-based power control can be formulated as a minimization problem [8, 32]:

$$\mathbf{p}^* = \arg\min_{\mathbf{p} \in P} F(\mathbf{p}), \qquad (6)$$

where $F : \mathbb{R}_+^K \to \mathbb{R}$ describes a (negative) network utility function given by

$$F(\mathbf{p}) := \sum_{1 \le k \le K} w_k \psi\big(\mathrm{SNR}_k(\mathbf{p})\big) \qquad (7)$$

Here $\mathbf{w} = (w_1, \ldots, w_K)$ is a given positive weight vector and $\psi(x)$, $x \ge 0$, a (negative) utility function. The function SNR_k describes the signal-to-interference ration of the k-th user, so that $\mathrm{SNR}_k(\mathbf{p}) := p_k / I_k(\mathbf{p})$ where $I_k : \mathbb{R}_+^K \to \mathbb{R}_+$ is an interference function and thus the special value $I_k(\mathbf{p})$ the effective interference power (including noise power) at the k-th receiver output when the transmission power outputs are given by $\mathbf{p}$. Here it should be noted that in general every SNR is influenced by all transmitted power outputs. In Fig. 9 the reader will find axiomatic definitions (A1)–(A3) of general interference functions.

Little is known about the solution $\mathbf{p}^*$ of problem (6) for general interference functions that only satisfy (A1)–(A3). Since a closed solution is only possible in a few trivial cases, recursive algorithms come into consideration. Due to the scarcity of resources in wireless networks, these algorithms should demonstrate global convergence and allow for an efficiently distributed implementation. The first condition is met if each local minimum to the problem (6) is at the same time a global minimum. The important question as to which functions ψ in (7) this requirement is fulfilled for has only been given a satisfactory answer for one class of interference functions so far. All local minimums of the problem (6) are simultaneously global if $\psi(e^x)$ is concave for real x and $I_k(e^s)$ with $\mathbf{s} = \log(\mathbf{p})$, $\mathbf{p} > \mathbf{0}$, is a log-convex function of the logarithmic power vector $\mathbf{s} \in \mathbb{R}^K$ [32]. In the meshed network with base stations the log-convexity requirement is met when the reception beamformers for given channel realizations are random but stable. In the case of the optimal reception beamformer the interference function is not concave rather than log-convex.

In [32] recursive algorithms for power control are presented and analyzed. For application in mobile networks both simple gradient algorithms as well as primal-dual algorithms come into consideration for the search for stationary points of a corresponding Lagrange function. In the latter case, alongside classical (linear) Lagrange functions, modified non-linear Lagrange functions are also being considered in order to accelerate the convergence to local or global minimums. An important requirement is that the algorithms must be able to be efficiently implemented in shared networks. In the example of application in Fig. 7 this means that every base station has to be able to calculate its iteration steps autonomously using local measurements and variables. The procedure presented in [32] based on the so-called adjugate network allows every user a very efficient estimation of the search direction using local measurements such as SNR. This procedure can be used in connection with gradient and primal-dual algorithms.

Fig. 8 Utility-based power control for radio networks

is maximized. This product function is an elementary mathematical interference function, which, however, does not allow for any originally physical interpretation. The resource allocation strategy based on the Nash bargaining solution is equivalent to the proportionally fair resource allocation strategy for wireless communication systems developed in telecommunications engineering. An important accomplishment of J. Nash (Nobel Prize for Economics, 1994) was to characterize his solution axiomatically [22]. This approach can also be expected to take on increasing importance for wireless communications systems in the future. For a further discussion of this see Sect. 5.4.

As in Fig. 8 $\mathbf{p} = (p_1, \ldots, p_K) \in \mathbb{R}_+^K$ describes the power vector the components of which equal the transmission power outputs of the users. In wireless communications it is assumed that every interference function $I : \mathbb{R}_+^K \to \mathbb{R}$, $K \geq 1$, satisfies the following conditions (axioms):

(A1) $\mathbf{p} > 0 \implies I(\mathbf{p}) > 0$
      (Positivity)
(A2) $\mathbf{p} > 0, \mu > 1 \implies I(\mu\mathbf{p}) < \mu I(\mathbf{p})$
      (Scalability)
(A3) $\mathbf{p}^{(1)} \leq \mathbf{p}^{(2)} \implies I(\mathbf{p}^{(1)}) \leq I(\mathbf{p}^{(2)})$
      (Monotony)

It can be shown [30, 32] that (A1)–(A3) implies the continuousness of interference functions. The functions from the example in Fig. 8 satisfy the axioms (A1)–(A3) when the transmission beamformers are chosen independently of the power vector and are thus fixed if the instantaneous channel realizations are known. In contrast, the reception beamformers can very much depend on transmission power outputs. The case that interests us here is the optimal reception beamformer that maximizes the corresponding SNR for a fixed power vector without influencing the other SNR values. In this special case the interference function is a continuous concave function.

The most important interference functions from the perspective of today's mobile networks are linear interference functions, which have the following form:

$$\tilde{I}(\mathbf{p}) = \sum_{l \in \mathcal{K}} a_l p_l + z, \qquad (8)$$

whereby $\mathcal{K} = \{1, \ldots, K\}$, $z > 0$ is a constant proportional to noise variance and $a_l \geq 0, l \in \mathcal{K}$, is a channel-dependent damping factor for the transmission power output of the l-th interferer. Consequently, the interference power resulting from this model at the output of each receiver is a linear combination of all transmission power outputs plus a noise-dependent constant. For the formulation of algorithms for power control, the following property of the function $\tilde{I}$ plays a decisive role (see also Fig. 8): $\tilde{I}(e^s)$ is a log-convex function of the logarithmic power vector $\mathbf{s} = \log(\mathbf{p}), \mathbf{p} > \mathbf{0}$ [30, 32].
We get another important class of interference functions if we consider that the constants

$a_l = a_l(u), l \in \mathcal{K}$, and $z = z(u)$ in (8) are generally influenced by the receiver structure. Here the receiver structure is represented with the help of an (abstract) variable $u \in U$. This variable belongs to a compact set U that reflects certain constraints concerning the receiver design. With these definitions, the interference functions among optimal receivers take on the following form:

$$\bar{I}(\mathbf{p}) = \min_{u \in U} \left(\sum_{l \in \mathcal{K}} a_l(u) p_l + z(u) \right) \qquad (9)$$

In contrast to the previous example, this function is not linear. Another important difference results from the fact that $\bar{I}(e^s)$ is not a log-convex function of the logarithmic power vector $\mathbf{s} \in \mathbb{R}^K$. Rather, this function is concave, which follows from the properties of the minimum operator.

The functions $\tilde{I}$ and $\bar{I}$ satisfy the axioms (A1)–(A3), so that they are interference functions in terms of the axiomatic definition. Sometimes, however, it can be advantageous to neglect noise in formulating strategies for power control or more generally for the allocation of resources—for example when the noise variance is unknown or very small relative to the entire interference power. For this reason [30] looks at interference functions that satisfy the following axioms:

(A1′) $\mathbf{p} > 0 \implies I(\mathbf{p}) > 0$
       (Positivity)
(A2′) $\mathbf{p} > 0, \mu > 0 \implies I(\mu\mathbf{p}) = \mu I(\mathbf{p})$
       (Homogeneity)
(A3′) $\mathbf{p}^{(1)} < \mathbf{p}^{(2)} \implies I(\mathbf{p}^{(1)}) \leq I(\mathbf{p}^{(2)})$
       (Monotony)

In comparison to (A1)–(A3), the scalability (A2) is replaced by homogeneity (A2′). Another advantage of the axioms (A1′)–(A3′) is that they capture functions that allow for the characterization of all supportable quality of service values in a multi-user network [32]. Thus these functions are interference functions in terms of the axioms (A1′)–(A3′).
At first glance the two axiomatic definitions seem incompatible. However, they can be translated into one another [32].

Fig. 9 Axiomatic theory of interference functions

3.4 Capacity Planning for UMTS Radio Networks

The coverage of UMTS networks depends on the capacity of the radio cells. (In contrast, with GSM these two factors can be determined in two independent steps.) This dependency results from the load-dependent interferences between radio connections and from the fact that the network controls the transmitted power outputs by means of the quality of the signals received.

Thus traditionally simulations are used in planning to assess UMTS networks—simulations that determine the load capacity of each individual cell for random realizations of user demands (position, mobile service). The realizations are set up according to a given distribution of demand intensity. Conclusions can then be drawn about the network quality following an analysis of a statistically significant number of demond instantiations and the network state determined in each case. The simulations computationally demanding. Moreover, they provide hardly any insight into what changes in the network configuration could lead to an improvement in the network quality.

A closed description of the behavior of complex systems represents a challenge, but is also an important contribution to the understanding of the system and comprises an essential foundation for the planning of cost-efficient networks [2]. Figure 10 describes a closed system model that allows for an assessment of large sections of a network with hundreds of cells in seconds, while at the same time its coupling matrix provides information about the effects of interferences between the cells. Thus the matrix forms the basis of several mathematical optimization models and automated network planning procedures [18, 35]. The underlying system model has to be adapted to the changed properties of the system in the course of the development of the system technology.

Automated network optimization procedures are of great significance in practice. Savings of 30% in terms of infrastructure cost were ascertained for a part of a network in Switzerland, provided that network planning was consistently executed with the use of optimization tools [12].

3.5 Hardware Development

The rapid development of wireless communication was also made possible by innovations in the semi-conductor industry. The technological progress in this industry is the basis of what is known as Moore's Law, namely the doubling of computing power for the same cost every 18 months [26]. In the meantime hardware platforms are being used for receiver implementation that can reach a computing power of several tera operations per second. The rapid development of hardware can be expected to continue in the coming years. Even today high-power mathematical techniques are being used to map telecommunications algorithms, such as optimal channel equalizers or the optimal multi-user receivers, onto high-power hardware

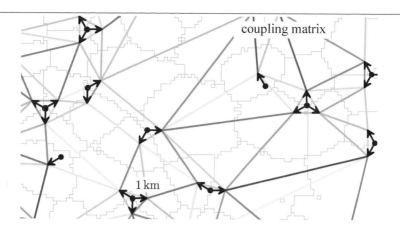

The coupling of interference between the cells of a UMTS radio network can be described mathematically using linear complementarity systems. These allow to calculate cell load and, as the case may be, overload.

The continual adaptation of transmission power outputs for the connection between base stations and their users is a characteristic of the UMTS system. The example of the downlink will show how a closed description can be given of the dynamic behavior of a network. The coupling system of the cells (e.g. [25]) describes the equilibrium that emerges between the transmission power outputs of the individual cells. This equilibrium is decisively determined by the traffic that has to be supported and the coupling of the transmitting power outputs between the cells by mutual interferences.

The SNR inequalities that state which minimum signal-to-noise ratio has to be achieved for each connection form the starting point for this modeling. Power control ensures that this minimum is just met. The assumption of perfect power control turns the inequalities into equations. In a second step these equations for each user can then be aggregated such that they result in a system of equations in which only the total transmitting power output of each cell appears (from which the transmitting power outputs for the individual connections can be derived):

$$p_i = C_{ii} p_i + \sum_{j \neq i} C_{ij} p_j + p^\eta{}_i + p^c{}_i \quad (10)$$

This equation expresses the dependence of the transmission power in cell i from the power of the other cells $j \neq i$. The users determine the inputs of the coupling matrix C through their geographical distribution and the services used. The above graphic illustrates a section of such a matrix. The color intensity of the connecting lines increases with the strength of the coupling between two cells (represented by the arrows pointing into the main direction of radiation of the sectors and their cell surfaces).

The value p^η_i states the power share per cell that would suffice to service the users if there were no interference, and p^c_i describes the constant transmitted power output used for the so-called pilot signals or for similar purposes. The pilot signal is a type of "beacon". Communication with the cell can only take place as long as the user sees this beacon.

The coupling system of all equations (10) can be solved very quickly with methods from numerics. This allows to determine (with sufficient accuracy) the transmission power per cell that is neccesary to service the users as well as to calculate the system load. However, this calculation only applies to the case where the system is not in overload.

In practice a UMTS network responds to overload with a partial or complete rejection of service requests. This behavior can be accounted for in an expanded model assuming perfect load control [14, 18]. Mathematically the linear system of equations becomes a complementarity system that can be solved with comparable effort.

Fig. 10 A system model for the power assessment of UMTS radio networks under load

platforms. Besides issues of the optimal scheduling of telecommunications algorithms, mathematics is just as crucial to innovations in the semi-conductor industry. For example, quantum physical effects will take on increasing importance in the continuing miniaturization of the semi-conductor industry, where, for example, the result of quantum information theory and operator algebra will have to be used in circuitry design.

4 Reinforcing Collaboration

The development of mobile communications continues to bring forth better and better systems. What leads to more efficient communication networks on the user side means a rapidly growing complexity in the entire system for the research and development side. This complexity can only be mastered when the means for its description and analysis advance correspondingly. There is no other option than to continue to push forward the use of mathematics as the medium of abstraction

Sections 2 and 3 provided an impression of the many applications of mathematics in the field of mobile communications. This partly stems from the educational system for telecommunications engineers, which traditionally imparts mathematics concepts and methods. However, the focus of this education is often on analysis. This limitation has to be overcome. The foundations of the mathematical fields of probability theory and graph theory, combinatorics and optimization are an important toolkit for today's prospective telecommunications engineer.

The progress described here is the result of the intensive collaboration of telecommunications engineers and mathematicians. This should continue to be expanded on a broad front. Section 5 will compile important fundamental questions for which the funding and support of joint projects through the German Research Foundation and/or the Federal Ministry of Education and Research can create optimal conditions.

The following will provide examples of recent developments for which greater support and reinforcement is also recommended.

4.1 Interdisciplinary Collaboration

The interdisciplinary collaboration between mathematicians and engineers is increasingly important. This can begin with the joint education of students, for example as happens as a part of the techno-mathematics course of study, and can be continued in interdisciplinary doctorate research programs. These doctoral research programs are particularly important, since here the essential foundations can be laid for research careers and for the transfer of methods and knowledge to industry for those students with sustained scientific interests. A broader use of this instrument as a part of academic education is very much to be desired.

Interdisciplinary fundamental research between mathematics and telecommunications engineering is also gaining significance outside of the doctorate research program. The German Research Foundation has provided important impulses to mobile communications as well with the research center MATHEON in Berlin and the Excellence Cluster High Speed Mobile Information and Communication at the RWTH Aachen.

4.2 International Collaboration

Interdisciplinary collaboration at the level of research and development is being promoted more and more in the European Union. An example of this is the European Cooperation in the Field of Scientific and Technical Research, COST for short [27]. The past and present collaboration of telecommunications engineers and mathematicians under the sequence of COST Actions 259, 273 and now 2100, for example, contributed to the effective network planning for each current generation of technology. An Action features three to four meetings per year over a period of several years (typically three or four). The participants include experts from academia and industry. The regular meetings help overcome initial "language barriers" between the disciplines and create a common foundation for work. They facilitate the mathematicians' access to research in mobile communications and introduce new mathematical methods to the telecommunications engineers. The instrument has proven its value.

The classical research and development projects under sponsorship of the European Union's Framework Programs are another example. The Seventh Framework Program, for example, is currently funding a project where engineers and mathematicians work together to research mechanisms for the self-organization of future mobile networks [9]. These mechanisms are intended to make networks based on the LTE standard considerably more manageable in practice than is the case for GSM and UMTS networks. This issue is of great economic interest [24]. The associated challenges are also scientifically very interesting and include questions of game theory, mechanism design and (online) optimization. The successful interdisciplinary collaboration of engineers and mathematicians within EU programs is a good argument to make considerable use of this instrument in the future as well.

4.3 Transfer

The great significance of technological innovation for the German economy is well known. The transfer of knowledge from research to practice is an essential source of innovation.

Alongside the transfer of knowledge through the transfer of personnel, joint project work between academic institutions and industrial partners also plays an important role. This collaboration is traditionally supported by the Federal Ministry of Education and Research (BMBF). Since small and mid-sized companies increasingly play a considerable role in the transfer of knowledge, the efforts of the BMBF to integrate these companies into its sponsorship are important. In view of the smaller amount of capital smaller companies have at their disposal compared to large industry, the funding needs to be delivered promptly. The BMBF and the EU are currently making this important adjustment.

The funding of newly founded technology-based companies arising from universities and research institutes has been improving for years now. However, there is still a need for action to help innovative technologies find their way into free enterprise more frequently and with greater success. For example, the reentry threshold for entrepreneurs coming from a research background is still too high when they want to resume their academic path after one or two years of entrepreneurial activity. Young researchers especially find it difficult, in these circumstances, to decide in favor of an entrepreneurial risk.

4.4 Sponsoring Research from Young Talent

New concepts are often devised and researched by young scientists. Thus support for independent fundamental research from this age group is particularly important. Junior professorships for those who are still young is one instrument; the well-established Emmy Noether Program from the German Research Foundation is another. Both should be used more intensively in the area of wireless communication systems in order to equip young talent with a kind of "venture capital" for their (university) research. The third-party funding of scientists from industry for industrially related research and development can only be helpful to this purpose.

5 Prospects

This section discusses important potential directions in which communication and information theory could develop. Although these questions have a long history, so far only small steps have been taken towards their solution. A complete solution of these problems would allow us to uncover new applications for wireless communication as well as achieve a better understanding of nature and of social processes.

5.1 Information Theory and the Digital World

The results of information theory discussed in Sects. 2 and 3 have without a doubt had an enormous influence on the development of mobile communications. However, important questions in information theory are still unresolved—for example, the capacity of continuous-time channels. Further development of information theory for more general channel models, i.e. a methodical connection between information and queueing theory as well as information and control theory, would be of great significance and would enable numerous new applications of wireless communications.

The information theory of discrete-time systems has developed well in the last 60 years. It was one of the starting points of the development that led to our "digital world" today. The Shannon sampling theorem forms the theoretical foundation for the digitalization of data (and of language and images as well), the digital processing of signals and their digital transmission. R.P. Feynman (Nobel Prize for Physics, 1965) honored the importance of this theorem as a basis of modern computation in [15] by saying "Consideration of such a problem will bring us on to consider the famous Sampling Theorem, another baby of Claude Shannon". The continuation of this trend to increased digitalization requires new mechanisms for signal reconstruction to be developed. The methodical cornerstone of this has to be found in an expansion of signal theory within foundational research.

5.2 Network Information Theory

Section 2 worked out the significance of information theory on multiple-access channels and broadcast channels for the development of cellular communications. It was found that the characterization of the optimal transmission strategies had decisively influenced the standardization of HSDPA, HSUPA and LTE. In general, however, central information-theoretical questions for ad hoc systems, wireless sensor networks, etc., are still unresolved. A deeper understanding of communication systems with many simultaneously active users is lacking, and for this reason the development of these systems has progressed relatively slowly thus far. We can expect the solution of the relevant information-theoretical questions to provide an essential contribution to our understanding of complex communication systems. The goal is the development of a network information theory. This would have great practical significance for many new applications in classical wireless communications, in the automotive industry, in machine construction and plant engineering, etc.

5.3 Quantum Information Theory

The questions described in Sects. 5.1 and 5.2 belong to the field of classical information theory. Quite recently the progress in experimental physics has led to the development of quantum information theory, based on quantum physics. The coding theorem for quantum channels represents a central result of this theory, showing, in analogy to Shannon information theory, that information can be transmitted over quantum channels as well. Quantum information theory is a very young discipline.

The quantum capacity regions for important multi-user communications scenarios are not yet known. A universal quantum coding that would allow for a robust transmission over quantum channels has to be developed in collaboration with experimental physics. We can expect quantum information theory to represent the key to understanding and mastering miniaturized systems. In particular quantum information theory will contribute to our understanding of the transmission and manipulation of classical information under the influence of quantum-mechanical noise, a scenario that will come up in on-chip communication in miniaturized chip architecture, for example. This also shows how pure and abstract fields of mathematics can become relevant to production through new areas of application. Thus convex geometry and the geometry of Banach spaces take on the role in quantum information theory that combinatorics does in classical information theory. The theory of operator algebra and the quantum probability theory offer a suitable framework to describe and examine the stochastic structure of the systems of quantum information theory.

5.4 New Approaches to Frequency Use

The above-mentioned concepts to retain a certain measure of fairness in the allocation of resources form the starting point of a development that could, in the end, produce entirely new concepts of frequency use and allocation. It is undisputed that the frequency spectrum is an important economic resource. The allocation of frequencies is a very complex problem requiring worldwide coordination, for example in the periodic world radio conferences. Typically legislation has to take into account a great variety of different standpoints in allocating wave bands. The necessary discussion is rarely conducted at a rational level. The mathematical disciplines such as social choice theory/social welfare theory and mechanism design can offer a more profound grasp of the frequency allocation problem. Within axiomatic theory, issues such as fairness, efficiency, robustness, security and social utility can be formalized and to some extent antagonistic contradictions can be uncovered. A very successful example of this is K.J. Arrow's "Impossible Result" (Nobel Prize for Economics, 1972) from the classical social choice theory [4, 5]. The formulation of an axiomatic theory for frequency use is crucial due to its social relevance as described above. This theory has to both come to terms with the very complex communication scenarios given in telecommunications technology and at the same time provide a formulation for different concepts of fairness, different legislative concepts and market mechanisms.

5.5 Self-Organizing Networks

Mobile radio networks such as GSM and UMTS are used over large surfaces as the infrastructure for mobile language and data services. The effort associated with the operation of one or several such networks has since come to surpass the acceptable level for the operators. Thus the standardization of successive systems, such as LTE, should fundamentally offer a system with fewer possibilities of configuration; and mechanisms of self-organization in the network should help considerably reduce the necessary manual intervention. The research on this is just in its beginning phase. The principles according to which a decentralized control of partly overlapping sets of parameters can be organized on different time scales have so far not been understood. Mathematically sound design principles for such control are not available, nor are any results known that would limit the conditions under which a stable regulation is possible in this field of application. Answers to these underlying questions would provide the urgently needed contributions for the creation of so-called self-organizing networks, since these could characterize the mechanisms with which a high automation of the operation of future mobile radio systems would be possible.

References

1. Aardal, K.I., van Hoesel, C.P.M., Koster, A.M.C.A., Mannino, C., Sassano, A.: Models and solution techniques for the frequency assignment problem. 4OR: Q. J. Belg. French Ital. Oper. Res. Soc. 1(4), 261–317 (2003)
2. Abiri, R., Neyman, Z., Geerdes, H.-F., Eisenblätter, A.: Automatic network design. In: Nawrocki, M., Aghvami, H., Dohler, M. (eds.) Understanding UMTS Radio Network Modelling, Planning and Automated Optimisation: Theory and Practice. Wiley, New York (2006). Chap. 14
3. Aigner, M., Ziegler, G.: Proofs from THE BOOK. Springer, Berlin (2004)
4. Arrow, K.J.: A difficulty in the concept of social welfare. J. Polit. Econ. 58(4), 328–346 (1950)
5. Arrow, K.J.: Social Choice & Individual Values. Yale University Press, New Haven (1963)
6. GSM Association: Press statement from 16.04.2008
7. Bertsekas, D.P., Gallager, R.: Data Networks. Prentice Hall, New York (1992)
8. Chiang, M.: Balancing transport and physical layers in wireless multihop networks: jointly optimal congestion control and power control. IEEE J. Sel. Areas Commun. 23(1), 104–116 (2005)
9. Socrates Project Consortium: Self-optimisation and self-configuration in wireless networks (2008)
10. Correia, L.M. (ed.): Wireless Flexible Personalized Communications—COST 259: European Co-operation in Mobile Radio Research. Wiley, New York (2001). COST Action 259—Final Report
11. Cover, T.M., Thomas, J.A.: Elements of Information Theory, 2nd edn. Wiley, New York (2006)
12. Dehghan, S.: A new approach. 3GSM Daily 2005 1, 44 (2005)
13. Eisenblätter, A., Grötschel, M., Koster, A.M.C.A.: Frequency assignment and ramifications of coloring. Discuss. Math., Graph Theory 22, 51–88 (2002)
14. Eisenblätter, A., Geerdes, H.-F., Koch, T., Martin, A., Wessäly, R.: UMTS radio network evaluation and optimization beyond snapshots. Math. Methods Oper. Res. 63, 1–29 (2006)
15. Feynman, R.P.: Feynman Lectures on Computation. Penguin Book, Baltimore (1999)
16. UMTS Forum: Press statement from 20.01.2008
17. Gamst, A.: Application of graph theoretical methods to GSM radio network planning. In: Proceedings of IEEE International Symposium on Circuits and Systems, vol. 2, pp. 942–945 (1991)
18. Geerdes, H.-F.: UMTS radio network planning: mastering cell coupling for capacity optimization. Doctoral thesis, Technische Universität Berlin (2008)

19. O₂ Germany: Telefónica investiert 3,5 Milliarden Euro in Deutschland. Press statement 11.10.2007 (2007)
20. Grötschel, M., Lovász, L., Schrijver, A.: Relaxations of vertex packing. J. Combin. Theory, Ser. B **40**, 330–343 (1968)
21. Hale, W.K.: Frequency assignment: theory and applications. Proc. IEEE **68**, 1497–1514 (1980)
22. Kuhn, H.W., Nasar, S.: The Essential John Nash. Princeton University Press, Princeton (2002)
23. Lovász, L.: On the Shannon capacity of a graph. IEEE Trans. Inf. Theory **25**, 1–7 (1979)
24. NGMN Ltd.: Next generation mobile networks: Beyond HSPA and EVDO. White Paper (2006)
25. Mendo, L., Hernando, J.M.: On dimension reduction for the power control problem. IEEE Trans. Commun. **49**, 243–248 (2001)
26. Moore, G.E.: Cramming more components onto integrated circuits. Electronics **38**(8) (1965)
27. European Science Foundation: COST Office. European cooperation in the field of scientific and technical research
28. Plehn, J.: Applied frequency assignment. In: Proceedings of the IEEE Vehicular Technology Conference. IEEE, New York (1994)
29. Annual report of the German Federal Network Agency 2007. Bundesnetzagentur für Elektrizität, Gas, Telekommunikation, Post und Eisenbahnen; Bonn (2008)
30. Schubert, M., Boche, H.: QoS-based resource allocation and transceiver optimization. Found. Trends Commun. Inf. Theory **2**(6) (2006)
31. Shannon, C.E.: A mathematical theory of communication. Bell Syst. Tech. J. **27**, 379–423, 623–656 (1948)
32. Stanczak, S., Wiczanowski, M., Boche, H.: Fundamentals of Resource Allocation in Wireless Networks. Foundations in Signal Processing, Communications and Networking, vol. 3. Springer, Berlin (2008)
33. Emre Telatar, I.: Capacity of multi-antenna Gaussian channels. Eur. Trans. Telecommun. **10**(6), 585–596 (1999)
34. Tse, D., Viswanath, P.: Fundamentals of Wireless Communication. Cambridge University Press, Cambridge (2005)
35. Türke, U.: Efficient methods for W-CDMA radio network planning and optimization. Doctoral thesis, Teubner Verlag (2007)
36. Wunder, G., Zhou, C.: Queueing analysis for the OFDMA downlink: throughput regions, delay and exponential backlog bounds. IEEE Trans. Wirel. Commun. (2009). doi:10.1109/TWC.2009.071244
37. Wunder, G., Zhou, C., Kaminski, S., Bakker, H.E.: Throughput maximization under rate requirements for LTE OFDMA Downlink with limited feedback. EURASIP J. Wirel. Commun. Netw. Special issue on Multicarrier (2007)
38. Xiao, M., Schroff, N.B., Chong, E.K.P.: A utility-based power control scheme in wireless cellular systems. IEEE Trans. Netw. **11**(2), 210–221 (2003)
39. Zhou, C., Wunder, G.: Throughput-optimal scheduling with low average delay for cellular broadcast systems. EURASIP J. Adv. Signal Process. (2009). doi:10.1155/2009/762050. Special issue on Cross-Layer Design for the Physical, MAC, and Link Layer in Wireless Systems

Mathematics of Chip Design

Jürgen Koehl, Bernhard Korte, and Jens Vygen

1 Executive Summary

Chips are probably the most complex structures ever designed and produced by man. On a small silicon chip of the size of a fingernail, one is today able to accommodate billions of transistors, linked by millions of connecting wires whose total length can exceed several kilometres. Figure 1, which was made with a scanning tunnelling microscope, shows a very small part (about one billionth) of a modern chip with two wiring levels.

The realization that these electronic midgets could never have been designed without the methods of discrete mathematics is a source of great satisfaction to mathematicians. It fills a mathematician with a very special feeling of happiness to have discovered a theorem which adds to the knowledge of mankind. When we look at a tiny chip on the palm of our hand and know that it was made possible by our mathematical algorithms, we feel a different but no less intense happiness.

No doubt the chip industry has, of all industries, experienced the greatest increase not only in the complexity of its products, but also in the accompanying technical challenges and the pressures of cost and time.

- *Complexity*: technological advances have made it possible to double the number of transistors on a chip every two years (Moore's law). The same holds for the

J. Koehl
IBM Deutschland Research & Development GmbH, Schönaicher Straße 220,
71032 Böblingen, Germany
e-mail: koehl@de.ibm.com

B. Korte (✉) · J. Vygen
Universität Bonn, Lennéstr. 2, 53113 Bonn, Germany
e-mail: dm@or.uni-bonn.de

J. Vygen
e-mail: vygen@or.uni-bonn.de

M. Grötschel et al. (eds.), *Production Factor Mathematics*,
DOI 10.1007/978-3-642-11248-5_10, © Springer-Verlag Berlin Heidelberg 2010

Fig. 1 A small excerpt
($\sim$ one billionth) of a chip
taken with a scanning
tunnelling microscope

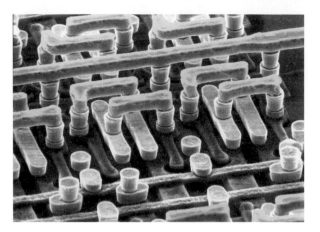

number of other components on a chip, as well as for the number of electric
connections that must be laid out by routing algorithms, the number of logic paths
for the optimization of the clock rate, and the size of the logical description which
must be turned into an optimized circuit by synthesis programs.

- *Technical challenges*: the most advanced production method today is the 45 nm
  technology. Here the gate oxide of a transistor consists of no more than five
  atomic layers. In spite of an immense technical effort, variations of one atomic
  layer more or less are unavoidable. This implies a 20% variation in the layer
  thickness, in spite of which every one of the billion transistors on the chip must
  keep working reliably. At the same time, the clock rate can rise to 5 gigahertz,
  while the various clock signals must be balanced out to an accuracy of only several picoseconds. Moreover, the power consumption of a new chip containing
  substantially more logic circuits must not be any greater, as this would make it
  impossible to cool the chip adequately.
- *Cost limits*: with every new technological generation it becomes possible to accommodate the functions of two chips on one, which, however, brings with it
  significantly greater technical challenges. In spite of the better performance, customers expect their new computer to cost no more than their old one. As increased
  prices are thus not possible, the development of the new chip must not cost more
  than that of the previous one.
- Figure 2 depicts the cost reduction in microelectronics over the last 60 years with
  respect to several different parameters. The cost reduction amounts to 35% per
  year. As far as we know, no other industry exhibits anything like such a cost
  reduction.
- *Time limits*: the new game console, for example, must be on the market in time for
  Christmas shopping. Every year customers expect new, faster laptops and mobile
  phones with even more functions. In spite of the exponentially increasing complexity, the turnaround time for the development of a new chip must not only not
  increase, but must, if possible, be decreased significantly.

These challenges cannot be met without a steady improvement in the mathematical
design tools. Even though the mathematical penetration of the design problems in-

Fig. 2 Cost reduction in microelectronics

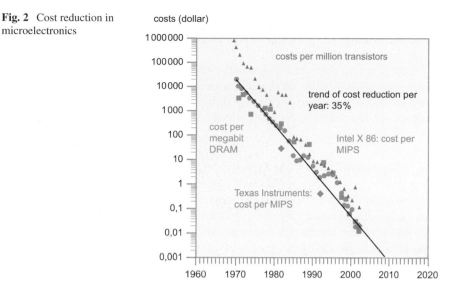

volved has reached a high standard, new approaches continue to be necessary, and this is a wide field of research with immediate technical and economical applications.

In this article we will report on some results of more than 20 years of cooperation between industry (IBM) and science (the Research Institute for Discrete Mathematics of the University of Bonn). This cooperation is widely considered to be one of the longest and most successful between science and industry. We restrict ourselves here to the application of combinatorial optimization in physical design and timing optimization of chips.

2 Success Stories

At a conference in 1985 we happened to meet a microelectronics design engineer, who described his problems in chip design to us. Although the design complexity at that time was several powers of ten less than today's, it was already great enough to render manual methods and simple heuristics useless for obtaining a high design quality. We realized immediately that here was an area of application *par excellence* for combinatorial optimization.

After a brief analysis we could state that the methods of combinatorial optimization would be able to improve results by about 30%. Engineers are, however, distrustful, in particular of mathematicians—perhaps justifiably so. Thus we had to prove our claim in a test case. We worked on a given chip and succeeded in improving it by more than 30%. This inaugurated an extremely successful cooperation between industry and science and led to a complete design system for the physical design of chips. The so-called BonnTools® are in the meantime being widely used in the chip industry. More than a thousand chips have been designed with these algorithms. The three following examples will serve to elucidate the progress made during the last twenty years.

Discrete optimization has significantly improved chip design and has, in some cases, even made possible otherwise impossible designs. On the other hand, new problems and questions arising in chip design have led to the extension of theory, methods and algorithms of discrete optimization.

The Telecommunication Chip ZORA

The first great challenge for the application of discrete optimization in chip design was a telecommunication chip designed in 1987 in the IBM Research Laboratory Zurich/Rüschlikon for the US telecommunication firm Rolm. The complexity of today's chips is greater by at least two or three powers of ten. Even so, *ZORA* was a real challenge at the time. The square chip had an edge-length of 12.7 mm and accommodated about a million transistors, 15 566 nets and 58 738 connecting pins. Its cycle time of 28 nanoseconds was not very critical, so that timing optimization was not necessary at all. It had HPCMOS technology, one of the earliest CMOS technologies to arise after the bipolar transistor technology. The main difficulty with this chip was its wiring, which at that time was only possible on two (mutually orthogonal) wiring planes. The chip incorporated a very large memory macro which could only be placed centrally (see Fig. 3), and all the I/O-pins were on the periphery of the chip. All earlier memory macros for data or instruction caches were smaller and were placed in the corners of chips. As the two-layer wiring could not cross the macro, a feasible wiring was considered impossible by all designers and experts.

However, our at that time still rudimentary routing algorithm immediately came up with a wiring of total length 14.9 m and 91 439 vias (inter-layer connections) (see Fig. 4).

The chip went into production (Fig. 5 gives a microscopic view of *ZORA*). We were for the first time able to admire a piece of hardware which exactly matched the designs produced by our algorithms (see Figs. 3 and 4).

The Microprocessor P2SC

In 1995 IBM produced the POWER2 chip in CMOS5X6S technology as the microprocessor for its POWER series. It was the first time that several units (processor, floating point, caches and storage controllers), usually on separate chips, were placed on a single chip. This chip required timing optimization to maintain its cycle time of 7.4 ns. It had 149 613 nets and a total wiring length of 166.3 m with about 1.4 million vias at all. By the way, this was the microprocessor in the computer Deep Blue which made history by winning in chess against Kasparov.

In hierarchical design, the logic part is subdivided into several hundred separate units, for each of which the timing behaviour is prescribed globally. Every unit must

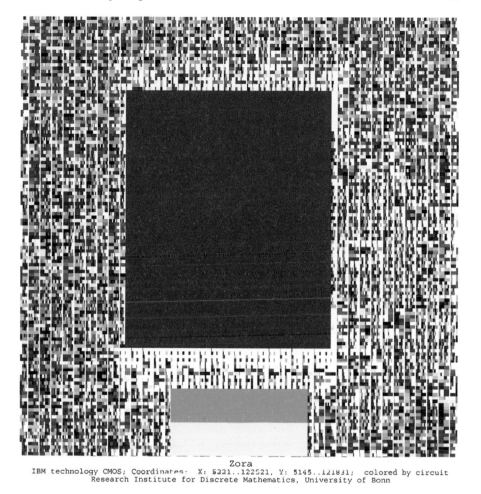

Zora

IBM technology CMOS; Coordinates· X: 5231..122521, Y: 5145..121831; colored by circuit
Research Institute for Discrete Mathematics, University of Bonn

Fig. 3 Placement of the ZORA chip

then maintain its preset running times. After the units of a hierarchical design have been designed, they are placed on the chip and interconnected (top level wiring). One advantage of such a hierarchical design method is that it allows a more complete control of the design process of the small segments, in particular so that one can quickly intervene locally in the case of later design alterations. Of course, such a segmentation can take one a long way from a global optimum. Hierarchical design involves an extremely great amount of development amounting to several hundred man-years; twenty man-years are required for the physical design alone.

The microprocessor P2SC gave us our first opportunity to prove the advantages of a flat design, in which no segmentation into smaller units takes place. Instead, the entire net list comprising all nets and circuits is placed flat and with no hierarchical restrictions on the chip surface. This, however, requires placement and routing algorithms which use methods of combinatorial optimization in an essential way.

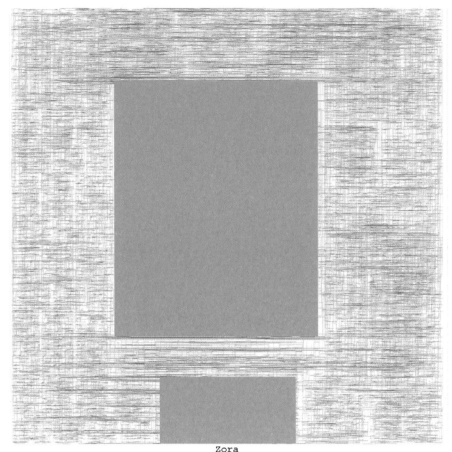

Zora
IBM technology CMOS; Coordinates: X: 5221..122521, Y: 5145..121831; colored by circuit
Research Institute for Discrete Mathematics, University of Bonn

Fig. 4 Routing of the ZORA chip

Our results were astonishing: a new layout could be designed in only six weeks and
only two team members were involved. The flat design made possible a reduction
of more than 20% in the net length and the number of vias. The cycle time could not
be improved because the critical path, i.e. the slowest path, which thus determines
the cycle time, lay in a macro which, even in a flat design, could not be altered.
We could, however, achieve a substantially better distribution of running times on
the various paths, which resulted in a much more robust chip which was also much
less sensitive to production tolerance fluctuations. The slack histogram which depicts the difference between expected and actual arrival times had a much better
distribution.

Figure 6 shows a placement for P2SC using hierarchical design, and Fig. 7 a
placement using a variant of our flat design method. It is evident that the latter is
substantially more compact. Large parts of the chip surface remain unused. One

Fig. 5 The ZORA chip in silicon, microscopically enlarged

can then use a smaller die size in the chip production process, which leads to a substantial reduction in cost.

The Systems Controller U3 for the Apple Power Mac

In the year 2002, Apple decided to design a new Power Mac P5 which, according to their CEO Steve Jobs, would become the fastest PC in the world. Its system architecture (Fig. 8) embodies a dual core microprocessor consisting of two Power PC 970 processors. The physical design of its systems controller, which was manufactured millions of times, was designed as a flat ASIC by an IBM engineer in Bonn during the winter months of 2002–2003, using BonnTools®. The chip had an edge-length of 9.3 mm and was based on 130 nm-Cu11-technology. A million circuits with more than a million nets had to be accommodated and wired up using four million pins. The total net length was around 260 m and used six wiring layers.

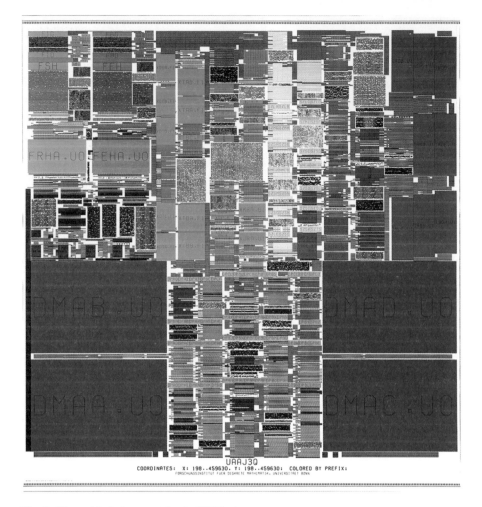

Fig. 6 Hierarchical placement for the P2SC microprocessor

The special issue arising in the design of the U3 was its cycle time. It was for the first time that an ASIC design had to be clocked with a frequency of more than one gigahertz. For the cycle time optimization we used a new method called BonnCycleOpt®, a brief description of which is given in the box of Fig. 9.

The success of BonnCycleOpt® was very convincing: we start with a chip whose placement and wiring has been completed and for which all further methods for running time improvement (gate sizing, buffering, timing-driven placement) have been used. The longest path between two registers determines the cycle time of the chip or its clock frequency. The question as to whether it is possible to improve the bottleneck cycle time further is answered clearly in the affirmative by BonnCycleOpt®: this is achieved by opening and closing registers in accordance with the optimal clock signals (clock skew) calculated for every register in turn.

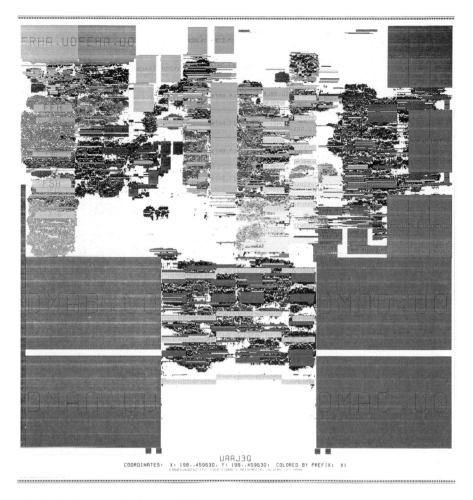

Fig. 7 Flat placement for the P2SC microprocessor

The chip U3 has more than a dozen different groups of clock nets. Figure 10 depicts a particularly critical high-speed clock net. Here, the application of BonnCycleOpt®, i.e. using only mathematical methods, led to an improved clock frequency of 29.6% and thus made possible a clock frequency of more than one gigahertz [9].

An improvement in chip performance or quality of this order of magnitude is in general only possible by employing a new technology (e.g. silicon-on-insulator, copper instead of aluminum wiring, strained silicon). Such technological advances entail an entirely new production process and thus a new silicon foundry. The cost of building a new silicon foundry is today in the region of 3.5 to 5 billion dollars. That is a very large investment. As mentioned above, performance improvements of

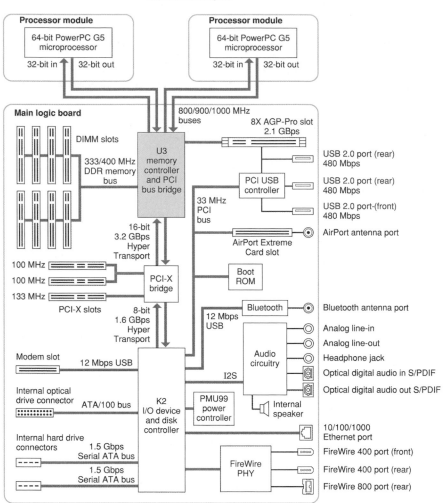

Block Diagram and Buses
Preliminary © Apple Computer, Inc. July 2, 2003

Fig. 8 Schematic diagram of the Apple Power Mac G5 chip architecture

similar magnitude can be achieved by means of mathematical methods, only. These incur no investment costs, one just has to think hard!

Figure 11 depicts the chip U3 before and after optimization with BonnCycleOpt®. Those components for which signals arrive on time or early are coloured blue and green, respectively. Yellow, orange and red indicate late signal arrival. The associ-

For cycle time optimization, one assumes that the physical design, in particular placement, has been completed. All logical operations on a chip are synchronized with memory elements (= registers). The registers are controlled by periodic clock signals. The signal flow between two registers travels over several steps (logical switching elements = circuits) and over many nets which connect these circuits. The signals enter the chip via primary inputs and leave it via primary outputs, which we can regard as special I/O registers.

Let R be the set of registers. On a modern chip there can be several hundred thousand of these. We construct a directed graph $G^R = (V(G) = R, E(G))$. Two registers $v, w \in R$ are joined by a directed edge (v, w) if there is a signal flow from v to w. Each edge (v, w) is assigned a delay $\delta(v, w)$, which is the longest time needed by a signal to go from v to w. These delays are made up of the signal propagation times through wires and the switching times of the circuits involved.

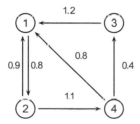

The above diagram depicts a register graph G^R with four vertices (and no primary I/Os). The numbers assigned to its edges are the signal propagation times (in ns, for example). If all the registers open and close simultaneously (zero skew), this circuit would have a clock time (= cycle time) T of 1.2 ns, corresponding to the longest delay on edge $(3, 1)$. However, if one lets the registers switch at different times, say

register 1

at the times $0; T; 2T; \ldots$

register 2

at the times $-0.1; T - 0.1; 2T - 0.1; \ldots$

register 3

at the times $-0.3; T - 0.3; 2T - 0.3; \ldots$

register 4

at the times $0.1; T + 0.1; 2T + 0.1; \ldots$

then one can, as is easily verified in the above graph, achieve a cycle time of $T = 0.9$ ns.

In register graphs with several hundred thousand vertices one cannot find the switching time of each register and the optimal cycle time by trial and error. Here a mathematical theorem is of help, which states that the optimal cycle time is equal to the greatest average delay on directed cycles in G^R. The register graph depicted above has three directed cycles, namely

first cycle:

$1 \rightarrow 2 \rightarrow 1$ with average delay

$(0.8 + 0.9)/2 = 0.85$

second cycle:

$1 \rightarrow 2 \rightarrow 4 \rightarrow 3 \rightarrow 1$ with average delay

$(0.8 + 1.1 + 0.4 + 1.2)/4 = 0.875$

third cycle

$1 \rightarrow 2 \rightarrow 4 \rightarrow 1$ with average delay

$(0.8 + 1.1 + 0.8)/3 = 0.9$

(cf. [1, 9, 13]).

This theorem is a cornerstone of BonnCycleOpt®. Of course, the practical problems involved in cycle time optimization are incomparably more complicated and difficult than in the above simple example. Here we have only taken the latest signal arrival time (late mode) into consideration. The signals must, however, also not arrive too early (early mode). Finally, it is not only a question of minimizing the cycle time determined by the directed cycle in G^R with the greatest average delay. The switching times at the registers of this cycle are tight, i.e. there is no slack between the signal arrival time and the switching times of the registers. All the other registers can have positive slack. A further problem thus arises, namely not only to minimize the worst slack, but also to optimize the slack distribution (slack balancing problem). This corresponds to balancing the potentials on the vertices of G^R. For all the problems associated with cycle time optimization we can design strongly polynomial algorithms.

Fig. 9 Outline of BonnCycleOpt®

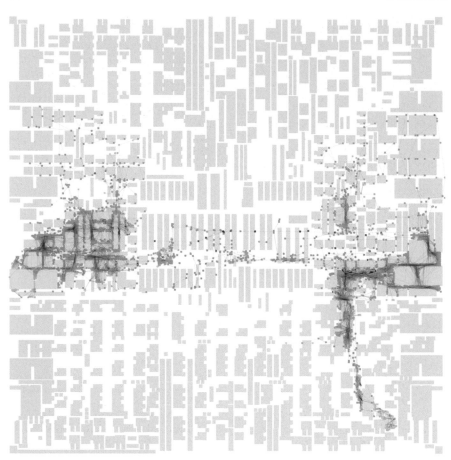

Fig. 10 Gigahertz clock net for the chip U3

ated slack histograms indicate the signal arrival time distribution at the registers. The colours have the same meaning as above. A direct comparison of the histograms or of the design pictures reveals the above-mentioned decrease in cycle time by nearly 30%.

3 Status Quo in Chip Design

As was mentioned earlier, the work at the Research Institute for Discrete Mathematics of the University of Bonn has since 1985 been centred on mathematical methods and algorithms for chip design. In the meantime, commercial enterprises selling tools for chip design, so-called EDA (Electronic Design Automation) companies, have started up. The three market leaders in this field (Cadence founded in 1988, Synopsys founded in 1986, and Magma founded in 1997) were founded

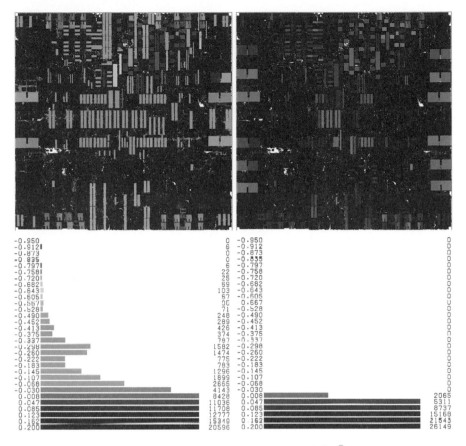

Left histogram:

Bin	Count
-0.950	0
-0.912	6
-0.873	0
-0.835	0
-0.797	6
-0.758	22
-0.720	25
-0.682	69
-0.643	103
-0.605	67
-0.567	0
-0.528	71
-0.490	248
-0.452	289
-0.413	426
-0.375	374
-0.337	787
-0.298	1582
-0.260	1474
-0.222	775
-0.183	783
-0.145	1296
-0.107	1899
-0.068	2665
-0.030	4143
0.008	8428
0.047	11036
0.085	11708
0.123	12777
0.162	15340
0.200	20596

Right histogram:

Bin	Count
-0.950	0
-0.912	0
-0.873	0
-0.835	0
-0.797	0
-0.758	0
-0.720	0
-0.682	0
-0.643	0
-0.605	0
-0.567	0
-0.528	0
-0.490	0
-0.452	0
-0.413	0
-0.375	0
-0.337	0
-0.298	0
-0.260	0
-0.222	0
-0.183	0
-0.145	0
-0.107	0
-0.068	0
-0.030	0
0.008	2065
0.047	5311
0.085	8737
0.123	15168
0.162	21545
0.200	26149

Fig. 11 The chip U3 before and after optimization with BonnCycleOpt®

later. The annual turnover of these three leading companies is approximately 4 billion dollars. We will now outline the development and status quo of chip design in terms of our BonnTools®. The major part of BonnTools® comprises nontrivial innovative (discrete) mathematics. The aim of chip design is to produce a functioning industrial product that will be manufactured in very large numbers. The design engineer expects to get a complete solution of his problem, not just the application of beautiful, complicated mathematical theorems. Thus we, too, must, in addition to complex mathematical methods, use some "quick-and-dirty" heuristics in order to achieve the final functional chip design. The ratio of mathematics to "quick-and-dirty" methods in BonnTools® is roughly 90% to 10%. This is the main advantage of BonnTools® over commercial products, for which the above-mentioned relation between mathematics and heuristics is inverse. A detailed description of BonnTools® is given in [10, 11].

Figure 12 gives a schematic description of the design steps from the initial HDL (hardware description language) formulation of the chip through to the chip production stage. The HDL is something like a higher programming language and is used

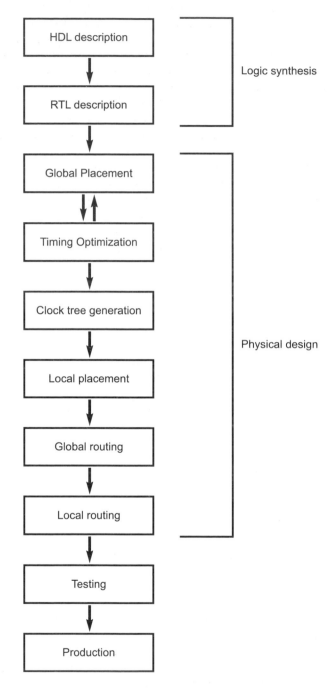

Fig. 12 Simplified schematic representation of the design process

to obtain a formal description of the function of the chip and its function blocks (without considering the problems of technical realization). The first step of the logic synthesis partitions this description down to the RTL (register transfer level). Registers are small memory elements that store interim results of steps in the combinational logic functions of the chip. The running time between the various registers determines the cycle time T or the clock frequency $1/T$ of the chip. In the next step of the logic synthesis, formal (technology-independent) Boolean expressions are generated, which are then represented as circuits (elementary switching units) in the corresponding technology. Apart from certain modifications in the logic made during the timing optimization, logic synthesis is not the domain of BonnTools®. Logic optimization is an extremely difficult, widely open field, for which an efficient mathematical approach is not yet known but very necessary.

The domain of BonnTools® is everything connected with the physical design of the chip, which comes after the logic synthesis. The first two steps of the physical design, namely global placement and timing optimization, are performed iteratively many times. They are today the main and probably hardest part of the physical design process. These iterations are also called "timing closure". At this stage of the design flow, the placement is completed up to small changes and movements, the optimal transistor size of each circuit is determined, the slack is optimized for each register, the topology of the repeater trees is determined, the signal strength by means of inverters and buffers (repeaters) is optimized, and the threshold voltage V_t is determined.

After that, the clock tree is constructed according to the calculated slacks at the registers by means of precisely determined clock signal arrival times. Modern chips can have hundreds of different clocks and thus forests of clock trees.

In the ensuing local placement, an overlap-free arrangement of all circuits is achieved by means of the smallest possible movements. During global as well as local placement, the routability of the chip must be preserved. The signals arising from logic operations in various circuits leave the output pin of a given circuit and go to the input pins of several other circuits. A modern chip can have more than 10 million such nets, all of which must be embedded vertex-disjointly in up to ten rectilinear wiring layers. This is achieved by global and local routing.

The problems of physical design can be simplifyingly described in the following way: consider several million small Lego blocks (placeholders for the circuits), with the same length but differing widths, as well as several larger ones (memory elements, macros, cores). All of these must be placed disjointly on a chip with a surface area of 1 to 3 cm^2. These Lego blocks communicate with one another by means of nets. The output pin of a given Lego block sends its signal through a net to several other Lego blocks. These nets must be accommodated vertex-disjointly in a three-dimensional grid-graph in such a way that, in particular, the signal-flows between the registers comply with the given cycle time. Many other technical conditions (e.g. noise, coupling) have to be fulfilled, which we cannot discuss here.

The various components of BonnTools® use a large number of different combinatorial optimization problems which we also cannot discuss in detail here. In order to demonstrate the mathematical diversity and depth of these applications in chip design, we list the most important: shortest paths, Steiner trees, maximum flows, transportation problems, minimum cost flows, multicommodity flows, disjoint paths and trees, discrete time-cost trade-off problems, minimum mean cycle problem, parametric shortest paths, minimum spanning trees, bin packing, knapsack problems, travelling salesman problems, Huffman codes, facility location.

As is to be expected, these combinatorial optimization problems cannot be utilized in their pure form as given in textbooks, but must almost always be modified for the particular application, requiring a proof of their correctness and efficiency.

As mentioned above, modern chip design would be impossible without significant use of discrete mathematics. On the other hand, these applications in chip design have resulted in new theoretical and methodological questions which have led to further development of theory and method. Because of the numerous problems which have arisen in connection with the newest advances in technology, this mutual stimulation will increase and last for a long time to come.

We will now briefly elucidate the separate steps of the physical design of a chip.

3.1 Placement

The placement algorithm BonnPlace® consists essentially of two components: the global and the local placement. The former yields an infeasible result in which the various objects may still overlap. However, the densities of the different regions of the chip permit an overlap-free placement using only small local movements. This is achieved with the local placement algorithm, which is also called legalization or detailed placement.

There exist three different approaches for global placement:

- simulated annealing
- min-cut method
- analytical placement

Simulated annealing is a heuristic which mimics a physical cooling process. We classify it as a "quick-and-dirty" method as it is not motivated by the special application problem (here chip design), nor does it have a mathematical justification. A simulated annealing algorithm applied to chips with 10 million movable objects would take astronomically long on even the fastest computers and, moreover, yield poor results. This criticism applies to other similar heuristics (e.g. genetic algorithms, neural networks etc.), which have acquired an interesting philosophy and perhaps even a meta-theory in other sciences which, however, are of no use in applications to chip design.

Min-cut methods were very popular up to 10 or 15 years ago. They aim to minimize the number of nets crossing a given cut by means of an exchange heuristic. The main criticism of this method, which yields useful results for some chip structures, is that the objective function gets minimized with respect to a one-dimensional object (the given cut), while the chip is obviously two-dimensional. Minimizing the number of nets crossing the cut can yield placements on both sides of the cut which are hard or even impossible to route.

Analytical placement methods minimize a linear or quadratic function which estimates the net length and leads to a nondisjoint placement that has to be made free of overlaps in a further step. BonnPlace® belongs to this category.

BonnPlace® minimizes the sum of the weighted squared lengths of all nets, where the estimator for the net length is obtained by modelling each net as a clique (called quadratic placement hereafter). Brenner and Vygen [5] have shown that the clique model is the topology-independent approximation of a Steiner tree of minimal length. There are several reasons for choosing a quadratic objective function. In general, quadratic placement leads to uniquely defined positions for the movable objects, it is stable, i.e. practically invariant under small changes in the net list [15], and it is very quickly solved in parallel by means of conjugate gradient methods with incomplete Cholesky decomposition. A quadratic placement of 5 million movable objects can be generated in only a few minutes of computer time. A further argument for the quadratic approach is that the signal delay increases quadratically with the wire length.

Quadratic placement usually has very many overlaps during its first steps, and these cannot be corrected locally. We therefore have to ensure that no region of the chip becomes too dense, i.e. that the space needed by the movable objects must not exceed the available space. A second algorithm called multisection takes care of this. The basic idea is to divide a region of the chip into several parts and to allot the movable objects to these parts in such a way that the density conditions are fulfilled and the total movement of the objects is minimized, i.e. the arrangement found by quadratic placement should be altered as little as possible. This allocation problem of objects to regions involves millions of objects but in general only very few regions (four or nine, say). Even so, the associated $(0, 1)$-allocation problem is NP-hard. But here we can work with the non-integer allocation problem which is a Hitchcock transportation problem, because Vygen [14] has shown that every fractional solution has only $k - 1$ non-integer allocations, where k is the (small) number of regions. Brenner [4] derived a new algorithm for this problem with $O(nk^2(\log n + k \log k))$ running time, which has turned out to be extremely fast in practice.

Figure 13 depicts the result of a multisection applied to nine regions. Each colour shows the allocation of objects to one region. The red objects belong to the upper left region, the light green ones to the upper middle region, etc. The allocation is optimal with respect to L_1-distance. Figure 14 shows the first six steps of the global placement using BonnPlace®. The global algorithm terminates when the height of a region corresponds to the height of the circuits. The overlaps are then removed by applying legalization (local placement).

We will describe legalization only briefly and in very simple terms. After the global placement has been completed, there will be regions of the chip that are too densely packed, i.e. they contain more cells than can be placed without overlap. These are the supply vertices. Regions having a surplus of space are the demand vertices. Edges are generated between these, along which flows can take place. A disjoint placement, i.e. the balancing by means of flows from supply to demand vertices, is achieved by solving a minimum cost flow problem. Determining the regions,

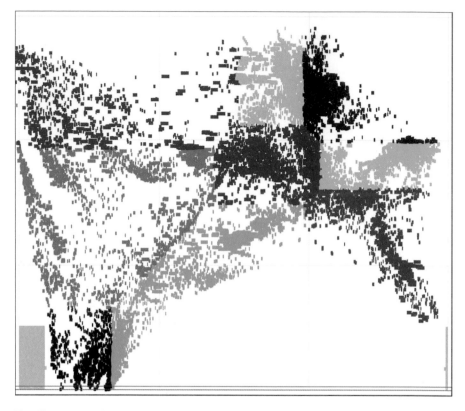

Fig. 13 An example of multisection with $3 \times 3 = 9$ regions

Fig. 14 The first six steps of global placement with partitioning into 1, 4, 16, 64, 256 and 1024 regions (i.e. 4^n regions with $n = 0, 1, 2, 3, 4, 5$)

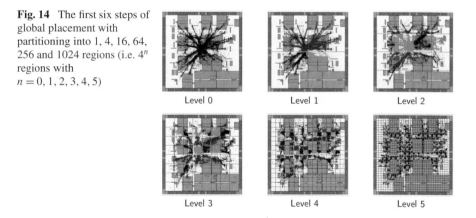

their size, as well as the flow edges, is very complex indeed. We will not discuss this topic here—Fig. 15 shows a "flow graph", i.e. the legalization of a small part of a chip. Supply regions are red and demand regions green. The blue edges represent the minimum cost flow, the width of the edges being proportional to the amount of

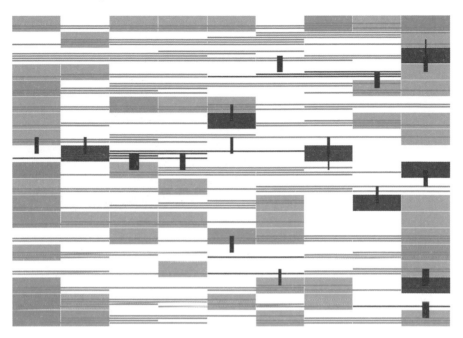

Fig. 15 Legalization on a small part of a chip

flow through them. Legalization is applied repeatedly in design flow, for example after the clock tree has been generated or after changing the circuit sizes in timing optimization.

3.2 Timing Optimization

Twenty years ago, timing optimization played practically no role in the physical design of a chip. After completion of the placement and wiring, the timing was calculated. The preset cycle time was always achieved and this completed the physical design process. One must remember that in those days one dealt with cycle times of 20 ns or more, corresponding to a clock frequency of 50 MHz or less, while today's processors have a clock frequency of 4 GHz, corresponding to a cycle time of only 250 ps. Several technological tricks and mathematical methods are necessary in order to achieve this.

The entire field of timing optimization, also called timing closure, comprises many different methods and algorithms. By means of these, new circuits (amplifier = buffer, repeater, inverter) may be added or the size of circuits changed. Moreover, these methods alternate iteratively with local placement. The main components of timing closure are fan-out tree (repeater tree) construction, fan-in tree optimization, gate sizing (circuit or transistor sizing), clock skew scheduling (cycle time optimization), slack balancing, and clock tree construction.

Fig. 16 Three different tree topologies of minimal length

Repeater trees are a good example for demonstrating how new technologies give rise to new challenges in design methodology and thus make new mathematical algorithms necessary. Earlier, signal delay was dominated by the active switching elements (circuits, transistors), while the wiring played a secondary role. In the latest technologies it is the other way round. It is foreseeable that in future technologies more than half the circuits will be used not for the actual switching elements but just for bridging the distances (as amplifiers = repeaters).

Repeater trees are needed in order to distribute the output signal (source) of a circuit optimally to the input pins (sinks) of a number of other circuits. Not only must the topology of the tree be optimal, but amplifiers (repeaters) must be inserted in order to bridge over-long distances. Figure 16 depicts three Steiner trees connecting the source r with seven sinks. All three trees have the same minimal length. If the path from r to the sink s is time critical, then the left-hand tree is the worst possible because the capacitive load from r to s is maximal. For the middle tree, the direct path from r to s is minimal, but the circuit s must also drive the capacitive loads of the edges and vertices following it, which results in signal delay. The right-hand tree is the optimal repeater tree. As was mentioned above, in addition to the topological generation of the tree, bridging repeaters will be necessary. Details of the very complicated method for constructing repeater trees can be found in Bartoschek et al. [2].

The construction of fan-in trees presents the reverse situation. Here several signals must be combined into one signal given by a logic function (Boolean expression). Such a fan-in tree can, however, have different shapes, in particular different depths. In general it is true that the shorter such a tree, the faster the signals are propagated. Basically fan-in tree optimization amounts to logic optimization (logic synthesis) which in this case, however, does not get done before the physical realization but after the physical layout is known. Figure 17 shows three logically equivalent physical designs of a Boolean expression consisting solely of logical AND-gates (red) and logical OR-gates (blue). This expression has eight Boolean variables (a)–(h). The reader can readily check that these three designs are logically equivalent, whereby the trees with shorter depth make possible an increased signal propagation. Fan-in tree optimization uses mainly dynamic programming. Further theoretical and practical details of this method can be found in [16]. Fan-in tree optimization is used very selectively towards the end of the timing closure iterations, and then only in the case of time-critical paths. Here, however, time savings of 100 ps or more can signify a large step towards achieving timing closure.

Gate or transistor sizing is a classic method of timing optimization. A technology library contains each circuit in several forms. A large circuit drives a signal very

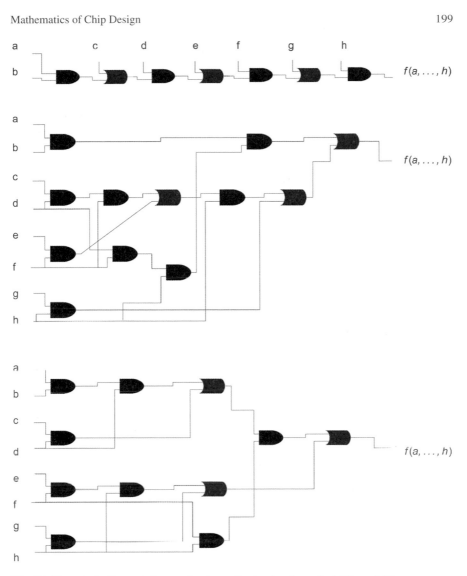

Fig. 17 Three logically equivalent realizations of a circuit using only AND- and OR-gates

quickly, but uses a lot of power and has a large capacitance. Smaller circuits work more slowly, use less power and have a smaller capacitance. Unfortunately, gate sizing produces opposing effects which must be taken into account in the optimization. Although a large circuit accelerates signal propagation, it has a large capacitive load leading to delayed operation of its predecessors which, of course, have to drive the whole of the following load. Thus gate sizing, which is oriented both forwards and backwards (driven load), is necessary. Classic gate sizing methods are based on convex/geometric optimization [3, 6, 7], where it is assumed that all functions that arise can be approximated by posynomials. Unfortunately these methods have the disadvantage that they can only deal with very small instances, which makes them practically useless. We use an iterative method which, when optimizing circuit size, takes into account how quickly the potential of a signal changes from 0 to 1 or vice versa (slew). This method is very fast, even for large practical instances. If one assumes that the delay function is a posynomial and the circuits can be scaled continuously (as required in geometric programming), one can prove that this iterative method yields minimal signal delay [8].

The problem of controlling the clock signals individually and optimally at each register (clock skew scheduling) was mentioned above in Sect. 2 (systems controller U3, see also Fig. 9). Here we can only add briefly that we have recently started using a new method for balancing the skews at all registers, which basically comprises a balancing of potentials at the vertices of a directed graph. Held [8] was the first to describe a strongly polynomial method for the potential balancing problem in non-acyclic graphs.

There are several other tuning parameters for the timing behaviour (e.g. plane assignment, V_t-level), which we cannot, however, discuss here. After all the tricks of timing optimization have been applied and "the last picosecond has been squeezed out", only the construction of the clock tree remains, which is a tree that performs the clock control at each register. Modern chips have several hundred different clock trees with a million or more registers. With a clock frequency in the gigahertz region one has to reserve 20–30% of the cycle time as a safety interval for manufacturing tolerances, so that only several hundred picoseconds remain. This implies that just a few picoseconds can be significant when optimizing the arrival times of the clock signals at the registers (clock skew).

We construct the clock tree bottom-up and assemble it from partial trees. Clock tree construction requires many sinks (the registers) with a time interval for each one, and also several sources (the roots of the partial trees). Furthermore, an (abstract) logically correct clock tree is required which supplies each register with clock signals. In order to ensure the arrival of the clock signal within the given time interval, one usually performs a balancing with respect to the wire length, i.e. the exact signal delay may be achieved by means of wiring detours. This approach has the disadvantage that substantial wiring resources are needed just for this balancing. This makes it difficult to design a fully defined clock tree wiring layout on a chip whose wiring is already fixed. We therefore use a different approach [9], in which the wiring is shortest possible and the signal delay balancing is achieved by means of the topology of the clock tree and a corresponding sizing of the repeaters. This method is similar to the construction of repeater trees described above. Signal amplification is achieved by means of inverters, where two inverters in series amount to a repeater (buffer). Figure 18 depicts different stages of the clock tree construction using BonnClock®. The coloured areas are regions in which the inverters (sinks) can be placed. The colours correspond to different arrival times in the clock tree: blue for signals near the source, green, yellow and red for later arrival times. During the bottom-up construction the coloured areas move slowly in the direction of the source which is somewhere near the middle of the chip in this example.

As we said, the clock tree is constructed bottom-up. Nearly 90% of the power consumption of the clock tree is required for the bottom stage. Thus, a good cluster-

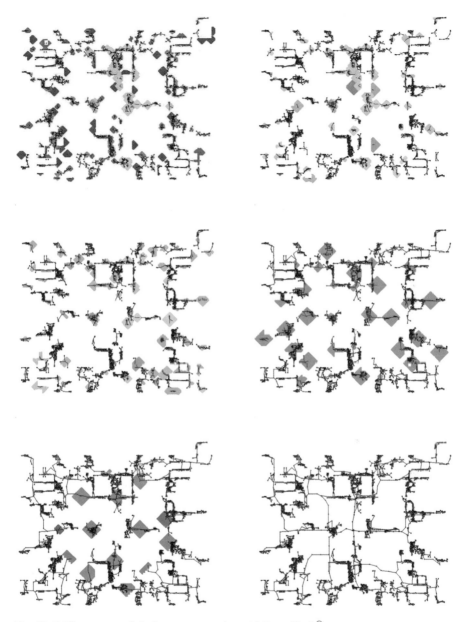

Fig. 18 Different steps of clock tree construction with BonnClock®

ing of the sinks is of paramount importance. We solve this problem with a modified facility location approach for which Maßberg and Vygen [12] were the first to describe an approximation algorithm with constant approximation factor. Figure 10, as shown earlier, depicts a gigahertz clock tree for the chip U3.

3.3 Routing

The disjoint layout of the nets of a chip within a three-dimensional grid-graph takes place in the routing step. Here the data volumes are gigantic. A very large chip has as many as ten million nets which must be wired up in the grid-graph, which has several hundred billion vertices and edges. For a flat routing, i.e. without hierarchical partitioning, this is an immense challenge which our routing algorithm BonnRoute® has mastered outstandingly thousands of times in many different design centres.

BonnRoute® consists of two main parts, namely global and local (detailed) routing. The global routing step allots to each net a corresponding area of the chip. This greatly reduces the searching in the local routing step. The global routing algorithm works with a substantially smaller graph obtained by contracting several hundred neighbouring vertices of a grid-graph to one super-vertex. The capacity of edges joining super-vertices is given by the number of edges of the grid-graph joining in the contracted vertices.

The global routing problem amounts to an edge-disjoint packing of Steiner trees corresponding to the nets, within this contracted grid-graph. A fractional relaxation of this problem can be solved with a modification of the multicommodity flow problem. The associated linear programming problem can then be extended by the addition of further constraints for timing, coupling, yield or power consumption, for example. In this way these objectives can be used in the optimizing process. The global router of BonnRoute® is the first algorithm with a provable performance guarantee and which incorporates constraints for timing, coupling, noise, yield and power consumption. The detailed solution of the relaxed linear program is then achieved using classic methods of randomized rounding. For a given placement, the global router determines a global routing if one exists, otherwise it terminates with a certificate stating that the placement cannot be routed.

The detailed router then places the individual nets within the original grid-graph, using the information provided by the global routing step. The edges of the global routing graph in which the Steiner trees of the different nets are already placed, define corridors in the grid-graph for the detailed placing of the nets. This substantially restricts the region of the grid-graph that has to be searched. This in fact makes it possible to route many millions of nets in reasonable computing time.

The Steiner trees of the nets are routed sequentially, for example in the order of their timing criticality, the Steiner trees themselves having been assembled sequentially as well, by means of two-point connections. These connections are then realized by means of shortest paths, for which Dijkstra's classic labelling algorithm is the natural choice. Unfortunately the usual implementation of this algorithm found in textbooks is of no use for such large problems. The theoretically fastest implementation of Dijkstra's algorithm with Fibonacci heaps has $O(m + n \log n)$ running time, where m and n are the number of vertices and edges of the graph. For our applications this is much too slow.

In order to speed up the algorithm, we modify it in essentially two ways. The first modification comprises a goal-oriented search, in which we determine for each vertex a lower bound of its shortest path to the sink. These lower bounds are taken to

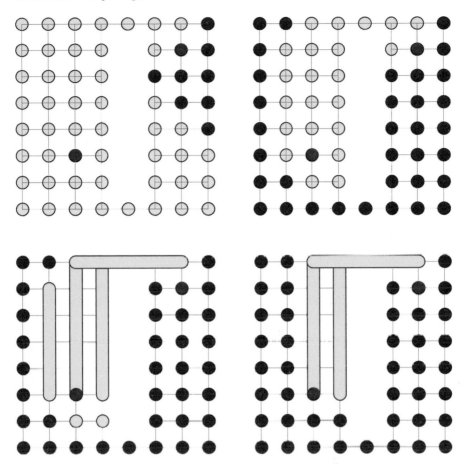

Fig. 19 Labelled vertices (yellow) in Dijkstra's algorithm using four different labelling methods

be vertex potentials of the vertices, and these are then used to modify the cost function of the edges. We are thus able to ensure that the labelling strategy of Dijkstra's algorithm does not move around the source in waves, but goes relatively directly from source to sink. The two upper pictures of Fig. 19 show the labelled vertices (yellow) of Dijkstra's algorithm applied without (left picture) and with (right picture) future cost. The goal is to find a shortest path from the bottom left red vertex to the top right red vertex. The classic form of Dijkstra's algorithm labels fifty vertices in order to achieve this, while the goal-oriented version marks only twenty-four.

A very substantial acceleration of Dijkstra's algorithm is obtained by labelling edge-paths (intervals) instead of individual vertices. The grid-graph used for the wiring of the chip is not complete—it has blockages and also holes. Now, one can label a whole edge-path up to the next obstruction (blockage or hole) in one step. Then the running time no longer grows with the number of vertices but with the number of obstructions, which is far smaller than the number of vertices. The two lower pictures of Fig. 19 show the labellings of an interval-based Dijkstra algorithm applied without (left picture) and with (right picture) goal orientation. The number of labellings in these two cases is only seven and four, respectively. As the running time of Dijkstra's algorithm is proportional to the number of labellings, the drastic improvement achieved by the interval-based version of Dijkstra's algorithm is strikingly apparent.

The newest chip technologies work with so-called gridless libraries. Here the connection points of the circuits no longer need to be located at grid-points. We cannot discuss this topic in detail here but just wish to point out that the data structures of BonnRoute® were chosen in such a way that grid-based as well as gridless circuits can be handled.

4 Analysis of Strengths / Weaknesses, Challenges

The design of processors with a clock frequency of several GHz still requires ten times the engineering effort needed for the design of an ASIC of one GHz. This efficiency gap can only be reduced by more powerful algorithms that can take account of the technical challenges of nanometer technologies:

- *Variability*: the layout must be designed in such a way that production variations in the electrical parameters of up to 30% still give correct timing behaviour. In the processor design process this is usually achieved by manually generating a highly symmetrical layout, which however is by no means a globally optimal solution. Any subsequent optimization must take account of the statistical variation in the electrical parameters and their correlation, and must yield a feasible layout in all situations.
- *Reliability*: the same factors which lead to an increase in variation also increase the failure rates of the transistors and the wiring. These risks are minimized by an increased design effort in DFM (design for manufacturability). A chip will, however, only function correctly if each one of its more than a billion transistors works faultlessly. A well-known method of achieving this in the manufacture of memory chips is to incorporate redundant memory cells. Logic chips and their wiring will in future also have to have redundant structures. Thus BonnRoute® is already now able to incorporate local loops in order to achieve a substantial increase in the robustness of vias. In future, the wiring must also be extensively protected against circuit failure by incorporating redundant connections. Redundant logic elements are also a further logical step for increasing reliability. This is avoided today for reasons of testability. In this respect we will thus see a paradigm shift which will require a completely new approach.
- *Power loss*: the reduction of power loss is becoming increasingly important and shows that simply increasing the clock frequency leads to a very unfavourable ratio of computing performance to power consumption. It makes far more sense to optimize computing performance per watt. The design programs developed in our cooperation ensure by means of timing optimization of the logic that the desired clock frequency is attained with as small a power loss as possible. Here not

only the dynamic power dissipation due to circuit operation is taken into account, but also the increasingly dominant passive energy consumption due to leakage currents. As we mentioned earlier, the energy consumption is determined mainly by the architecture and the logic description. Optimization methods must therefore already begin at the behaviour description level and must permit a quick comparison of different implementations.

5 Visions of, and Recommendations for the Future

Many of the problems arising out of future technologies and products can only be attacked by optimizing the layout and the logic structure simultaneously. For example, power dissipation is determined to 80% by the logic structure, and the best possible optimization of the remaining 20% has essentially been achieved already. The logic architecture must involve the layout possibilities at an early stage: it is a simple matter to double the number of entries and exits of a crossbar with each new technology, but the resulting quadrupling of the number of connections is impossible to realize from a certain size onwards. Recognizing such problems at an early stage requires close cooperation between logic design and layout and the chance to test a given logic structure efficiently with respect to realizability and the existence of problems. These considerations also show how important it is to be able to implement different architectures quickly and in a highly automated way, in order to obtain exact comparisons of the final performance parameters rather than having to rely on rough estimates.

On the other hand, the challenges inherent in the physical layout must be met in order to be able to produce structures that are significantly smaller that the wavelength of the light used. The entire lithographical and manufacturing process is today modelled and optimized mathematically (virtual fab). This model is then used to modify structures on the chip and to compensate for the inevitable inaccuracies inherent in the manufacturing process.

The complexity of the design process and of the technologies involved has reached a point where only a holistic view all the way from the logic structure to the manufacturing process can guarantee a successful product.

The ultimate goal of a silicon compiler that will, for every given logic description, yield a circuit automatically which works with a clock frequency of several GHz, will probably not be available in the foreseeable future. Even so, the automation of chip design must be driven forward. An important recipe for the success of our cooperation is the close collaboration on actual product designs using the newest available technologies currently with structural widths of 45 nm, and also the early discussion of challenges arising in future technologies with structural widths down to 22 nm. A successful cooperation is only possible with an open exchange of information and joint work on challenging new projects.

References

1. Albrecht, C., Korte, B., Schietke, J., Vygen, J.: Cycle time and slack optimization for VLSI-chips. Discrete Appl. Math. **123**, 103–127 (2002)
2. Bartoschek, C., Held, S., Rautenbach, D., Vygen, J.: Efficient generation of short and fast repeater tree topologies. In: Proceedings of the International Symposium on Physical Design, pp. 120–127 (2006)
3. Boyd, S., Kim, S.-J., Patil, D., Horowitz, M.: Digital circuit optimization via geometric programming. Oper. Res. **53**, 899–932 (2005)
4. Brenner, U.: A faster polynomial algorithm for the unbalanced Hitchcock transportation problem. Oper. Res. Lett. **36**, 408–413 (2008)
5. Brenner, U., Vygen, J.: Worst-case ratios of nets in the rectilinear plane. Nets **38**, 126–139 (2001)
6. Chen, C.-P., Chu, C.N.N., Wong, D.F.: Fast and exact simultaneous gate and wire sizing by Lagrangian relaxation. IEEE Trans. Comput. Aided Des. Integr. Circuits Syst. **18**, 1014–1025 (1999)
7. Fishburn, J., Dunlop, A.: TILOS: A posynomial programming approach to transistor sizing. In: Proceedings of the IEEE International Conference on Computer-Aided Design, pp. 326–328 (1985)
8. Held, S.: Timing-closure in chip design. PhD Thesis, University of Bonn (2008)
9. Held, S., Korte, B., Maßberg, J., Ringe, M., Vygen, J.: Clock scheduling and clocktree construction for high performance ASICs. In: Proceedings of the IEEE International Conference on Computer-Aided Design, pp. 232–239 (2003)
10. Held, S., Korte, B., Rautenbach, D., Vygen, J.: Combinatorial Optimization in VLSI Design. In: Chvatal, V., Sbihi, N. (eds.) Combinatorial Optimization: Methods and Applications. IOS Press, Amsterdam (2010, to appear)
11. Korte, B., Rautenbach, D., Vygen, J.: BonnTools: Mathematical innovation for layout and timing closure of systems on a chip. Proc. IEEE **95**, 555–572 (2007)
12. Maßberg, J., Vygen, J.: Approximation algorithms for a facility location problem with service capacities. ACM Trans. Algorithms 4, Article 50 (2008). Preliminary version in APPROX (2005)
13. Szymanski, T.: Computing optimal clock schedules. In: Proceedings of the ACM/IEEE Design Automation Conference, pp. 399–404 (1992)
14. Vygen, J.: Geometric quadrisection in linear time, with application to VLSI placement. Discrete Optim. **2**, 362–390 (2005)
15. Vygen, J.: New theoretical results on quadratic placement. Integration, VLSI J. **40**, 305–314 (2007)
16. Werber, J., Rautenbach, D., Szegedy, C.: Timing optimization by restructuring long combinatorial paths. In: Proceedings of the IEEE/ACM International Conference on Computer-Aided Design, pp. 536–543 (2007)

Materials and Mechanics

Chances and Visions of Advanced Mechanics

Wolfgang Ehlers and Peter Wriggers

1 Executive Summary

In coming years, the society of the 21st century will continuously increase its requirements on the quality of life and environmental standards. This concerns the quality and prediction accuracy in the computation of complex problems and, in particular, involves the overall design of products in our immediate surroundings i.e., architectural and industrial buildings, but also industrial products that we use in our daily life.

The requirements on the design, the construction, and the material strength of buildings make it necessary to overcome the current state of an empirically dominated material description and replace it with a simulation-based design of new materials with customised high-tech properties. What is true for buildings is also true for industrial-made products, especially within an industrially dominated society like ours. In such a society, product quality ensures the good world-wide reputation of the industrial site Germany and of its society. Moreover, it also secures our high gross national product and our living standard.

Besides research in materials science that contributes to an improvement and advancement of materials and their mechanical properties, an overall quality improvement of products can only be achieved if the solution of increasingly more complex engineering problems is advanced by new and enhanced strategies of modelling and simulation methods. First developments of new simulation tools influences already large parts of our society. Such developments are beyond usual engineering applications. This is particularly true for the fields of medical engineering and many areas in life sciences, which gains an increasing importance within an aging society.

W. Ehlers (✉)
Universität Stuttgart, Lehrstuhl II, Pfaffenwaldring 7, 70569 Stuttgart, Germany
e-mail: ehlers@mechbau.uni-stuttgart.de

P. Wriggers
Universität Hannover, Appelstraße 9a, 30167 Hannover, Germany
e-mail: wriggers@ibnm.uni-hannover.de

M. Grötschel et al. (eds.), *Production Factor Mathematics*,
DOI 10.1007/978-3-642-11248-5_11, © Springer-Verlag Berlin Heidelberg 2010

What can mechanics do for science, economy and for the society of the 21st century? Today, the usage of high-tech materials and the systematic investigation of scale-overlapping problems embrace, in the same way as the modelling of multi-physical relations, formulations in which coupled systems of equations from multiple physical fields have to be solved. So-called multi-scale methods are becoming increasingly more popular to represent more accurately processes in space and time. These methods allow, by zooming into smaller spatial and temporal scales, a detailed analysis of the given structures and thus contribute to a better understanding of material behaviour.

In general, the simulation of three-dimensional problems spanning multiple scales and embracing multiple physical properties leads to very large systems of coupled partial differential equations that can only be solved numerically via the use of state-of-the-art computers. For this, cooperations between scientists working in the fields of mechanical engineering and mathematics are essential. In principle, this conclusion is not new and can be traced back to the tradition of the Gesellschaft für Angewandte Mathematik und Mechanik (GAMM), which has been established in Leipzig on September 21st, 1922. This tradition has to be continued and strengthened for the benefit of our society, its science, and economy. By combining actions between mechanical engineers and mathematicians, we will succeed in advancing the already developed simulation tools. The new generation simulation tools will be expanded by scientists from computer and information sciences to an integrated tool of simulation engineering, see also the Cluster of Excellence "Simulation Technology" (SimTech) at the University of Stuttgart (http://www.simtech.uni-stuttgart.de).

2 Success Stories

There is still a long way to go on the path to a more detailed analysis and a better comprehension of materials and structures. In recent years, the use of advanced computers and fast algorithms developed within the field of numerical analysis has formed the basis for a new era in mechanics. Today, the treatment of mechanical problems within a complex mechanical modelling framework, the development of the respective mathematical models, and their numerical solution strategy is often referred to by the term *Computational Mechanics*. The tools developed within this context will be the driving forces for many technical developments. In the following, we present some examples to highlight applications as well as the chances and visions of advanced mechanics.

2.1 Continuum Mechanics of Multi-Field and Multi-Physical Materials

Foundation of Buildings—A Millennium-Old Problem of Architecture

Frequently, it is sufficient to treat the foundation of buildings as a solid mechanical problem regarding the load on the subsoil due to a foundation. However, increas-

Fig. 1 Schematic description of the consolidation problem

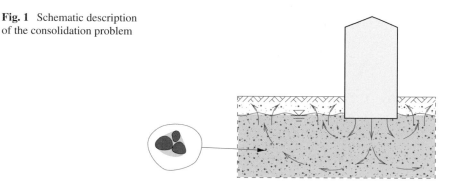

ingly more problems arise where this approach is no longer adequate. If one assumes that a building with its foundation has to be founded in water-saturated soil, then it is not sufficient to settle the structure load via the foundation plate on the soil and to treat it like a solid. Instead, a consolidation problem (Fig. 1) has to be computed in which, after imposing the external load, the load is carried at first entirely by the pore-water. But, the load on the water-saturated soil causes a pressure gradient in the pore water that, in turn, results in a flow process. The water is drained from the pore-space of the soil, which subsequently deforms, and hence yields through the relation between deformation and stress an equilibrium between the external load and the foundation plate.

Obviously, one needs to solve a coupled problem of fluid-structure interaction i.e., the characteristic of the solid and the fluid interaction and the influences on each other. The mathematical description of this problem within the Theory of Porous Media (TPM) [1], which is a continuum-mechanical approach to describe the interaction of bodies whose micro-structures are homogenised in the common domain, results in a complex non-linear system of coupled partial differential equations. Such a system of equations can generally only be solved numerically, e.g. with the help of the finite element method (FEM). Modern computers are increasingly able to master large and largest systems of equations. In order to use such computers in an optimal way, it is necessary to appeal to methods from the fields of civil, environmental, and mechanical engineering as well as applied mathematics to analyse and treat such coupled problems successfully using a computer.

CO_2 Sequestration—A Challenge for the Mechanics of Environmentally Relevant Systems

In the context of global warming, the storage of greenhouse gases (Fig. 2) increasingly gains importance. In particular, power plants that are driven by fossil fuels emit large amounts of carbon dioxide (CO_2). Modern flue gas purification plants are able to isolate CO_2 and therefore make it accessible for storage. Taking into account that fossil fuels, which produce CO_2 during the combustion process, have been stored in georeservoirs, it seems obvious to sequestrate the isolated CO_2 again

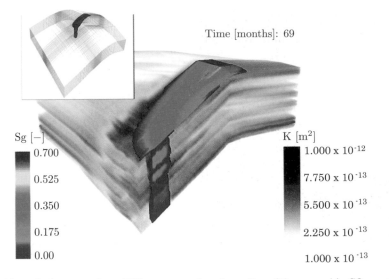

Time [months]: 69

Sg [−]

0.700

0.525

0.350

0.175

0.00

K [m²]

1.000×10^{-12}

7.750×10^{-13}

5.500×10^{-13}

2.250×10^{-13}

1.000×10^{-13}

Fig. 2 Numerical computation of CO_2 sequestration. Ascending of the pressed-in CO_2 up to the impervious cap layer (source: http://www.simtech.uni-stuttgart.de)

in georeservoirs. For this purpose, one chooses water-bearing soil layers (aquifers) that lie under an impermeable cap layer.

Comparable to the consolidation problem, the computation of the CO_2 sequestration process yields a coupled problem. In addition to the soil skeleton, which holds the aquifer and the water of the aquifer, there is a need to additionally consider the CO_2, which is injected under high pressure and in supercritical state into the aquifer. Thereby, the soil skeleton as well as the cap layer deform. If the fluid pressure decreases, the CO_2 changes from the supercritical state into the gaseous phase as a result of a phase transition. Hence, besides the coupled problem of fluid-structure interaction (solid deformation vs. filter flow), there are further chemo-physically motivated couplings that come into play. A highly complex system of strongly coupled partial differential equations is generated that can be solved numerically using the finite element method. However, the size of the numerical problem generally requires additional adaptive algorithms (adaptivity in space and time) on massively parallel systems, cf. also [1, 2].

The Biomechanical Challenge

Biomechanics, and in particular the biomechanics of soft and hard tissue, constitutes a relatively young and ambitious discipline of advanced continuum mechanics. The vision of the scientists associated with this field is the generation of a *virtual human model*, in which scale-overlapping information is available that range from the coarse-scale of the *crash-test dummy* over the fine-scale of tissues and organs to the micro-scale on the cellular level. Such a model could assist with analysing more precisely the consequences of accidents or by planning the course of surgeries virtually. But also the estimation of the aging processes of the musculo-skeletal system

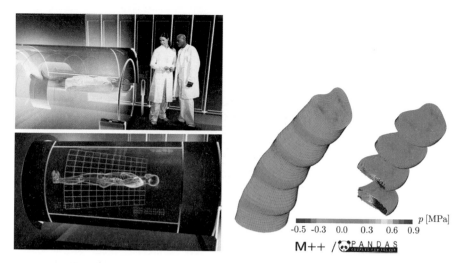

Fig. 3 *Left*: Recording of a patient and automatic generation of individual data (fiction). *Right*: Numerical computation of the flection of the lumbar spine and presentation of the induced pressure buildup in the separate intervertebral discs (source: http://www.simtech.uni-stuttgart.de)

or the planning of patient-specific implants as substitutes for worn joints or parts of the spine are conceivable.

Parts of the human body can already be simulated within the scope of advanced continuum mechanics of porous media [1]. If we consider the spine as the essential part of our musculo-skeletal system, then, the mechanical behaviour of single vertebrae as well as single intervertebral discs can be described, cf. [3].

In particular, the mechanical modelling of intervertebral discs with their gelatinous tissue from the nucleus pulposus (interior) and the annulus fibrosus (outer ring) poses a great challenge since the material is not only inhomogeneous, but also deforms anisotropically. In addition, electro-chemically driven swelling and shrinking processes have to be taken into account by which the intervertebral disc, as an avascular tissue, is provided with nutrients. We personally experience this effect, e.g., due to the fact that our body is in the morning about 2 cm taller than in the evening. During the day, the spine is under a load that leads to an internal pressure. That pressure contributes to extrude a part of the interstitial fluid from the tissue and creates a chemical imbalance that adjusts during the nightly recovery by reabsorbing the extruded amount of interstitial fluid.

Besides the coupled equations for the description of the complex mechanical behaviour of the solid material, which consists of collagen fibres and proteoglycans with adhesive electrical charges, and the behaviour of the interstitial fluid, in which positive and negative charged ions are located, diffusion equations for the description of the process of swelling and shrinking have to be considered, i.e., the mechanical behaviour and the electro-chemical behaviour are coupled and cannot be examined separately from each other. In the same way as in the prior discussed examples, a highly complex system of strongly coupled partial differential equations arises that can be solved numerically using the finite element method. Again, the size of the numerical problem in general requires the application of adaptive algorithms (space and time adaptivity) on massively parallel systems.

2.2 Simulation Techniques for the Description of Heterogeneous Materials (Multi-Scale Modelling)

The classical approach to gain a deeper understanding for the constitutive behaviour of materials that are composed of different ingredients on different length scales— e.g. concrete, a material that consists of cement, sand, and gravel—is the use of experimental techniques. Based on the experimental results, material models are developed that describe the different behaviour on the different length scales. These days, numerical simulation techniques can be utilised to incorporate the behaviour on the different length scales. Such an approach necessarily leads to multi-scale methods. This technique is new and poses a real challenge, since simulations, which yield quantitative results, have to be carried out in three dimensions.

Within a multi-scale simulation, different three-dimensional (3-d) mechanical models are used on each scale to take into account the material behaviour of the ge-ometrical and constitutive characteristics on each scale. For this purpose, so-called representative volume elements (RVE) are frequently used to characterise the ma-terial behaviour on the considered scale with sufficient accuracy. The RVE are then subjected to mechanical loads that are known from a coarser scale. This leads to a material response that, after averaging, is returned back to the coarser scale. This process is called homogenisation, see also [4, 5].

A numerical multi-scale analysis takes the complex 3-d geometry of the micro-structure on the respective scale into account and therefore requires an elaborate numerical model that often leads to a system of equations with more than a million unknowns. Thus, besides fast computers, there is also a need for fast solvers using modern numerical analysis techniques in order to use such methodologies within engineering applications. The result is a powerful tool that can dissolve the internal mechanical properties of a heterogeneous material and thus allows for new insights that, in this form, are usually not available from experiments.

Multi-Scale Analysis of Concrete

As an example, the new method for the analysis of heterogeneous materials pre-sented above is applied to the material concrete. This complex material and its re-spective scales are depicted in Fig. 4. Starting from the left, Fig. 4 depicts first the macro-scale of the building, which presents the meter range. On the meso-scale, which follows next, one finds the aggregates of the concrete (gravel) and the mortar. On the micro-scale, one considers the components of the mortar, e.g. the hardened cement paste. In this example, the nano-scale is not resolved, although some of its processes are certainly of importance to concrete.

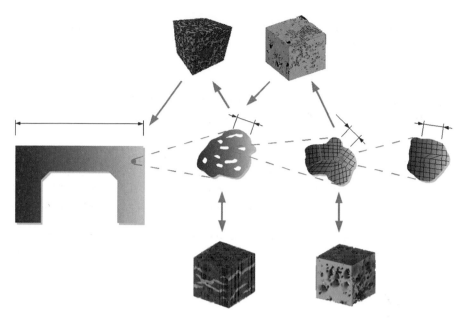

Fig. 4 Different scales that are to be considered in the material description of concrete

From the described scales depicted in Fig. 4, the micro-scale, in particular the hardened cement paste, is considered in more detail in the following. For this purpose, the knowledge of the micro-structural geometry, as well as the distribution of the different material phases is required. In the hardened cement paste, the geometrical model is determined via CT recordings of the cement paste using a resolution of one micrometer, see Fig. 5.

Fig. 5 Micro-structure of hardened cement paste and local stresses in the micro-structure [6]

At this level, no detailed statements on the internal structure of the single hydrati-sation products in the hardened cement paste can be made. But one can distinguish between hydrated and non-hydrated regions as well as micro-pores.

Now, different simulations are possible:

1. A homogenisation of the stresses and strains obtained on the micro-scale leads to a material model on the meso-scale. These homogenisations can be determined for the elastic part as well as for the inelastic material responses. Due to the fact that micro-structural geometry of the RVE is randomly distributed, different RVE need to be incorporated in order to obtain quantitative results. The corresponding computations are very costly, as up to 10 000 RVE are to be considered within a Monte-Carlo method.
2. Exchanging information between the micro-structure and the meso-structure to obtain a true multi-scale computations exchanged. With this, a micro-structure simulation, e.g. at every integration point of the meso-scale, has to be carried out. This exceeds today's computer capacities in the industrial field. Nevertheless, with growing computer capacity, it will be possible to carry out these simulations successfully in the future. This requires the development of adaptive methods that specify regions requiring multi-scale computations.

Homogenisation of Hardened Cement Paste

The method of homogenisation mentioned under item 1. is further explained using the example of the hardened cement paste.

Based on the CT scans given in Fig. 4, a 3-d finite element (FE) model is developed. The discretisation of an RVE, which has an edge length of 64 µm, leads to a FE mesh with about 300 000 elements and 820 000 unknowns. Since the elastic properties of the hydrated and non-hydrated regions are known from micro-indentation experiments, they can be applied within the model on the micro-scale and hence be used for the homogenisation process. After evaluation, a computation leads to the averaged or effective values for the Young's modulus and the Poisson's ratio of the hardened cement paste.

The use of 8200 different RVE within a Monte-Carlo simulation requires the processing of large amounts of data and the computation of 8200 micro-structures, each with about 800 000 unknowns. This can be accomplished in parallel on PC clusters, where each separate computation requires approximately six minutes on present PCs. The results for the effective values of the Young's modulus and Poisson's ratio are given in Fig. 6 together with the corresponding Gaussian distribution. These results have been compared to experiments carried out at the Institute of Materials Sciences at the RWTH Aachen. Thereby, it could be shown that the effective material constants obtained by the numerical simulation are very accurate (deviation approx. 2%) concluding that the presented strategy is already very well applicable for the prediction of material constants, see [6].

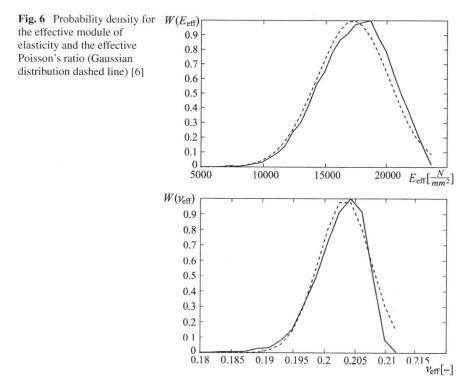

Fig. 6 Probability density for the effective module of elasticity and the effective Poisson's ratio (Gaussian distribution dashed line) [6]

Damage Due to Frost in Hardened Cement Paste

The method of homogenisation can also be extended to materials that react inelastically. As an example, we consider damage due to frost within hardened cement paste. Such damage can occur due to the filling of the pores with water, see also [7]. The mechanical behaviour of the examined hardened cement paste can be characterised by brittle damage and plastic deformations due to microcracks. The microstructural material model has to be verified on the basis of experimental results. Thereupon, the damage due to frost, i.e., a damage of the micro-structure due to the increase in volume of water-filled pores, can be numerically simulated. A statistical evaluation for different temperatures and moistures and subsequent homogenisation finally leads to a micro-structurally-based, effective material model that can be used in structural simulations on the macro-scale.

For the numerical simulation of damage due to frost on the micro-level, a numerical model for the freezing process of water is developed that describes an increase in volume of 9%. Obviously, the freezing process is temperature-dependent and, therefore, the micro-structural temperature gradient has to be determined by a thermo-mechanically coupled simulation. However, the micro-structure is so small such that an existing temperature gradient can be neglected. In consequence, one can control the temperature of the micro-structure by an external parameter that depends on the actual load step. Accordingly, no thermo-mechanical coupling on micro-level is necessary, and it is sufficient to carry out a purely mechanical analysis.

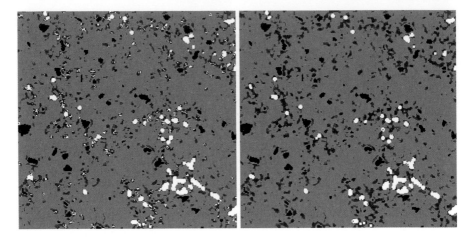

Fig. 7 Cross-sectional view through the water-filled hardened cement paste (256×256 µm) for the moistures $w_h = 0.70$ and $w_h = 0.80$

In experiments, an absorption of water occurs in thermal load cycles that causes a continuous increase in moisture. The water absorption is not simulated, instead, a nearly saturated micro-structure is used that has been generated numerically using the medial axes. Figure 7 depicts cross-sectional views of micro-structures with different degrees of moisture content, w_h. As expected, the water depicted in blue covers first the edge of a pore and then the interior of the pore volume. In the figures, the unhydrated clinker phases are depicted in black, the hydratisation products in gray, and the pores in white.

An increase in volume of the water-filled pores occurs as soon as the temperature drops below a certain freezing temperature θ_f, which depends on the current radius of the pore. In the CT-imaged micro-structures, the pore radii vary from 0.5 µm to 2.0 µm. Due to the limited resolution, small pores with radii less than 10 nm cannot be resolved. Hence, simulations appealing to the FE method neglected the influence of small pores. For the numerical analysis, the water-filled pores are described using a linear-elastic material law including temperature-induced strains.

Then, the damage within an RVE can be determined based on constitutive relations. This is depicted in Fig. 8 due to the cooling of an RVE. The lighter areas hereby indicate the regions of damage. Based on these results, one can deduce an effective damage law on the macro-scale with the help of Monte-Carlo simulations of many different RVE and homogenisation techniques. The homogenisation allows to prescribe the damage due to frost on the next coarser scale.

Multi-Scale Methods for 3-d Crack Propagation

Many materials contain micro-cracks that grow under constantly varying loads and, thus, can essentially lead to macro-cracks. The knowledge of the crack growth is essential for gaining new insights in the field of fatigue strength of materials. Here,

Fig. 8 Damage of an RVE
due to frost attack in the
micro-structure [7]

Fig. 9 Vertical normal stress
in an RVE with nine
micro-cracks and a crack
front as a result of a
multi-scale analysis

micro-macro simulations under dynamical loading conditions ensure progress that
assists engineers to improve the construction design. But such computations are still
extremely time-consuming as they are based on the analysis of FE discretisation
that possess several millions of unknowns. Promising are new approaches like the
eXtended Finite Element Method (XFEM). This method incorporates the influence
of cracks by introducing additional degrees of freedom, such that one does not have
to adapt the existing FE mesh in order to include crack propagation.

Figure 9 depicts nine randomly distributed elliptic cracks within a part of a solid,
which consists of a linear-elastic material. In vertical direction, this part is subjected
to a traction force which causes the normal stress with respect to the vertical direction as depicted in Fig. 9.

The coupling of the XFEM with a multi-scale analysis then allows the analysis
of finer structures, in which macro- and micro-cracks can be resolved, see Fig. 10
for an example of a straight and a curved crack.

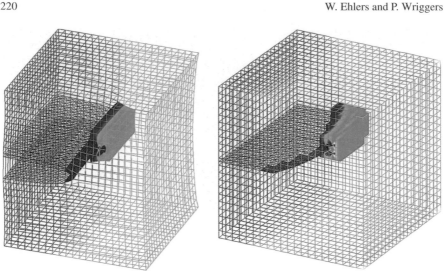

Fig. 10 Crack front with macro- and micro-crack in a multi-scale analysis

The numerical simulation of crack formations, which can lead to the growth of a crack on the macro-scale, has been computed using a projection method within a multi-scale analysis. The macro-cube is stretched in the vertical direction. The colour spectrum resembles the normal stresses in the vertical direction. With such simulations, one can examine crack amplification as well as shielding and thus can clarify the influence of micro-cracks on the growth of a crack in a 3-d structure.

3 Synopsis and Status Quo

3.1 Facts

Continuum Mechanics of Multi-Component and Multi-Phase Materials (Multi-Field and Multi-Physics Materials)

Continuum mechanics describes the behaviour of materials (solids, fluids, and gases) on the macro-scale. This also holds for the Theory of Porous Media (TPM) and the underlying Theory of Mixtures (TM). While the TM has been developed as a continuum-mechanical modelling tool for miscible and reactive substances, the TPM is a tool that describes immiscible materials (e.g., solid skeletons and pore fluids) in the sense of superimposed and interacting continua, i.e., each part of a body, which occupies only a portion of the entire domain, is smeared over the entire domain such that all incorporated fields are now defined simultaneously on the same total domain. This requires an actual or a virtual homogenisation of the partial bodies. While the TPM generally assumes that a homogenisation has been already performed, a multi-scale modelling approach provides the opportunity to not only

Fig. 11 A sample made of polyurethane foam (*left*) with an open-cell micro-structure (*right*)

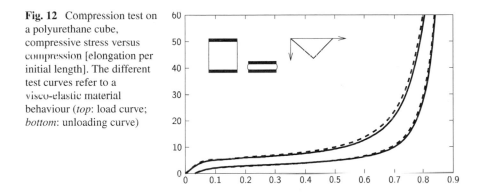

Fig. 12 Compression test on a polyurethane cube, compressive stress versus compression [elongation per initial length]. The different test curves refer to a visco-elastic material behaviour (*top*: load curve; *bottom*: unloading curve)

consider the homogenisation as a virtual process, but to carry this process out in a real environment by numerical techniques. This can have great advantages.

Heterogeneous materials or materials built from a heterogeneous composition of different immiscible materials possess a micro-structure that has great influence on the representation of macroscopic processes. This can be illustrated by a simple example: A sponge that consists of an open-cell foam can exhibit for the same porosity (pore volume per total volume) completely different permeabilities depending on whether the structure of the foam consists of either large or small pores or pore channels. In the laboratory, one would determine for both sponges a hydraulic conductivity that presents on the macro-scale a value that is essential within engineering calculations of coupled deformation and flow processes. The micro-structure of the sponges influences not only the permeability properties, but also the material properties of the sponge itself. If we assume for this example that an open-cell foam with 90-percent porosity (Fig. 11) and visco-elastic material properties is compressed, then, on the macro-scale, large deformations occur in the direction of compression, while the deformations on the micro-scale are mainly governed by local bending and buckling, cf. Fig. 12.

Fig. 13 Compression test of a polyurethane cube. From left to right: bending region, buckling region, and compression region

In the stress-strain diagram (Fig. 12), one observes an initial region that is related to bending modes on the micro-scale (Fig. 13). This is followed by a region that is characterised by local buckling, i.e., the foam ligaments buckle on the micro-scale and the foam layers *pile up* on top of each other. Adjacent to this region is the compression domain. Within this domain, the material increasingly hardens until finally no further deformation occurs, despite an additional load increase (vertical tangent). The reason for this macroscopically peculiar behaviour is again to be found on the micro-scale. Polyurethane is an incompressible material, i.e., the material does not reduces its volume, if one applies a pressure to its entire surface. While the polyurethane foam seems to be soft at first (bending and buckling region), the foam hardens increasingly due to closing of the pores until all foam layers are *piled up*, and the pore space is closed. The initially highly compressible foam becomes incompressible, and the behaviour on the micro-scale corresponds to the behaviour on the macro-scale.

The example of the open-cell polyurethane foam shows that on the macro-scale, every change in porosity and every change in pore geometry (large pores vs. fine pores) results in another macroscopic material behaviour, for which separate experimental tests and test evaluations are required to determine the macroscopic material parameters. Assuming that each foam consists of the same polyurethane with the same intrinsic material properties, i.e., the material properties on the micro-scale are the same for each case, then the macroscopic material properties can be obtained within the scope of a numerical homogenisation process using appropriate RVE. This assumes the knowledge of the internal geometry of the material, i.e., the knowledge of the geometric data on the micro-scale. Thus, the optimisation process for the generation of the material parameters on the macro-scale mentioned in Sect. 3.1 is obsolete; however, the numerical costs associated with the homogenisation process are at least comparable.

Given sufficient computational resources, it would also be feasible to use so-called FE^2 calculations instead of numerical computations on the micro-scale in order to compute macroscopic material properties. In this approach, a macroscopic

initial-boundary-value problem (IBVP) is solved without using a macroscopic constitutive law. This means within the framework of the FEM that one does not evaluate a constitutive law at the Gauss points of the finite elements, but instead embeds
a further FE simulation at the Gauss point level in the sense of an RVE. The homogenised result of RVE calculations replaces then the evaluation of a macroscopic
constitutive law. Although this approach at first sight seems to be too costly for the
solution of a single IBVP, it has a decisive advantage. A numerical laboratory is generated that can be used in the sense of an optimisation process to embed arbitrarily
many micro-structures into the macro-structure in order to generate an optimal solution to a complex problem or to find the optimal composition of an inhomogeneous
material in the sense of so-called *tailored materials*.

Here, one could give many further examples that show the forward-looking possibilities of continuum mechanics of multi-component and multi-phase materials
as well as simulation techniques to describe heterogeneous materials (multi-scale
modelling).

All presented examples reveal that problems with multi-physical properties always lead to a system of strongly coupled partial differential equations (PDE),
which, in general, has to be solved monolithically. However, there exist strategies
that check the degree of coupling of the overall system in order to decide whether
it is reasonable to use an operator-splitting method. Such methods reduce the total
costs for the solution of the system of equations without compromising the convergence properties of the entire system. Beyond that, volumetrically coupled fluid-
structure problems generally require the use of Taylor-Hood elements such that stability criteria like the Ladysenskaja-Babuška-Brezzi condition (LBB condition) are
fulfilled. Taylor-Hood elements appeal to quadratic ansatz functions approximating
the displacement of the solid and linear ansatz functions for the pore pressure. This
element type further increases the degrees of freedom of the overall problem. This
yields, in particular for geometrically 3-d problems, very large systems of equations,
which, even today, pose a great challenge for their solution.

The analysis of complex material behaviour confronts the modeller with further
problems. The material parameters contained in the complex constitutive laws often
elude a direct determination, i.e., they cannot be derived directly, but only indirectly.
Thus, optimisation strategies and the solution of inverse problems are required. Alternatively to this approach, a multi-scale analysis can be used. This is further expressed in the following.

Simulation Techniques for the Description of Heterogeneous Materials (Multi-Scale Modelling)

The consideration of small scales within a numerical simulation model leads to a
very costly discretisation, since the prediction of quantitative results always requires
a 3-d micro-structure. This leads to high-dimensional systems of equations that can
only be solved using specific mathematical methods. These include iterative solvers
that have to be adapted to a specific problem in order to obtain optimal simulation

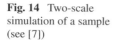

Fig. 14 Two-scale simulation of a sample (see [7])

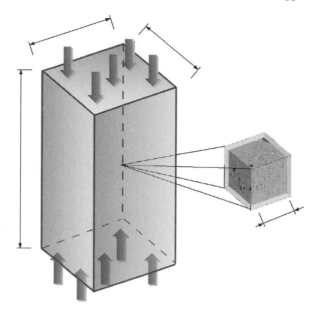

times. Due to the size of the systems, often parallel multigrid methods have to be used, whereas algebraic multigrid methods are advantageous for engineers because of their independence of the mesh.

The local deformations arising in the substructures of the RVE are frequently very large such that robust finite elements have to be applied. For this purpose, there is still a need to develop new robust elements, in particular, in the field of nonlinear deformations.

In the same way as for multi-physical problems, parameter identifications of not directly measurable constitutive parameters of the micro-structure are still required for the multi-scale analysis. This leads to inverse problems, for which large systems of equations have to be solved causing high computational costs due to a large number of function evaluations within the underlying optimisation problem. Again, parallel computation methods are indispensable if the solution needs to be determined within a computing time reasonable for engineers.

The disadvantage and the associated challenge of this approach are the extremely high requirements on computing power. This results from the dimension of the micro-structural model that leads to systems of equations with about 1 million degrees of freedom. The micro-systems included in this model have to be analysed at all integration points of the macro-model, where it is assumed that the internal structure of the micro-models is randomly distributed such that no micro-system equals another. Since the macro-model generally exhibits more than one thousand integration points, it follows that over thousand micro-systems have to be evaluated at every computational step of the analysis of the entire sample. Since these are used for nonlinear material behaviour, this also necessitates a nonlinear simulation of the micro-systems. If we imagine that the macro-system within a relevant computation does not only consist of one thousand finite elements, but of one million finite elements, then, it is clear that presently even state-of-the-art computer architectures are not sufficient to carry out such simulations within a reasonable time. Thus, there is great need for research in parallel computing, in systems solvers, and in the development of robust finite elements for automated nonlinear calculations.

3.2 Summary

The preceding discussion has shown that advanced mechanics can contribute to the description of complex problems of material modelling under consideration of multi-physical properties on all scales. Advanced mechanics is also able to formulate large initial-boundary-value problems and to solve them numerically. The resulting problems have been mentioned and refer mainly to questions of the analysis of the system of equations and the construction of suitable system solvers including the required preconditioners. Moreover, the problem of parameter optimisation and the associated inverse problems become increasingly important. This makes clear once more, that the problems of advanced mechanics cannot be handled by engineers alone, but collaboratively with mathematicians. Therefore, engineers and mathematicians have to proceed further on the way that has been taken by Ludwig Prandtl and Richard von Mises when they founded the GAMM in order to solve the problems of the future together.

4 Strengths-Weaknesses Analysis, Challenges

On the basis of comprehensible facts, the preceding sections have illustrated the potential of advanced mechanics for the challenges of the future, in particular, in cooperation with numerical mathematics. This is further explained in the following. Thereby, also strengths and weaknesses are mentioned and the challenges of the future are pointed out.

4.1 Basis and Visions

Today, there are very powerful computers on which multi-component materials with multi-physical properties as well as multi-scale problems can be analysed. This has been totally different 20 years ago. Indeed, the Theory of Mixtures and the Theory of Porous Media have been elements of extended continuum mechanics for a long time (the TM since approx. 1960, the TPM since approx. 1980). However, in those days, there have been besides the fundamental equations (kinematics and balance relations of superimposed continua) no noteworthy material models that could have been included into a relevant numerical simulation. A reasonable numerical analysis of complex material models was simply not possible. The same holds for multi-scale models, where the physical properties are numerically analysed on multiple scales and compared and related to each other with the help of different homogenisation methods.

In advanced mechanics, both elements, continuum mechanics of multi-field and multi-physical problems and the analysis of multi-scale problems, take an essential role. These elements, which can be easily combined as described above, hold a huge potential for the tasks lying ahead of us. A consequent design of complex problems with the methods of the TPM combined with multi-scale analyses offers the possibility to solve future problems that seem to be unsolvable today. In the following, this is emphasised by an example.

Mankind has always been interested to know what "holds the world together in its innermost". Therefore, it is not astonishing that biomechanics enjoys high attention, and this not only among colleagues, but in the whole society. Looking visionary at future possibilities of biomechanics and comparing them with what is feasible and possible today, then it becomes clear that today merely isolated problems within the field of biomechanics are treated. On the one hand, there are groups that examine the human musculo-skeletal system using rigid body dynamics. In particular, these groups can be found in the field of sports sciences or sports medicine. On the other hand, there are groups that examine and describe numerically parts of the human body such as bones, cartilage, organs, or the vascular system. In general, these groups belong to the field of *continuum mechanics* and can be found mostly in civil engineering and mechanical engineering faculties. If the expertises presented in the different fields can be bundled and combined to a common multi-scale structural model, then, we are on the way to generate a *virtual human* that describes the human body at large with all its mechanical properties. This model could use rigid body mechanics on the coarse scale of the musculo-skeletal system, it could describe the bones, cartilage, and organs under consideration of the vascular system with the methods of continuum mechanics, and could use the TPM in particular on the meso-scale, furthermore performing cell-mechanical examinations on the micro-scale. All scales could be combined into a multi-scale analysis.

With a virtual human at hand, one could, for example, analyse the consequences of accidents (Fig. 15) by carrying out an initial analysis on the coarse scale using a crash-test dummy and then continuing the analysis on the meso-scale in areas where certain critical loads are exceeded. Numerical investigations of this kind

Fig. 15 Crash test with dummy (source: http://www.euroncap.com)

could also be used in the construction of the external shell of vehicles in order to reduce possible consequences of accidents with pedestrians, bicyclist, and other traffic participants. These days, patient-specific implants of the hip, knee, or spine can be produced for which shape optimisations algorithms minimise, for example, luxation probabilities. Furthermore, surgical interventions could be planned and even carried out virtually for training or testing purposes. For this, a stored virtual human model on the computer would have to be adjusted to individual patient data, cf. Fig. 3. The vision of the computer-oriented virtual human is also supported by the scientists at the Cluster of Excellence for Simulation Technology (http://www.simtech.uni-stuttgart.de).

If such computations have to be carried out as part of medical engineering applications accompanying a surgery, then, complex simulations and visualisations in real-time are required. However, this is not possible on present computers.

4.2 How Does Mathematics Come into Play?

Complex continuum mechanical problems always lead to a system of coupled partial differential equations. Such a system can be solved analytically only in special cases: hence, approximate solutions are required. A suitable tool is available with the finite element method (FEM). Numerical analysis and computational mathematics are necessary if multi-scale and multi-physical problems have to be handled numerically. This incorporates the quantification of the uncertainties contained in the models as well as a self-adaptive choice of scales and of the knowledge of the basic physics in order to treat dynamical and coupled processes in real or virtual reality.

Real problems are three-dimensional in space. This leads to a huge and nearly unmanageable amount of degrees of freedom and corresponding equations, especially, in the treatment of coupled problems within the scope of FEM. Still, large and largest problems cannot be solved on single-processor machines but only on massively parallel systems. Furthermore, adaptive strategies are required in time as well as in space, cf. [1, 2]. In addition, domain decomposition methods have to be applied in order to segment the entire problem and feed it into a massively parallel computation. Moreover, fast iterative solvers with suitable preconditioners have to be found in order to treat large problems in an acceptable time.

In general, complex mechanical problems are based on complex material models, where the required material parameters cannot be determined directly, but only by means of inverse computations and by the use of optimisation methods. This is, for example, the case if shear band localisations [8] have to be handled. This frequently leads to mathematically ill-posed problems in the framework of a standard continuum (Boltzmann continuum). For example, this applies to the base failure problem (Fig. 16), where an elasto-plastic soil starts to form shear bands due to the load resulting from a building. This can lead to an instability of the whole foundation such that the construction collapses. By application of an extended continuum, the ill-posed problem can be turned back into a well-posed one without changing

Fig. 16 Shear band formation with shear failure (FE computation). Regularisation of the ill-posed problem with the help of a continuum extended by rotational degrees of freedom (Cosserat continuum)

the elasto-plastic constitutive law. If the shear failure problem is described, for example, not with a Boltzmann but with a Cosserat continuum, then, one obtains a well-posed problem, however, at the cost that one has to determine additional material parameters (Cosserat parameters).

The Cosserat continuum [1] is an extension of the Boltzmann continuum by the additional inclusion of independent rotational degrees of freedom. In addition, couple stresses corresponding to these rotations occur and have to be determined by additional constitutive equations. Numerical investigations furthermore exhibit that the Cosserat rotations and the corresponding couple stresses, as a result of local microstructural effects, are only active in the localisation zone and not in the reminder of the entire domain [9]. As a consequence, only the material parameters of the Boltzmann continuum can be determined using the usual optimisation methods. The Cosserat parameters have to be determined by inverse computations, i.e., by a back-analysis of an IBVP describing a related experiment, cf. e.g. [9]. In this case, the solution of the problem, which is characterised by the shear band initiation and the shear band direction and width, is taken as given and the problem is solved for the unknown Cosserat parameters based on the given material parameters of the Boltzmann part of the continuum.

All above mentioned problems, like many more, can only be treated successfully in collaboration with engineers and mathematicians. Besides this direct linkage between complex problems and mathematical solution strategies, there are also huge amounts of data that have to be processed and visualised. Here, besides mathematics, computer science and data processing also plays a crucial role.

4.3 Strengths, Weaknesses and Challenges

From the present point of view, it can be stated that advanced mechanics can cope with the challenges of the future. However, the models for capturing realistic problems are getting more and more complex. This leads inevitably to even larger problems that have to be handled with methods from computational mechanics and numerical analysis.

The computations belonging to the applications and examples cited above require fast iterative solvers that are suitable for large 3-d systems. The elements used for the discretisation have to be robust in nonlinear problems and have to possess the required stability. This field is well-investigated within the linear theory, but it is still open if nonlinear simulations are encountered. Furthermore, adaptive algorithms for the coupling of multi-scale computations have to be developed, cf., e.g., [10], in order to be able to carry out computations for practical applications as is shown in Fig. 14. Here, on each level, an optimal solution with respect to computational time (costs) has to be aspired. Therefore, considerable efforts are still necessary on the part of engineers and mathematicians in order to advance modelling, validation, and theoretical foundations.

In contrast to current models, which are almost deterministic, reality requires the inclusion of stochastic components, as otherwise all statements made by deterministic methods can only appear real within certain uncertainties. Therefore, the involvement of uncertainties in our models is in many cases absolutely necessary. If we consider again the example of CO_2 sequestration (Fig. 2), it is absolutely clear that the available data of the sequestration region are only available at selected locations. The data of the regions lying between the measuring points are not recorded, instead, assumptions on the data are made. In such cases, statistical methods have to be used in order to deal with uncertain data that lead to uncertainties in the geometry data and material parameters of our models.

In the future, virtual worlds are created on the computer that are used as prediction tool for real actions. The present and future challenges for the realisation of such computations and visualisations lie in the reduction of development and computational times as well as in the promotion of young academics in engineering and mathematics. After all, there cannot only be the *user*. As now and in the future, someone has to be able to understand and to advance complex models in order to carry out even more complex computations.

Only in this way, a numerical simulation environment can be created and constantly maintained and extended in the future that allows reliable predictions of complex engineering models.

5 Recommendations on Possible Actions

Advanced mechanics will make its contribution to the challenges of the future. In particular, in cooperation with mathematics and computer science, mechanics will be able to deal with these tasks. Visionary solution approaches are given in sufficient quantities as for example the *virtual human* as described in Sect. 4.1, see also http://www.simtech.uni-stuttgart.de. Yet, the realisation of these visions requires not only time, but also committed and hard-working scientists. Insofar, it is necessary to inspire young people for mathematics and mechanics, to arouse their motivation and to shape the future with them. For this purpose, advertising efforts at schools and other facilities, where young people can be reached, are also necessary in order to make the fields of mechanics and mathematics and their scientific, industrial, economic, and, last but not least, their social dimension accessible to them. The GAMM and its technical committees will contribute everything that is necessary for this.

References

1. Ehlers, W.: Foundations of multiphasic and porous materials. In: Ehlers, W., Bluhm, J. (eds.) Porous Media: Theory, Experiments and Numerical Applications, pp. 3–86. Springer, Berlin (2002)
2. Wieners, C., Ammann, M., Graf, T., Ehlers, W.: Parallel Krylov methods and the application to 3-d simulations of a triphasic porous media model in soil mechanics. Comput. Mech. **36**, 409–420 (2005)
3. Ehlers, W., Karajan, N., Markert, B.: A porous media model describing the inhomogeneous behaviour of the human intervertebral disc. Mater. Sci. Mater. Technol. (Materialwissenschaft und Werkstofftechnik) **37**, 546–551 (2006)
4. Zohdi, T.I., Wriggers, P.: Introduction to Computational Micromechanics. Springer, Berlin (2005)
5. Löhnert, S., Wriggers, P.: Effective behaviour of elastic heterogeneous thin structures at finite deformations. Comput. Mech. **41**, 595–606 (2008)
6. Hain, M., Wriggers, P.: On the numerical homogenization of hardened cement paste. Comput. Mech. **42**, 197–212 (2008)
7. Hain, M., Wriggers, P.: Computational homogenization of micro-structural damage due to frost in hardened cement paste. Finite Elem. Anal. Des. **44**, 233–244 (2008)
8. Ehlers, W., Graf, T., Ammann, M.: Deformation and localization analysis in partially saturated soil. Comput. Methods Appl. Mech. Eng. **193**, 2885–2910 (2004)
9. Ehlers, W., Scholz, B.: An inverse algorithm for the identification and the sensitivity analysis of the parameters governing micropolar elasto-plastic granular material. Arch. Appl. Mech. **77**, 911–931 (2007)
10. Temizer, I., Wriggers, P.: An adaptive method for homogenization in orthotropic nonlinear elasticity. Comput. Methods Appl. Mech. Eng. **196**, 3409–3423 (2007)

Mathematics for Machine Tools and Factory Automation

Berend Denkena, Dietmar Hömberg,
and Eckart Uhlmann

1 Executive Summary

The article describes the state and perspective of the application of mathematical methods in the simulation and design of machine tools and factory automation. Successful examples for the application of mathematics can be found mainly in the modeling of machine tools, of multibody systems, and in the simulation of individual production processes.

The authors discuss important current mathematical concepts for production engineering. They show that a consistent mathematical modeling is necessary for the increasingly important description of entire process chains. For the numerical simulation adaptive algorithms are often an essential tool. They lead to a drastic reduction of the number of unknowns without increasing the approximation error, and thus frequently a numerical simulation for complex workpiece geometries becomes possible in the first place. The ultimate goal of modeling and simulation is usually the computation of optimal process parameters. Mathematically, this means the solution of a so-called optimal control problem. Optimal control can be particularly successful when applied in combination with process control of the machine.

Major challenges for the collaboration of production engineering and mathematics lie in the examination of the interactions between process, machine, and work-

B. Denkena
Institute of Production Engineering and Machine Tools (IFW), Leibniz Universität Hannover,
An der Universität 2, 30823 Garbsen, Germany
e-mail: Denkena@ifw.uni-hannover.de

D. Hömberg (✉)
Weierstrass Institute for Applied Analysis and Stochastics, Mohrenstr. 39, 10117 Berlin, Germany
e-mail: hoemberg@wias-berlin.de

E. Uhlmann
Institute for Machine Tools and Factory Management, Technische Universität Berlin,
Pascalstraße 8-9, 10587 Berlin, Germany
e-mail: uhlmann@iwf.tu-berlin.de

M. Grötschel et al. (eds.), *Production Factor Mathematics*,
DOI 10.1007/978-3-642-11248-5_12, © Springer-Verlag Berlin Heidelberg 2010

piece. Using the examples of milling and laser material processing this problem is explained in more detail.

Due to a continuously growing model variety with simultaneously growing pressure of shorter development times, from the industrial point of view there is great need for the development of tools for an automated replanning or reconfiguration of complex production facilities. For that purpose the development of a multi-scale mathematical standard model for factory automation is required. This could be jointly worked out in an interdisciplinary priority program of production engineers, computer scientists, and mathematicians.

2 Success Stories

2.1 Mathematics for the Development of New Machine Concepts: Parallel Kinematics

In recent years, the machine tools branch was increasingly concerned with the search of new machine concepts due to the cost and quality pressure of their clients. Thereby, machines with parallel and hybrid kinematics recently attracted particular attention. Compared to conventional machines, parallel kinematics can have advantages in stiffness and dynamics. Furthermore, no superposition of the errors arising in the axes takes place. By the use of common parts the production costs are reduced.

Despite these advantages reluctance still prevails with respect to their industrial application. A fundamental restraint consists in the difficulties in governing the complexity of the coupled kinematical positioning and motion behavior. Geometric errors (due to production inaccuracies of the machine elements or assembly errors, respectively), gravitational and inertia influences, as well as thermal deformations of the structure elements yield deviations between the kinematic transition behavior theoretically given by construction and the actual behavior, see [18]. The identification, computational mapping and compensation of these negative effects is the task of calibration.

Within the research project *Calibration methods for hybrid parallel kinematics*[1] a general method for the parametric calibration of parallel and hybrid kinematics has been developed (cf. Fig. 1). The basis is a model of the calibrated machine that should consider all error impacts acting on the end-effector. For this, a general approach for kinematic modeling has been developed, see [10]. The method is based on homogeneous transformations and the Denavit-Hartenberg-convention that has been extended corresponding to the requirements of the modeling of hybrid parallel kinematics. The standardised modeling has been automated and integrated in the calibration process. The developed method has been included in the calibration software and is generally applicable on parallel or serial structures [8].

---

[1] Supported within the DFG Priority Program 1099 "Production machines with parallel kinematics" (duration: 2000–2006).

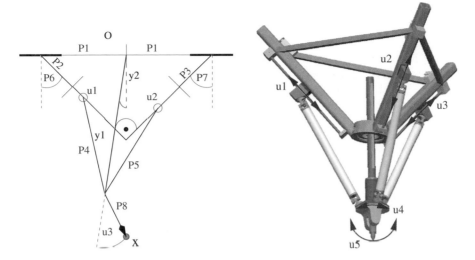

Fig. 1 Two and three dimensional variants of the hybrid parallel kinematic *Georg V*

2.2 *Mathematics for the Layout of Machine Tools: Cutting Processes*

The mathematical modeling and simulation of cutting processes with geometrically defined cutting edges such as planing, drilling and milling is a current field of research. Thereby, besides the consideration of chip formation, the load on the tool and the analysis of the surface layer of the workpiece are the focus of the examinations. The vast majority of the examinations, which rely on numerical computations using finite element methods, model the chip formation process with the three elements tool, workpiece and chip. Thereby, particular attention is paid to the deformation processes that take place in the primary and secondary shear planes. The examinations involve on the one hand analyses of the chip separation and of the chip flow, on the other hand the simulation of chip breakage by suitable breakage criteria.

In recent years, the consideration of the workpiece material has come increasingly to the forefront in cutting simulation. Thereby, constitutive laws have been developed for different materials that consider besides the dependence of the flow stress on the strain also the strain rate and the thermal behavior.

At the Institute for Machine Tools and Factory Management of the TU Berlin (IWF) the flow behavior of high-alloyed steels, e.g., nickel-based alloys, has been analysed and modeled under consideration of the damage mechanisms occurring during the stress of the material [23].

Figure 2 shows the actual chip formation in the cutting process as well as a simulation of this chip formation mechanism. An analysis of the specific cutting force shows good agreement between simulation and experiment. Even minor changes in the element distribution of the cutting edge lead to significant changes in the specific cutting edge. Using this results, the cutting edge geometry could be adapted to the workpiece material, which led to a considerable reduction of tool wear [20, 22, 24].

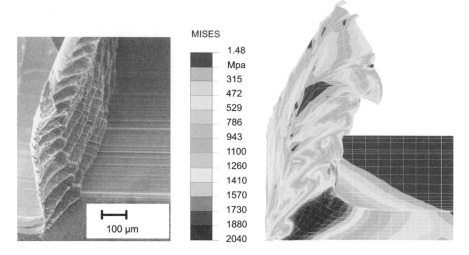

Fig. 2 Simulation of chip formation in outside longitudinal cylindrical turning

Fig. 3 Face-grinding with
lapping kinematics

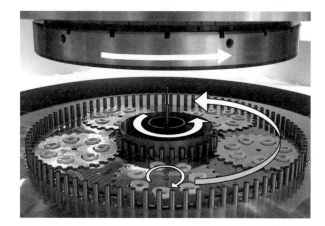

In comparison to the cutting simulation with geometrically defined cutting edge, the simulation of the undefined cutting edge is a particular challenge with respect to process understanding and optimisation. This is due to the high complexity of the effective kinematics, the existence of material separation in the microscopic range, the statistical nature of the cutting body, as well as the continuous cooling of the process with large amounts of cooling lubricant.

An additional difficulty in face grinding with lapping kinematics is the simultaneous movement of many workpieces between two grinding wheels. Thereby, the components are fixed in workpiece holders that rotate between two pin crowns like the planets in a planetary gear set (Fig. 3). A simulation for the prediction of the wear behavior of these cost intensive tools is indispensable for the economic application of this process.

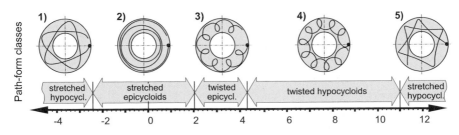

Fig. 4 Path-form classes in dependence of the rotation speed ratio

At the IWF the relative movement between the effect partners has been examined (see Fig. 4), thus a computation of the position probability of the components at different radial sections of the tool and therefore also the computation of the load on the grinding wheels could be carried out.

In order to avoid inhomogeneous wear of the grinding wheels, the design of the grinding wheels has been optimised based on simulations [2]. Furthermore, it could be shown how the cutting volume can be maximised without experimental effort by a suitable choice of the kinematic process parameters [21].

3 Mathematical Concepts in Production Engineering

3.1 Modeling of Process Chains

In recent years significant progress has been made in the simulation of individual processes and their effects on the workpiece. Meanwhile, the continuum-mechanical material behavior can also be simulated with a number of commercial software packages that are essentially based on finite element methods. Additionally, commercial tools for the setup and solution of the corresponding equations of motion are available for the description of the machine tool dynamics.

It turns out, however, that the significance of the simulation results is limited if one tries to improve the machine structure on the basis of these results. In the same way, this approach reaches its limit if by coupling of commercial software packages the interaction of different effects, for example between machine structure and process, is to be described. Frequently, the different packages can only be controlled via graphical user interfaces, which prevents an incremental coupling. Even if the packages possess scripting languages and clearly defined interfaces, that allow such an incremental coupling, this approach often leads to unacceptably small stepsizes for a stable numerical computation.

A reliable simulation of the interactions between machine, process and workpiece, or even of a whole process chain, requires a consistent modeling of the whole process. Of special importance is the definition of interfaces between the different interacting process chain components that can be described mathematically as boundary or initial conditions for the relevant field equations. For illustration this is

Multibody model machine (MBS):

- equation of motion of the cutter

$$\ddot{q} + D\dot{q} + Kq = \frac{1}{m}F$$

- force on the cutter

$$F = -\sum_{j=1}^{N_z} g(\phi_j) \underbrace{O(\phi_j)\hat{F}}_{F^j}$$

- cutting force $\hat{F}$ in dependence on chipping thickness h

$$\hat{F} = a_p \hat{K} h$$

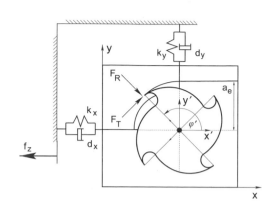

Field equations workpiece (PDEs):

- momentum balance

$$\rho_0 u_{tt} = \nabla \cdot \sigma,$$

$$\sigma = \lambda(\varepsilon : I)I + 2\mu\varepsilon$$
$$- 3K\alpha(T - T_0),$$

$$\varepsilon = \frac{1}{2}(\nabla u + \nabla^T u)$$

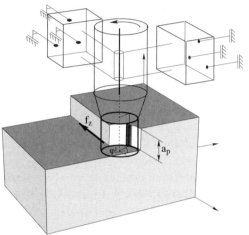

- energy balance

$$\rho_0 c_v T_t = \kappa \nabla T - 3K\alpha T_0 \nabla \cdot u_t$$

Boundary condition on the contact surface $\Gamma(t)$

- momentum balance

$$-\sigma \cdot n = \frac{F^j}{|\Gamma(t)|}$$

- energy balance

$$-\kappa \partial_n T = W(F^j, v_T, h)$$

- coupling between MBS and PDEs via chipping thickness h

$$h = \Big[\underbrace{R(t) - R(t-\tau)}_{\text{feed}}$$
$$+ \underbrace{q(t) - q(t-\tau)}_{\text{machine def.}}$$
$$- \underbrace{u(t) - u(t-\tau)}_{\text{workpiece def.}} \Big] e_r^j$$

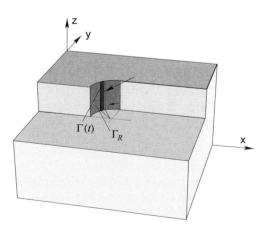

Fig. 5 Coupled model for the system cutter-workpiece

shown in Fig. 5 for the coupled modeling of the system cutter-workpiece. For simplicity, here the machine dynamics are only described as a double-mass oscillator (for further information, see [16]).

3.2 Adaptive Numerics

An efficient analysis of the entire system of production processes requires the application of modern algorithms of numerical mathematics. The numerical treatment of partial differential equations, today also called scientific computing, is the constant endeavor for optimal and reliable methods regarding the task of effective grid generation and the fast and, if necessary, parallel solution of the occurring high-dimensional discrete problems.

For efficiency reasons a grid should be as coarse as possible, but on the other hand it should be as fine as necessary to gain the desired accuracy in the computation. In order to meet both requirements, adaptive algorithms have been developed in recent years. These generate a series of grids and corresponding finite element spaces that do not refine uniformly, but only where an error indicator shows that the error is too large.

Even though many mathematical questions remain open in the context of adaptive algorithms, it is nevertheless the generally accepted state of the art that adaptive algorithms are preferred compared to uniform refinement methods. Adaptive algorithms reduce the number of unknowns dramatically and lead to a distinctively smaller discrete problem that then has to be solved effectively. Besides, a grid hierarchy is provided that allows a fast exchange of information within the fast solution algorithm and therefore enables solution methods of optimal complexity [3, 15].

Figure 6 shows two time points of an adaptive simulation of laser surface hardening. The laser light typically has a penetration depth of less than 1 mm, therefore the surface layer has to be strongly refined to spatially resolve the metallurgical effects that arise due to heating. A spatially and temporally adaptive simulation allows local grid refinement close to the laser focus. The upper picture shows the corresponding temperature profile and the lower one the grid. It can be seen how the refinement moves along with the laser focus.

Also the numerical simulation of the interaction of tool and workpiece in cutting processes requires the application of powerful numerical approximation methods. Typically, time-variant elastic and plastic regions occur and because of friction and deformation the superposition of thermal effects also appears. Figure 7 shows the adaptive numerical simulation of the NC form grinding of freeforms [25]. Thereby, the system spindle-grinding wheel (top left) is discretised by a FE-model (top right). In addition, there is a contact condition for the workpiece. The lower picture shows the adaptive linking of the contact area as well as the size of the error.

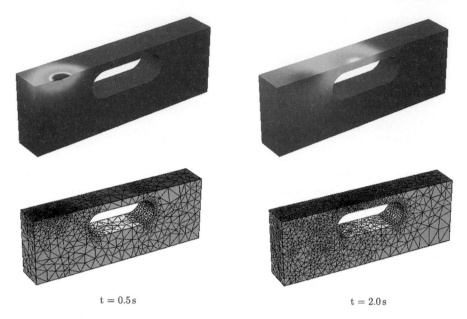

t = 0.5 s   t = 2.0 s

Fig. 6 Temperature distribution and adaptive grids during laser hardening at two different time points

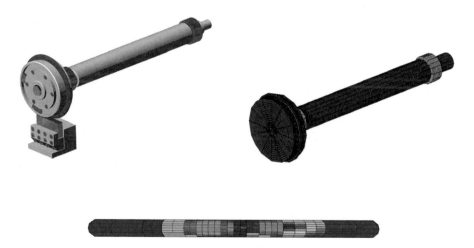

Fig. 7 Adaptive simulation of the NC form grinding [25]. System spindle-grinding wheel (*top left*), FE-linking (*top right*) and adaptive linking of the contact area (*red*: large, *blue*: small error)

3.3 Optimal Control

The ultimate goal of every process simulation is the optimisation of the process run. Typical goals thereby are the improvement of product quality, the reduction of cycle

- Minimize

$$J(T, z; p) = \int |z(x, t_E) - z_d|^2 dV$$

- under the constraints
  - temperature T fulfills the heat equation;
  - phase fractions z satisfy a rate law modeling the phase transitions in steel.

Fig. 8 Formulation of an optimal control problem for the heat treatment of steel

time or the lowering of energy costs.[2] Mathematically, this requires the solution of an optimal control problem. To this end, an objective function or a quality functional is defined allowing for the evaluation of the process result. Then, one tries to minimise this functional under the constraint that the optimal state is a solution of the corresponding field equation. The mathematical theory of optimal control of partial differential equations is an active research field in Germany. By a number of new appointments in recent years this field, which is extremely important for applicants, is by now represented at almost every technical university. A good overview of the mathematical theory can be found in a recently released textbook [19].

Figure 8 shows the formulation of an optimal control problem for the heat treatment of steel. The aim of a heat treatment is to obtain certain microstructure conditions in the workpiece that are described by the vector of the desired phase fractions z_d. At the end time t_F it is attempted to minimise the distance of the actual computed phase fractions $z(x, t_E)$ to z_d, under the constraint that the temperature field T and the vector of the phase fractions z fulfill the corresponding model equations.

The choice of the control variables p depends on the respective heat treatment method. In induction hardening this can be the generator frequency or the coupling distance between coil and workpiece, in laser hardening the feed rate and/or the power of the laser. Frequently, the underlying model is based on simplifying assumptions or the material parameters are not known exactly. In laser hardening, for example, the absorption coefficient that describes which fraction of the laser power is coupled into the workpiece is a hard to determine value as it depends not only on the surface roughness but also on other hardly predictable factors such as colour and staining.

Therefore, optimal control yields the best and also for practice most relevant results in connection with a machine control. In [1, 12] it has been shown that a sole machine control with constant surface temperature does not yield the desired constant hardening depth. On the other hand, by numerical simulation or by the solution of an optimal control problem, respectively, the power of the laser that yields the desired hardening depth can be computed. Because of the mentioned inaccuracy in the identification of the absorption coefficient, the direct application of the computed

---

[2] A recent study on behalf of the Federal Ministry of Education and Research (BMBF) shows the need for action and research in the field of energy efficiency in production [14]. Here, a wide field for applications of optimal control opens up.

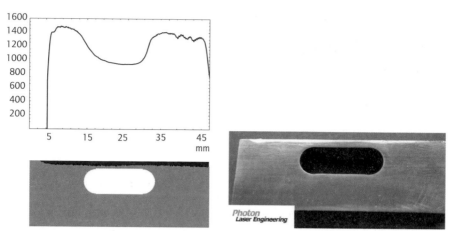

Fig. 9 Optimal computed temperature in the laser focus and resulting simulated hardening depth (*left*), experimentally reached hardening depth with numerically computed optimal focus temperature as set-point for the machine control (*right*)

optimal laser power would lead to a poor result. However, a resulting optimal focus temperature can also be computed from the optimal laser power. This temperature has to be reached independently of the respective absorption coefficient. Therefore, the best result is obtained when the computed optimal focus temperature is used as set-point for the machine control (cf. Fig. 9).

4 Challenges

4.1 Interaction Between Structure and Process

In recent years machine tool manufacturers evolve more and more from machine suppliers to system suppliers that offer besides the actual machine tool also tools, measuring devices, and services to the end user. Customer specified criteria such as component quality, processing time and delivery date have to be fulfilled. Thus, after fixing a suitable machine concept, already in the planning phase the determination of suitable tools, process parameters, and tool paths for the machining task belongs to the duties of the machine tool manufacturer. Depending on the complexity of the produced component, the NC[3] programs for the control of the machine are manually developed or CAM[4] software systems for the generation of the tool paths are used. CAM systems provide the possibility to generate the tool paths for the production of the complete workpiece, for a given workpiece geometry model and after the

---

[3]Numerical control.

[4]Computer aided manufacturing.

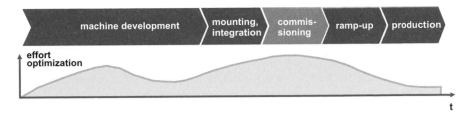

Fig. 10 Effort in the startup of machine tools [6]

determination of the required cutting tools, and to translate them by using a post processor into an NC code that is executable on the machine.

After completion of the planning phase, the commissioning and afterwards the acceptance of the machine at the end user takes place. The effort in the individual phases from the machine development to its application is schematically depicted in Fig. 10, whereby the commissioning phase requires the largest effort [6].

The so-called start-up of the NC programs follows the mounting of the production facility. This 'start-up' occurs in the 'commissioning' phase of the machine tool development. In this phase, the raw material is loaded in the machine and the NC program that has been developed during the planning is carried out to produce sample workpieces. After processing, the produced workpieces are measured, e.g. in a 3D coordinate measuring machine, and a protocol regarding the existing dimensional deviations on the verifying functional surfaces is written. Since it has to be verified that all relevant dimensions of the reference workpiece lie within the tolerances in order for the machine to be accepted, the causes that led to the dimensional deviations must be determined and rectified. A usual approach for the remedy of dimension faults is the adaption of the traverse paths of the tools by the programmer responsible for the start-up of the machine until the dimensions at the functional surfaces lie in the middle of the tolerance band. The adaption of the NC programs and the corresponding additional test series can lead to a delay of several days in the commissioning phase [6]. In general, in the planning and compilation of NC programs ideal conditions are considered. It is assumed that the machine with its components as well as tool and workpiece meet in their dimensions the construction specifications and that during the processing no deformations occur. However, batch fluctuations may lead to varying material properties and raw part dimensions, clamping conditions may not be exactly repeatable or the tool may be pushed aside by process forces. All these effects can lead to deviations of the produced workpieces in reality.

A further difficulty may arise due to badly adjusted process parameters, such as spindle speed and cutting depth. These lead to instable processes and therefore to quality losses and a lower process safety. If unstable states occur, due to which characteristical marks on the workpiece surface can result, then the process parameters must be adapted such that stable processing is attained. Therefore, for the corresponding processing step, e.g., the cutting depth is increased or the spindle speed is adapted until the process runs stable. A reduction of the cutting depth however, is directly connected with a decrease of the material removal rate and leads to a non-optimal utilisation of machine power. Moreover, the process lasts longer than originally planned, such that the processing times stipulated by the customer cannot be met reliably anymore.

In order to specify tool paths and process parameters in the planning phase, for which in the commissioning phase only marginal adaptions are necessary, it is mandatory to know and take into account the mechanisms that lead to dimension deviations or instabilities. In this context, the important mechanisms between the production process and the properties of the machine structure can be summarised under the term interactions between structure and process and are further explained in the following subsection using the example of milling.[5]

4.2 Interactions in Milling

Vibrations caused by cutting lead to a reduction of tool endurance and machine lifetime as well as to poor surface qualities and a high noise development. Particularly critical in this context are vibrations that lead to instable states with sudden strongly rising amplitudes ("chattering"). The occurrence of machine vibrations depends on several parameters. For example, long overhanging tools, large cutting depths, or thin-walled components increase the probability that vibrations occur. The main goal of the investigation of interactions in cutting processes is to be able to predict these vibrations in order to configure milling processes such that a preferably uncritical excitation of vibrations takes place.

The stability properties of a machine tool are frequently determined via stability maps. On these diagrams which are also called "stability lobe diagram" the cutting depths are displayed as a function of the current spindle speed up to which stable processing is possible. On the stability map depicted in Fig. 11 operating points above the stability border would lead to instable processes, and vice versa. Besides the stability border in the configuration of milling processes also the performance limit of the machine has to be considered.

In order to obtain a high-quality and reliable stability map, nowadays a very high experimental effort is required. Thereby, cutting tests with systematically varying cutting depths and rotation speeds are carried out. If the parameter combination yields a stable or instable process, then this is marked in the stability map. Thereby, the considered stability criterion influences also the quality of the stability map. To

---

[5]In order to better understand and to predict the relations between the production process, the machine behavior and the resulting workpiece properties, in 2005 the Priority Program 1180 "Prediction and Manipulation of Interactions between Structure and Process" by the German Research Foundation (DFG) has been established (http://prowesp.ifw.uni-hannover.de). In this research program, designed for a total of six years, approx. 50 researchers throughout Germany from the fields of production engineering, mechanics, mathematics and materials sciences work together on innovative approaches and methods for the prediction and systematic manipulation of interactions, cf. [7].

Fig. 11 Stability map of a cutting process [7]

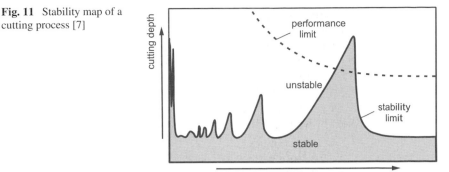

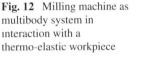

Fig. 12 Milling machine as multibody system in interaction with a thermo-elastic workpiece

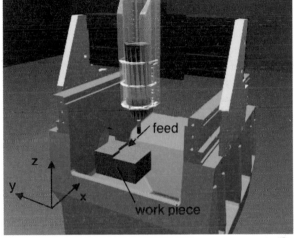

restrict the parameter domains to be examined, usually further investigations are made before the stability analysis. The experimental modal analysis provides the possibility to determine the eigenfrequencies of a machine. As the eigenfrequencies usually correspond to the chatter frequencies, in this case only spindle speeds that lie near to a machine eigenfrequency have to be examined.

In order to reduce the experimental complexity in the generation of stability maps, simulations are increasingly employed. They are able to reproduce the principle course of the actually measured stability borders. However, deviations between experimentally determined and simulated stability maps can occur that heavily reduce the model's significance and applicability. In order to increase the quality of these simulation-based stability investigations, in the Priority Program 1180 different approaches are investigated. One is based on a coupled mathematical model (cf. Fig. 5) representing the dynamics of the machine tool as a multibody system and describing the workpiece as a thermo-elastic continuum. Figure 12 shows the reduced depiction of machine tool and workpiece.

Even if some questions are still open with regard to parameter identification for the machine model and the modeling of the joints in the multibody system, still a deeper comprehension of the interaction of machine and component dynamics could already be gained [16].

4.3 Reconfiguration of Production Facilities

Figure 13 shows the rapidly growing model variety in the automotive industry. With every model generation the pressure to shorten the time to market is increased. This has not only strong effects on the component development, but the time for development and realisation of production systems is also permanently reduced. The growing variance of models, the shorter life cycles, and the request for more flexibility form enormous goal conflicts.

Given this situation, the virtual simulation of complex production facilities is of growing significance. Figure 14 shows a typical production scenario. A group of robots that work at different geostationary devices can be observed, and in addition there are conveyor belts for material transport and further devices for supply and removal as well as intermediate bearing of components. In recent years, a number of commercial tools have been developed for the virtual production that are able to represent production processes of high complexity. Thereby, whole production facilities can be assembled and simulated on the computer. However, they are far from enabling an at least semi-automatic planning of new or a reconfiguration of existing facilities, respectively.

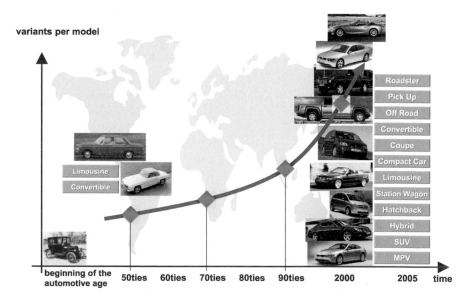

Fig. 13 Variants development in automotive production (by courtesy of Rücker EKS, Weingarten)

Fig. 14 Virtual simulation of a complex production facility (by courtesy of Rücker EKS, Weingarten)

For the reconfiguration of such a facility at least three levels can be distinguished: In the easiest case only the program of a robot has to be changed, e.g., as it has to attach a further weld spot. In a reconfiguration of mean complexity, e.g., the geometry of a component could be changed, such that the assignment of operating points and the path-planning for all robots of a group has to be modified. A complex reconfiguration would require an additional hardware modification.

Here, a wide field for cooperations between production engineering and mathematics opens up. The modification of a robot program due to mathematical path-planning has belonged to the state of the art since the nineties [11, 17]. But, one has to state that the development has slowed down. There are many open questions, e.g., in path-planning for cooperating robots [4, 13].

A thrilling challenge on the interface of mixed-integer programming, optimal control and production engineering is for example the development of a planning tool for a robot welding cell. Based on the CAD data of a component and the given coordinates of the welding spots, this tool would have to carry out an automatic assignment of the welding points to the robots, fix an optimal sequencing and finally automatically generate optimal and collision-free robot paths.

5 Perspectives

From the industrial point of view there is great need for planning and optimisation tools that allow an accelerated planning and reconfiguration of complex production facilities.

The modeling and simulation of fundamental production processes such as milling, grinding and forming and of joining processes such as soldering, welding and bonding has been an active research field of production engineering and of applied mathematics for more than twenty years. The same holds true for the multibody simulation of machine tools and robots. In contrast, the examination of interactions between process, machine and workpiece has only been tackled in recent years, driven in particular by the establishment of a DFG Priority Program for this topic.

In addition, over the past years significant progress has been made in the description of global production and material flow. The investigation of supply chains as well as the discrete-event simulation have been established as active research fields in applied mathematics and computer sciences [5, 9].

The next step on the way to automatic tools for the planning and reconfiguration of complex production processes is the development of a multi-scale mathematical standard model. For this, a common effort of production engineering, mathematics and computer sciences is required:

- On the lowest level this concerns the modeling of elementary production processes such as welding, drilling, or milling under consideration of the executing machine tool or the corresponding manipulator. Mathematically this means models of partial differential equations and multibody models.
- On the mean level these processes are combined to work cells consisting of a group of robots, machine tools, holding equipment and conveyor belts. Here, mathematical tasks are in particular mixed-integer optimisation as well as path-planning.
- Finally, these work cells are joined to form complex production systems. Here, the main tasks are the description of the workflow and the development of a corresponding language, i.e., the tasks that lie in the field of applied computer sciences.

These research tasks could be the scope of an new interdisciplinary DFG Priority Program.

The advantage in the development of such a concept lies in the fact that it allows for the consideration of improved simulation and optimisation techniques on the different scales. In connection with suitable sensors the factory model could be used for the control and monitoring of the entire production process. It would yield an intelligent error control and could be used for the controlled ramp-up and ramp-down of the facility. By the coupling with corresponding stock-keeping software a business-to-manufacturing integration of the production process could be reached. In other words: The much-trumpeted digital enterprise can become reality.

References

1. Alder, H., Hömberg, D., Weiss, W.: Simulationsbasierte Regelung der Laserhärtung von Stahl. HTM Z. Werkst. Wärmebeh. Fertigung **61**, 103–108 (2006)
2. Ardelt, T.: Einfluss der Relativbewegung auf den Prozess und das Arbeitsergebnis beim Plan-schleifen mit Planetenkinematik. Dissertation, TU Berlin (2001)
3. Brenner, S., Carstensen, C.: Finite element methods. In: Stein, E., de Borst, R., Hughes, T.J.R. (eds.) Encyclopedia of Computational Mechanics (2004), Chap. 3
4. Caccavale, F., Villani, L.: Impedance control of cooperative manipulators. Mach. Intell. Robot. Control **2**, 51–57 (2000)

5. Degond, P., Göttlich, S., Herty, M., Klar, A.: A network model for supply chains with multiple policies. SIAM Multiscale Model. Simul. (MMS) **6**(3), 820–837 (2007)
6. Denkena, B., Brecher, C.: Ramp-Up/2: Anlaufoptimierung durch Einsatz virtueller Fertigungssysteme. VDMA Verlag, Frankfurt (2007)
7. Denkena, B., Deichmüller, M.: Wechselwirkungen zwischen Prozess und Maschine. In: Biermann, D. (ed.) Spanende Fertigung: Prozesse Innovationen Werkstoffe. Vulkan Verlag, Essen (2008)
8. Denkena, B., Günther, G., Mehrmann, V., Möhring, H.-C., Steinbrecher, A.: Kalibrierverfahren für hybride Parallelkinematiken. In: Heisel, U., Weule, H. (eds.) Fertigungsmaschinen mit Parallelkinematiken – Forschung in Deutschland, Shaker, to appear
9. Fowler, J.W., Rose, O.: Grand challenges in modeling and simulation of complex manufacturing systems. Simulation **80**(9), 469–476 (2004)
10. Günther, G., Steinbrecher, A.: Strukturausnutzung hybrider Parallelkinematiken hinsichtlich ihrer Kalibrierung. In: Proceedings of 3. Dresdner WZM-Fachseminar, Technische Universität Dresden (2001)
11. Heim, A., von Stryk, O.: Trajectory optimization of industrial robots with application to computer-aided robotics and robot controllers. Optimization **47**, 407–420 (2000)
12. Hömberg, D., Weiss, W.: PID control of laser surface hardening of steel. IEEE Trans. Control Syst. Technol. **14**, 896–904 (2006)
13. Lippiello, V., Siciliano, B., Villani, L.: An open architecture for sensory feedback control of a dual-arm industrial robotic cell. Ind. Robot **34**, 46–53 (2007)
14. Neugebauer, R.: Energieeffizienz in der Produktion. Studie der Fraunhofer Gesellschaft im Auftrag des BMBF (2007)
15. Rannacher, R., Suttmeier, F.-T.: Error estimation and adaptive mesh design for FE models in elasto-plasticity theory. In: Stein, E. (ed.) Error-Controlled Adaptive Finite Elements in Solid Mechanics. Wiley, New York (2002)
16. Rott, O., Rasper, P., Hömberg, D., Uhlmann, E.: A milling model with thermal effects including the dynamics of machine and work piece. WIAS Preprint No. 1338 (2008)
17. Steinbach, M.C., Bock, H.G., Kostin, G.V., Longman, R.W.: Mathematical optimization in robotics: towards automated highspeed motion planning. Surv. Math. Ind. **7**(4), 303–340 (1998)
18. Tönshoff, H.K., Günther, G., Grendel, H.: Influence of manufacturing and assembly errors on the pose accuracy of hybrid kinematics. In: Proceedings of the 2000 Parallel Kinematic Machines—International Conference, September 13–15, 2000, Ann Arbor, Michigan, USA, pp. 255–262 (2000)
19. Tröltzsch, F.: Optimal Control of Partial Differential Equations. AMS (2010)
20. Uhlmann, E., Mattes, A., Zettier, R., Graf von der Schulenburg, M.: Investigations on the adjustment of the modeling reaction in 2D simulation of milling processes. In: Proceedings of the 10th CIRP Int. Workshop on Modeling in Machining Operations, August 27–28, 2007, Calabria, Italy, pp. 157–164 (2007)
21. Uhlmann, E., Paesler, C., Ardelt, T.: Qualitäts- und Leistungssteigerung beim Planschleifen mit Planetenkinematik. Talk, 10. Internationales Braunschweiger Feinbearbeitungskolloquium, Braunschweig, October 7–9, 2002
22. Uhlmann, E., Zettier, R.: 3D FE simulation of turning processes. In: CIRP Workshop on "FE Simulation of Cutting and Forging Processes", January 29, Paris, France (2003)
23. Uhlmann, E., Zettier, R.: Experimentelle und numerische Untersuchungen zur Spanbildung beim Hochgeschwindigkeitsspanen einer Nickelbasislegierung. In: Tönshoff, H.K., Hollmann, F. (eds.) Hochgeschwindigkeitsspanen metallischer Werkstoffe, pp. 404–425. Wiley-VCH, Weinheim (2005)
24. Uhlmann, E., Zettier, R., Sievert, R., Clos, R.: FE simulation of high-speed turning of Inconel 718. In: Moisan, A., Poulachon, G. (eds.) Proceedings of the 7th CIRP International Workshop on Modeling of Machining Operations, May 4–5, 2004, Cluny, France, pp. 67–74 (2004)
25. Weinert, K., Blum, H., Jansen, T., Rademacher, A.: Simulation based optimization of the NC-shape grinding process with toroid grinding wheels. Prod. Eng., Res. Dev. **1**(3), 245–252 (2007)

Production and Use of Novel Materials

Wolfgang Dreyer

1 Executive Summary

For a long time now there is a rapidly rising application of new materials for key technologies. Steels with novel properties are designed. The tremendous demand for solar cells involve new processes to produce the semiconductors needed to this. In order to achieve higher clock rates in computers, or likewise to enhance the power of solar cells, the pure silicon technology is abandoned, novel semiconductors are planed. In connection with the miniaturization of chips, the semiinsulator becomes of great importance.

Here to win the market leadership requires a deep and, in particular, interdisciplinary understanding of the physical phenomena within the technology chain: *characterization, production and life time of novel materials*. This can only be achieved by use of applied mathematics with its methods *Modeling, Analysis and Simulation*. The artful application of mathematical methods namely can

- shorten the time off development,
- even lead to patents and finally
- by means of subtle methods to prevent the engineer to produce numerical errors and other artifacts.

This is illustrated here with selected aspects within the topics: Production of crystals, enhancement of the properties of the semi-insulator gallium arsenide, fracture tests of a wafer, aging and damage of solder joints in chips. The essential mathematical context are initial and boundary value problems for coupled systems of nonlinear partial differential equations.

W. Dreyer (✉)
Weierstraß-Institut für Angewandte Analysis und Stochastik, Mohrenstraße 39,
10117 Berlin, Germany
e-mail: dreyer@wias-berlin.de

M. Grötschel et al. (eds.), *Production Factor Mathematics*,
DOI 10.1007/978-3-642-11248-5_13, © Springer-Verlag Berlin Heidelberg 2010

Fig. 1 Cut out of a crystal growth device with magnetic coils and some calculated temperatures (from [10])

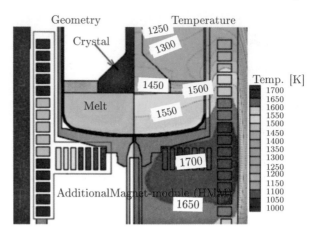

2 Success Stories

Crystal Growth in a Traveling Magnetic Field

The essential ingredients of electronic devices are based on growing crystals. Their production process starts at high temperature with a melt and then the crystal arises by controlled cooling. The so called Czochralski method pulls the crystal from a rotating melt.

Figure 1 shows a cutout of a typical Czochralski device inclusively a calculated temperature distribution [10]. In order to achieve a homogeneous crystal, three things must be guaranteed, viz. preferably a spatial and temporal homogeneous temperature, a laminar flow within the crucible, and a convex interface between melt and crystal. This requires long-standing experience and great art of experimentation. If now every parameter is optimal adjusted, to grow a crystal from a melt of given diameter and melt height, and if it is decided then to produce larger crystals, so that a larger crucible diameter and a larger melt height is involved, usually this will lead to a crystal of inferior quality. Once more tedious changes of the growth device and new experiments to determine the new optimal parameter becomes necessary. Parallel to the industrial production of crystals this cannot be accomplished.

In this context a consortium was founded in Berlin, that consists of two industrial companies, physicists, engineers, and mathematicians to carry out the interdisciplinary project Krist$\widetilde{MAG}$.[1] It should be discovered if a traveling magnetic, that is generated by the same coils that produce the induction heating, may favorably control temperature, flow and interface boundary in the above sense. The project, that recently has been successfully terminated, has impressively demonstrated that objectives: several pilot devices have been manufactured, there are five given patents

[1]Funded by Zukunftsfonds der Stadt Berlin and by Technologiestiftung Berlin and cofinanced by EU (EFRE).

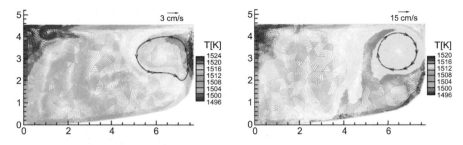

Fig. 2 Snap shot of the temperature and flow velocity distribution within the melt, from [10]. *Left*: without magnetic field, *Right*: with magnetic field. Note the different scales of velocity and temperatures

From the mathematical point of view, the problem to control temperature, flow and interface leads to a coupled system of nonlinear partial differentials equations for a non-smooth domain with free boundaries, see [6]. For example, the variables in the melt are the temperature T, the flow velocity $v = (v_i)_{i=1,2,3}$ and the magnetic induction $B = (B_i)_{i=1,2,3}$. The differential equations contain the Navier-Stokes equations in the setting of the Boussinesq approximation.

$$\rho\left(\frac{\partial v}{\partial t} + v \cdot \nabla v\right) + \nabla p$$

$$+ \mathrm{div}(\eta(T)(\nabla v + (\nabla v)^{\mathsf{T}}))$$

$$= \rho(1 + \alpha(T - T_{\mathrm{R}}))g + \frac{1}{\mu}(\mathrm{rot}B \times B),$$

$$\mathrm{div}\, v = 0.$$

On the right hand side we have the buoyancy force, which induces to a great extent unwanted eddies, as well as the Lorentz force, whereby a magnetic filed may control the flow. The magnetic field is calculated by means of reduced Maxwell equations, viz.

$$\frac{\partial B}{\partial t} + \mathrm{rot}\left(\frac{1}{\mu\sigma(T)}\mathrm{rot}B\right) = 0$$

and finally there is the heat conduction equation with convection,

$$\rho c_{\mathrm{V}}\left(\frac{\partial T}{\partial t} + v \cdot \nabla T\right)$$

$$= \mathrm{div}(\kappa(T)\nabla T) + \eta(T)\nabla v \cdot \cdot \nabla v$$

$$+ \frac{\mathrm{rot}\,B \cdot \mathrm{rot}\,B}{\mu^2\sigma(T)}.$$

In other parts of the system additionally occurs radiation. Hereupon as well as on boundary conditions is not committed here.

Fig. 3 Mathematical context to crystal growth within a traveling magnetic field

and a further proposal for a patent that is in particular initialized by the mathematicians of the project.

The combined heating-magnet- module is sketched in the Czochralski-device off Fig. 1. It consists of three cylindrical coils, which are flowed through by phase shifted currents that generate the traveling magnetic field. On its left hand side, the Fig. 2 gives within the melt and without the magnetic field a snap shot of a calculated temperature and flow velocity distribution. The right hand side of Fig. 2 shows the influence of a magnetic field that travels downwards, which apparently changes the temperature distribution and additionally reverses and changes the shape of the right eddy.

Moreover there are two additional coils below the melt crucible, which generate a magnet field traveling outwards, and this is the subject of the new patent proposal [7]. That arrangement was motivated by means of various simulations, that have proved: Properly adjusted parameter lead to a temporal damping of unwanted temperature oscillation, and moreover also the shape of the crystal-melt interface may be varied. This important finding could also experimentally be observed. However, even slightest modifications of a crystal growth installation, for example, the system of magnetic coils, need weeks or sometimes months. During this time there are neither experiments nor growth of crystals possible. For this reason the virtual crystal growth device, as it is represented by a mathematical model, gives the grower a competitive advantage with respect to time as well as to effort.

Unwanted Formation of Precipitate in Gallium Arsenide Crystals

From the crystal, see Fig. 4, so called wafer are fabricated, thin single-crystal plates, which are further processed to chips, laser or photovoltaic cells. In order to homogenize its electrical, optical and mechanical properties, the wafer must be subjected to a heat treatment above a certain temperature. In this connection unwanted precipitates occur. In semi-insulated gallium arsenide crystals we meet liquid droplets with an arsenic content of 90%. In this state, use in micro- and opto-electronic applications is not possible, so that a procedure to dissolve the precipitates must be found, and also in this case methods of applied mathematics make important contributions.

Before an illustration of the problem, a short excursion on the crystal structure is given [4]. Figure 5 shows the basic structure of the crystal lattice of gallium arsenide, consisting of three cubic sublattices with equal edge length, which is periodically continued. The ideal crystal consists of equal numbers of gallium and arsenic atoms, that are distributed on the lattice sites as follows. The first sublattice is completely occupied with gallium atoms (blue balls), the second exclusively with arsenic atoms (red balls), and the third sublattice is empty (white balls). They say it is occupied

Fig. 4 Crystals and wafer from gallium arsenide

Fig. 5 Sublattice structure of a gallium arsenide crystal

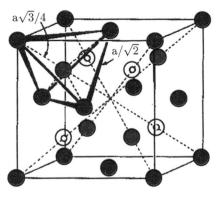

with vacancies. Certainly this crystal is no semi-insulator as it is needed for the applications. Such crystal emerge from a melt with slightly added concentration of arsenic, for example 0.500082 at Freiberger Compound Materials (FCM), one of the world leading producers of semi-insulating gallium arsenide. This concentration, however, leads to a different distribution of the atoms over the sublattices. The site of the first sublattice are occupied by the actual gallium atoms, some arsenic atoms and by some vacancies, whereas the second and third sublattice are occupied by the remaining arsenic atoms and by vacancies. However, that distribution lead to diffusion during the heat treatment, which finally is the origin of the formation of unwanted precipitates.

In order to describe the diffusion phenomena including the formation of precipitates, up to now we have already seven variables, viz. the particle densities on the particular sublattices. Furthermore there are N radii of N precipitates, where $N \approx 10^6$, and the displacements of the surrounding crystal lattice with respect to the three spatial directions. These displacements result because the surfaces of the precipitates are equipped with energy, and, due to the difference of mass densities between crystal and liquid, the precipitates need more space than the crystal. Unfortunately, above all the actual situation is much more complicated as the described situation. The crystal that are produced by FCM contain further eleven barely detectable impurities, that nevertheless have to be taken into account, for example, silicon occurs with a concentration of 10^{-9}.

Thus it is evident, that the reduction of the complete mathematical model is mandatory. However, despite many experimental dissertations, the regime of precipitate evolution is undetermined up to now. Here mathematical modeling sets in to assemble a hierarchy of verifiable models.

The simplest model considers the diffusion problem in the vicinity of a single precipitate within a crystal. Two possible classes of boundary conditions at the melt-crystal interface were studied, and a comparison of the results with the few available experimental data has already excluded one class, while the other one is still in the race.

The simplest many-precipitate model is the Becker-Döring model. Originally it was conceived to simulate nucleation of fog in air. Dreyer & Duderstadt have extended that model to describe nucleation of liquid precipitates within a crystal, and

The Becker-Döring model describes the evolution of the number of precipitates $Z(t, \alpha)$ with α atoms by the system

$$\frac{\partial Z(t, \alpha)}{\partial t} = J_{\alpha-1} - J_\alpha$$

with $J_\alpha = \Gamma_\alpha^C Z(t, \alpha) - \Gamma_{\alpha+1}^E Z(t, \alpha+1)$,

$\alpha \in \{2, 3, \ldots, \infty\}$,

whereby $Z(t, 1)$ is determined by the conservation of the total particle number

$$N = \sum_{\alpha=1}^\infty \alpha Z(t, \alpha).$$

The flux J_α gives the rate of produced precipitates with $\alpha + 1$ atoms, and Γ_α^C, Γ_α^E denote the condensation rate, respectively the evaporation rate. Both the quantities, according to the case at hand, depend on α and on the precipitate number $Z(t, \alpha)$ in a complicated manner, so that there results a non-linear and sometimes a non-local system of ordinary differential equations.

Fig. 6 Mathematical context of the Becker-Döring model

applied to the gallium arsenide problem [3]. The model consists of a large system of ordinary differential equations and calculates at any time the size distribution of precipitates in the system. From the point of view of mathematical physics, the Becker-Döring model is of most interest. However, there are some indications, that the industrial problem is not properly presented here.

A further many precipitate model reduces the complete, obvious not solvable, diffusion problem with many precipitates to a homogenized model, that exclusively gives the evolution of a mean precipitate radius. Currently there are hints that the reality is well represented here.

However, what is the practical value of these studies? What is the benefit for an industrial user to recognize that the reduction of the complete model by homogenization is near to the observed phenomena? Well, each of the introduced models identifies different quantities as the crucial parameter, that control the unwanted process of precipitate formation. If, for example, it is predicted, that a variation of the pressure has no influence upon the unwanted precipitates, but very well a variation of the temperature history during the heat treatment, consequently the experimental effort to identify the origin of the phenomenon considerably reduces.

The Bending Test for Gallium Arsenide Wafer

During further processing a wafer is exposed to various processes, that may lead even to fracture due to diverse mechanical loading. For this reason the wafer producer has to guarantee his customer a certain fracture toughness, and its determination requires besides a fracture test subtle mathematical methods to calculate not directly measurable stresses that occur within the wafer during the test.

The Fig. 7 illustrates the principle of the test device [5]. The circular wafer with diameter 15 cm and thickness 0.5 mm concentrically lies on the thrust ring. By means of a steel ball a force is applied, and brings a bending of the wafer. The measured quantity for a given force is the corresponding central deflection.

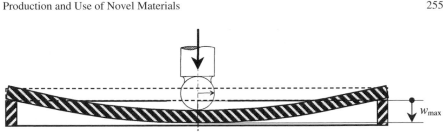

Fig. 7 Cross section of a ring-plate-load system (from [5])

Obviously this test also allows to identify the force needed for fracture of the wafer. The resulting value is a limited characteristic number only, because it does not exclusively evaluate the constitutive property, but also depends on the wafer dimensions and on the properties of the test. A purely constitutive dependent critical number can be obtained from the stresses that occur in the wafer and these cannot be measured but must be calculated.

The calculation of the stresses relies on the observation that we meet brittle fracture in gallium arsenide wafer, what means that purely *elastic* deformation precedes the abrupt fracture. Thus the stresses may be calculated by means of the equations of elasticity. Due to the large difference between the diameter and the thickness of the wafer, a finite element approximation of the three-dimensional equations of elasticity is out of the question. At first a reduction of those equations to a two-dimensional plate model must be carried out.

A careful analysis of the various number of possible model reductions motivates for the case at hand the von Kármán model as the appropriate plate model, and for that a finite-element approximation was carried out. A typical mesh [5] is drawn in Fig. 8. Note the mesh condensation within the very small load surface. Moreover the

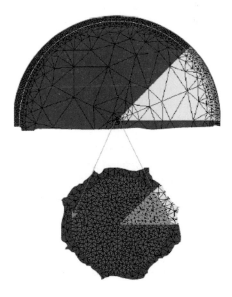

Fig. 8 Finite-element net of the plate problem [5]. *Above*: Half of complete mesh. *Below*: Mesh of the load surface with 350-fold zoom

Magenta: Static friction, Green: Slip, Red: Linear theory

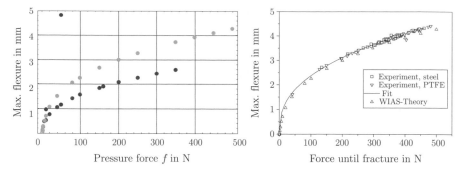

Fig. 9 Maximal flexure vs. force. *Left*: comparison of various models. *Right*: comparison of theory and experiment

cubic symmetry of the crystal is lightly recognized here. The foot print of the load sphere shows no circle on the wafer, but exhibits the crystal symmetry with respect to 90° rotations.

Incidentally, a delicate issue are the boundary conditions on the contact line waver-thrust ring. Namely from the experiment it cannot be extracted whether the wafer glides off or remains fixed here during the experiment. Therefore the mathematical model simply simulates both the possibilities by appropriate boundary conditions. The left Fig. 9 shows the difference, and the comparison with experimental data alongside implies, that the mathematically more involved case of gliding is realized in the experiment. Interestingly is to observe the complete failure of the linear plate model.

Finally a subtle phenomenon is reported here, whose ignoring may lead to calculational errors within order of magnitude of 40% [1]. It is the *Babuška paradox*, which is known among mathematicians, however, surprisingly it is ignored by engineers due to lack of knowledge. The paradox depends upon the discretization of the circle, that is generated by the contact line wafer-thrust ring. Obviously, a polygon will better approximate a circle, the more traverses there are. In spite of this, the boundary value problem of the circle will be worsen approximated by the boundary value problem of the polygon for increasing refinement of the finite element mesh, at least this happens if rectilineal elements are used. This is the Babuška-paradox. However, it is possible to develop methods so that it does not occur, even for rectilineal elements, but obviously only if the problem as such is recognized.

Aging-Damage-Fracture

At the end of the production chain of semi-conductors stand their use within electronic devices. Here occurs phenomena, that are typical for an entirely different issue. It is a matter of aging, damage and fracture, leading to total breakdown in the end. Because fracture is preceded by large variances of the properties of the involved materials, these *age*, it is not an unpredictable event, but, based on detailed

The system of plate equations according von Kármán consists of three coupled nonlinear partial differential equations of elliptic type for the displacements $U = (U_1(X_1, X_2), U_2(X_1, X_2))$ of the middle surface of the wafer and for its vertical displacement $W(X_1, X_2)$, whereby (X_1, X_2) are the Lagrange coordinates of the middle surface. For a given external load $p(X_1, X_2)$, the equations

$$\frac{\partial N_{\alpha\beta}}{\partial X_\beta} = 0, \qquad \frac{\partial^2 M_{\alpha\beta}}{\partial X_\alpha \partial X_\beta} + \frac{\partial^2 W}{\partial X_\alpha \partial X_\beta} N_{\alpha\beta} = -p, \quad \alpha, \beta \in \{1, 2\},$$

guarantee mechanical equilibrium. These are completed by stress-strain relations fur cubic anisotropic gallium arsenide with three Lamé constants λ, μ and μ':

$$N_{\alpha\beta} = h\left(\lambda \frac{2\mu + \mu'}{2\mu + \mu' + \lambda} G_{\gamma\gamma}\delta_{\alpha\beta} + 2\mu G_{\alpha\beta} + \mu'\delta_{\alpha\beta} G_{\beta\beta}\right) \quad \text{with } G_{\alpha\beta} = \frac{1}{2}\left(\frac{\partial U_\alpha}{\partial X_\beta} + \frac{\partial U_\beta}{\partial X_\alpha}\right),$$

$$M_{\alpha\beta} = -\frac{h^3}{12}\left(\lambda \frac{2\mu + \mu'}{2\mu + \mu' + \lambda} \frac{\partial^2 W}{\partial X_\gamma \partial X_\gamma}\delta_{\alpha\beta} + 2\mu \frac{\partial^2 W}{\partial X_\alpha \partial X_\beta} + \mu'\delta_{\alpha\beta} \frac{\partial^2 W}{\partial X_\beta \partial X_\beta}\right).$$

To this system appears boundary conditions, that, however, are not indicated here.

Fig. 10 Mathematical context of the bending test for gallium arsenide wafer [5]

Fig. 11 Solder joints on a chip

knowledge of the underlying mechanisms, can be predicted. However, the technologist often misses time to understand in detail the aging process, rather he prefers pragmatic-phenomenological solutions, despite the fact that a deep understanding of the problem would lead to better results.

In order to illustrate the problem, the production and use of solder joints, see Fig. 11, which connect the chip with the external world, is paradigmatic considered. Up to know the most popular solder consists of a tin-lead alloy, because two necessary, but intrinsically contradicting assumptions are meet here to some extent. For production of the electronic device, the solder must have a low melting point. However, due to the high service temperature, in use the device needs a high melting point.

Based on its toxicity, since 2006 the lead-tin solder is banned from the large scale production, so that a substitute must be developed. The last two years revealed that this is not a simple undertaking. Incidentally, the failure of a chip, of a photo-voltaic

Fig. 12 Whereto deposit the
electronic waste?

Fig. 13 Decomposition at high temperature (from [2]): *Left*: after solidification. *Middle*: after 5 hours. *Right*: after 40 hours

cell or of an airbag is by 90% due to the used solder joints. The economic loss is considerably. According to an estimate of the German environmental organization BUND the yearly electronic waste fills four Cheops pyramids (height 140 m, see Fig. 12).

However, what are the causes of the failure of solder joints? Roughly speaking one meets two different problems. Due to the high service temperature, which may be 70% of the melting temperature, there is a decomposition of the participating substances. Whereas the other problem already appears during the production of the contact, for example between the solder and a copper substrate.

Right away after solidification there is a decomposition of the chemical constituents [2]. For the example from Fig. 13 silver-rich islands within a copper-rich vicinity are formed. It is important to know the evolution, because the experience shows that the coarsening of the solder morphology is accompanied by an intensified formation of small cracks along the interfaces. For this reason it is a great success that this evolution can realistically be simulated by means of a mathematical model. To this end a so called phase field model is used, that models the encountered sharp interfaces from Fig. 13 between the islands and the surrounding weak diffusively. This trick considerably accounts for the capability to calculate such complex morphologies at all.

Currently the most promising lead-free solder consists of the three chemical constituents tin, silver and copper. When this solder in the liquid state is brought into contact with the solid copper substrate, an unwanted and extreme brittle layer with the chemical name Cu_6Sn_5 is formed, see Fig. 14. In particular, the geometric shape of the boundary between that layer and the interior of the solder is assumed to excite

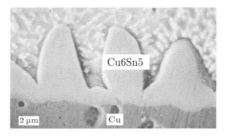

Fig. 14 Unwanted inter layer between substrate and solder (from [9])

fracture. One of the objectives is the identification of the crucial parameter, so that this boundaries becomes flat. Likewise here there is a mathematical model, however, not with diffuse but rather with sharp interfaces.

3 Status Quo

The demonstrated success stories concern the use of applied mathematics in the context of *novel materials*. The hereby implied working method is illustrated now for two examples at hand.

There are some crystal growth techniques where the melt-crystal system recides in crucible, whose wall is furnished with a thin polymer coating. Where melt and crystal simultaneously meets the polymer coating, a so called free triple line with the three participating substances is induced. This line is a priori not given rather it is part of the solution of the mathematical model. The problem now is that the position and shape of the triple line is of essential influence on the crystal quality, unfortunately the relevant equations that control this line are not completely known. In particular this is due to the complicated mechanical deformations, that are induced by the elasticity of the participating substances and that interact with the flow of matter as well as with the conduction of heat. Thus before applied mathematics can properly contribute, there is a subtle and often tedious process, which is accompanied by several iterations between physical modeling and numerical tests.

A further example concerns the availability of material data. About fifty characteristic numbers are needed to simulate the evolution of unwanted precipitates in semi-insulating gallium arsenide, and because this is a *novel material*, many of these quantities are not available. Partly resulting from the phenomenon, that many characteristic numbers are not directly measurable, rather for their determination a theoretical model is already required. In this context the analysis is of large importance, because by clever application a classification in important and unimportant characteristic numbers is possible.

These are typical problems of the mathematical modeling of novel materials, and they should illustrate that a strong interdisciplinary interlocking is indispensable for a successful treatment of the subject matter. However, currently this interlocking exists only in first attempts. Usually the industry does not arrange interdisciplinary pools to solve its problems. Instead funding mostly is given to disjunct areas of

The evolution in time and space of the interior micro structure of a solder with two constituents is described by a single concentration $c(t, X)$ and the displacement of the crystal lattice $u = (u_1(t, X), u_2(t, X), u_3(t, X))$. The phase field model consists of a parabolic-elliptic system of differential equations of the form

$$\rho_0 \frac{dc}{dt} + \frac{\partial J^i}{\partial X^i} = 0, \qquad \frac{\partial \sigma^{ij}}{\partial X^i} = 0,$$

that describe the conservation of mass and mechanical equilibrium. To this appears two newly equations for the diffusion flux and the stress tensor, that take into account the specific circumstances within a solder:

$$J^i = -\rho_0 \mathcal{M}^{ij}(T) \frac{\partial}{\partial X^j} \left(\frac{\partial \psi}{\partial c} - 2A^{kl} \frac{\partial^2 c}{\partial X^k \partial X^l} - \frac{\partial A^{kl}}{\partial c} \frac{\partial c}{\partial X^k} \frac{\partial c}{\partial X^l} \right.$$

$$\left. - \frac{\partial A^{kl}}{\partial E^{mn}} \frac{\partial c}{\partial X^k} \frac{\partial E^{mn}}{\partial X^l} - \frac{\partial a^{kl}}{\partial E^{mn}} \frac{\partial^2 E^{mn}}{\partial X^k \partial X^l} - \frac{\partial^2 a^{kl}}{\partial E^{op} \partial E^{mn}} \frac{\partial E^{op}}{\partial X^k} \frac{\partial E^{mn}}{\partial X^l} \right),$$

$$\sigma^{ij} = K^{ijkl}(E^{kl} - E_*^{kl}) - \frac{\partial a^{kl}}{\partial E^{ij}} \frac{\partial^2 c}{\partial X^k \partial X^l} - \frac{\partial b^{kl}}{\partial E^{ij}} \frac{\partial c}{\partial X^k} \frac{\partial c}{\partial X^l}.$$

Furthermore there are boundary conditions, however, exclusively along the outer boundary of the solder. If a model with sharp interfaces were used, the differential equations would have a simpler structure, rather conditions for each interface point were needed.

Fig. 15 Mathematical context to the topic aging-damage-fracture (from [2])

expertise. The above example for crystal growth within a magnetic field still represents an exceptional case. Even the German science foundation currently has too few instruments to fund interdisciplinary projects. For example, the special research fields with interdisciplinary problems are to large and cumbersome to treat industrial problems. Its treatment is better placed in small research groups with only two or three different partners, though their funding seems to be too elaborate for the administration. In this context the duration of an interdisciplinary project should also be discussed. Currently it is by far too long for the industrial user of the results! Many examples to the following circumstance could be cited here: The mathematical modeling solves a posed problem in a distinguished manner; however, nobody no longer is interested on the solution, because meanwhile the industry could dispose the problem pragmatically by another view on the subject.

4 Analysis of Strengths/Weaknesses, Challenges

The described set of problems encounters the treatment of novel materials in a much stronger manner than other topics, where applied mathematics exhibits excellent success almost always. This too is once again illustrated by an example.

Active flow control of vehicles, see the corresponding chapter, uses as the mathematical model the Navier-Stokes equations for water and air, and moreover the possible boundary conditions are known since long time. Even so this holds if the occurring turbulence is described by a reduced model. In other words: The prob-lcm to translate the technical process into a mathematical model happens according to established and simple rules. In addition, the involved substances water and air requires only a few and known characteristic numbers. These are at most three quantities, namely the viscosity, the thermal expansion and eventually the dynamic friction coefficient, which enters the boundary conditions.

Concerning novel materials, the circumstance is different. It is not the problem to efficiently solve existing model equations of high complexity within incredibly short time. Rather the present main challenge of mathematics is its use at modeling in the sense of the above examples.

Currently, at the authors opinion the necessary techniques are hardly or not appropriately imparted at German universities. Modeling in mechanics ignores most of the thermodynamic aspects, thermodynamics usually discusses mechanical aspects in an oversimplified version, and in order to obtain as much as possible explicit solutions of the model, physics often simplifies the real geometry in an unrealistic manner. Likewise the modeling of mathematicians leave many wishes to be desired. Namely in this setting too often it is the objective to end with a model that allows its treatment by means of already existing mathematical methods. However, this should be no criterion for realistic modeling. Conclusion: Interdisciplinarity must be considerably strengthened.

5 Visions and Recommendations

Problem solutions in the area *Use and production of novel materials* consist of the following articles:

- Mathematical modeling inclusively providing the material data;
- Analysis of the mathematical properties of the model equations;
- Development of numerical algorithms inclusively reliable error estimates;
- Numerical simulation, visualization and interpretation of the solutions.

Usually that chain of tasks must be passed through several times and thus the conclusion is: Exclusively an interdisciplinary composed study group will be successful within this area.

References

1. Babuška, I., Pitkäranta, J.: The plate paradox for hard and soft simple support. SIAM J. Math. Anal. **29**(5), 1261–1293 (1992)
2. Böhme, Th.: Investigations of microstructural changes in lead-free solder alloys by means of phase field theories. PhD thesis, Technical University of Berlin (2008)
3. Dreyer, W., Duderstadt, F.: On the Becker-Döring theory of nucleation of liquid droplets in solids. J. Stat. Phys. **123**, 55–87 (2006)
4. Dreyer, W., Duderstadt, F.: On the modelling of semi-insulating GaAs including surface tension and bulk stresses. Proc. R. Soc. Lond. Ser. A (2008). doi:10.1098/rspa.2007.0205

5. Dreyer, W., Duderstadt, F., Eichler, St., Jurisch, M.: Stress analysis and bending tests for GaAs wafer. Microlectron. Reliab. **46**, 822–835 (2006)
6. Druet, P.E.: On weak solutions to the stationary mhd-equations coupled to heat transfer with nonlocal radiation boundary conditions. WIAS-Preprint **29**(1290), 1–44 (2008)
7. Frank-Rotsch, Ch., Rudolph, P., Lange, P., Klein, O., Nacke, B.: Vorrichtung und Verfahren zur Herstellung von Kristallen aus elektrisch leitenden Schmelzen (Anordnung und Verfahren zur optimierten Schmelzzüchtung von Kristallen). Patent Application DE 10 2007 028 548.7. Submitted June 18th 2007
8. Herrmann, M., Naldzhieva, M., Niethammer, B.: On a thermodynamically consistent modification of the Becker-Döring equation. Phys. D **222**, 116–130 (2006)
9. Kim, H.K., Tu, K.N.: Kinetic analysis of the soldering reaction between eutectic SnPb alloy and Cu accompanied by ripening. Phys. Rev. B **53**, 16027–16034 (1996)
10. Klein, O., Lechner, Ch., Druet, P.-É., Philip, P., Sprekels, J., Frank-Rotsch, Ch., Kießling, F.-M., Miller, W., Rehse, U., Rudolph, P.: Numerical simulations of the influence of a travelling magnetic field, generated by an internal heater-magnetic module, on Czochralski crystal growth. In: Proceedings of the International Scientific Colloquium Modelling for Electromagnetic Processing (MEP 2008), Hannover, 27–29 October 2008
11. Rudolph, P.: Travelling magnetic fields applied to bulk crystal growth from the melt: The step from basic research to industrial scale. J. Crystal Growth **310**(7–9), 1298–1306 (2008)

Topology and Dynamic Networks: Optimization with Application in Future Technologies

Günter Leugering, Alexander Martin,
and Michael Stingl

1 Executive Summary

The optimal design and control of infrastructures, e.g. in traffic control, water-supply, sewer-systems and gas-pipelines, the optimization of structures, form and formation of materials, e.g. in lightweight structures, play a predominant role in modern fundamental and applied research. However, until very recently, simulation-based optimization has been employed in the sense that parameters are being adjusted in a forward simulation using either 'trial-and-error' or a few steps of a rudimentary unconstrained derivative-free and mostly stochastic optimization code. It has become clear by now that instead often a model-based and more systematic constrained optimization that exploits the structure of the problem under consideration may outperform the former more naive approaches. Thus, modern mathematical optimization methods respecting constraints in state and design variables can be seen as a catalyst for recent and future technologies. More and more success stories can be detected in the literature and even in the public press which underline the role of optimization as a key future technology. In particular, optimization with partial differential equations (PDEs) as constraints, or in other words 'PDE-constrained optimization' has become research topic of great influence. A DFG-Priority-Program (PP) has been established by the German Science Foundation (DFG) in 2006 in

G. Leugering (✉)
Universität Erlangen–Nürnberg, Martensstr. 3, 91058 Erlangen, Germany
e-mail: leugering@am.uni-erlangen.de

A. Martin
Department Mathematik, Universität Erlangen–Nürnberg, Am Weischselgarten 9,
64289 Darmstadt, Germany
e-mail: Alexander.Martin@math.uni-erlangen.de

M. Stingl
Universität Erlangen–Nürnberg, Exzellenzcluster EAM, Nägelsbachstraße 49b, 91052 Erlangen,
Germany
e-mail: stingl@am.uni-erlangen.de

M. Grötschel et al. (eds.), *Production Factor Mathematics*,
DOI 10.1007/978-3-642-11248-5_14, © Springer-Verlag Berlin Heidelberg 2010

which well over 25 project are funded throughout Germany. The PP focuses on the interlocking of fundamental research in optimization, modern adaptive, hierarchical and structure exploiting algorithms, as well as visualization and validation. With similar goals in mind, a European network within the European Science Foundation (ESF) 'PDE-constrained Optimization' has been recently established that provides a European platform for this cutting edge technology. In this report the authors dwell on exemplary areas of their expertise within applications that are already important and will increasingly dominate future developments in mechanical and civil engineering. These applications are concerned with optimal material and design in material sciences and light-weight structures as well as real-time capable optimal control of flows in transportation systems such as gas-pipeline networks. 'Advanced Materials', 'Energy-Efficiency' and 'Transport' are key problems for the future society which definitely deserve public funding by national and international agencies.

2 From Properties to Optimal Structures

'Structure formation' and its goal-oriented control and optimization is a central focus of the Cluster of Excellence (CE): 'Engineering of Advanced Materials: Hierarchical Structure Formation of Functional Devices' funded by the DFG at the Friedrich-Alexander-University of Erlangen-Nuremberg since November 2007. The notions associated with this cluster are e.g. 'new materials', 'meta-materials', 'material foams', 'printable electronics' etc. and belong to the vocabulary of DOE (Department of Energy) white-papers and have made their way even into the public press.

Light-weight structures, such as airplanes or micro-mechanical devices, will revolutionize future technologies. Topology optimization combined with material- and shape- optimization can save well over 30% of material while keeping the global strength of the structure.

The supply with water, power and gas provides a crucial problem of future societies. The design and maintenance of traffic and more general communication networks is the base for its mobility.

In order to be able understand the catalytic role of mathematics, and in particular of mathematical optimization, for key future technologies, one has to step back from a direct cost-merit analysis. It soon becomes clear then that the automated model-based design, simulation, optimization and visualization, so to speak the anticipation of a product on the screen, has already succeeded the classical prototyping. The crucial step is now to revert the process: from the functionality to the structural properties—to the design. 'Given a certain brilliance of a color, what is the structure of the nano-particles constituting the corresponding paint?', 'Given a certain demand of gas-consumers and a number of gas-providers that use a given pipeline-network, what is the right portfolio of inputs, valve positions and compressor activities, in order to satisfy the consumer's demands on pressure, amount and quality of gas while minimizing the costs of the network-flow-management?' These typical questions are examples of what has become to be known as 'reverse engineering' or even 'inverse optimization' and what in mathematical terms is just 'mathematical optimization'!

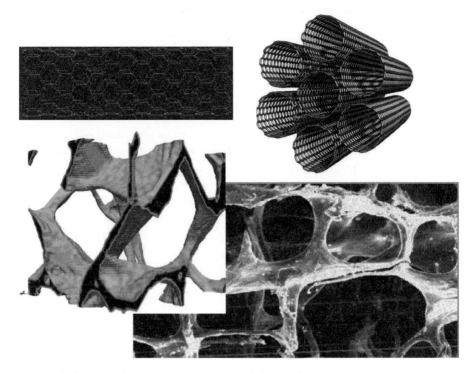

Fig. 1 Multiscale modeling: from nanotubes to cellular materials

In order to make optimization in that sense applicable one has to transform the engineering problem into the context of mathematical modeling which is to say into the language of algebraic-ordinary- or partial-differential-algebraic or integro-differential-equations or inequalities. The models are then interpreted as constraints together with other technical or legal constraints on the state- and design-variables that have to be complied with during the process of optimization with respect to a given merit-function.

During the phase of modeling, mathematics, as the science of structures or as a 'think-tank', offers structural models. The question may be: 'what is the structure behind hierarchical cellular materials?', 'how can I describe particle-structures that are build in a reactor?' or 'what is the right description of a transportation process along carbon nano-tubes?'. A mathematical answer will be in terms of mathematical graph modeling. In a graph, there are nodes (vertices) and connection lines (edges). The connection lines can have a mechanical reality as strings, elastic structures, roads, pipelines etc. or they can be a realization of interrelations such as binding energies, flow of information etc., thus the concept is extremely flexible, but still mathematics (graph theory) has a lot to tell even on this abstract level that can be extremely useful in a concrete modeling project. See some striking examples in Fig. 1. In a concrete situation the network topology is reflected in edge-degree, the number of edges joining a vertex. The properties of the process on such a graph will have local effects on individual edges and global effects while passing across multiple nodes. Mathematical graph theory is a well established subject of research since centuries beginning with the celebrated 'Königsberger Brückenproblem' of Leonard Euler (1736) which asks for a round-trip through the city of Königsberg in which each of the seven bridges (edges) that connect certain parts of the city (nodes) have to be trespassed exactly once such that one returns to the start position. The answer that Euler gave was in the negative, as the nodes had an odd degree while in order to have an affirmative answer they should have possessed an even edge-degree. This fundamental result of abstract mathematical research has an enormous consequence even for modern engineering applications!

Once the mathematical object of a graph had been established as a potential resource for modern modeling one started asking for modeling processes on such objects. The processes may be connected with transmission of light and waves in general, transport of masses, electrons or products etc. or deformations. Such processes are determined in part by the network structure of the graph underlying the model or, the other way round, the process determines the graph. For example, the vascular system on a leaf has always a particular bifurcation structure: nature optimizes topologies! [5]. In order to sustain forces in an optimal fashion, truss-structures have to obey certain optimality principles that are reflected in proper connectivity conditions at multiple nodes [8, 12]. Learning from these examples, one my ask now: 'what is the right topology for a transportation network?' and 'how about optimal topologies for ceramic foams in order to exhibit maximal strength?'.

A topology is not necessarily a static object, as seen in particular in transportation networks: closing a valve means 'deleting' an edge, a 'red' in a traffic light closes the road for a while (also cf. 'ramp metering'). Thus, topology optimization may have an overlap with active real-time optimal control, and sensitivities in order to update the strategy may be taken with respect to topology as well as with respect to running profiles for the compressors. The interplay between these different sensitivities and their coordination is a grand challenge in discrete-continuous dynamical systems (cf. Fig. 2). Robustness issues become more and more important in process of automating processes. 'When has a bridge to be repaired or reconstructed in order for the traffic flow to remain operative', and 'which pipeline has to be replaced in a gas-network under safety conditions', or 'is the minimal weight design for an airplane safe?'. These are questions pointing to robustness.

All of the problems mentioned can be reduced to mathematically very similar questions. If one has at hand a mathematical sound theory and numerically real-time capable simulation software together with a tool for robust topology- and material optimization, then one is in the position of a key technology applicable to a vast number of different problems that only upon stepping back reveal a structural paradigm.

While topology optimization appears more intuitive and easier to understand, free material optimization by itself seems to provoke fundamental criticism. With 'Free Material Optimization' (FMO) we understand the more or less free variation of material properties in every material point—an obvious absurdity! We are therefore asking nothing less than: 'how would nature perform, if there were no limits

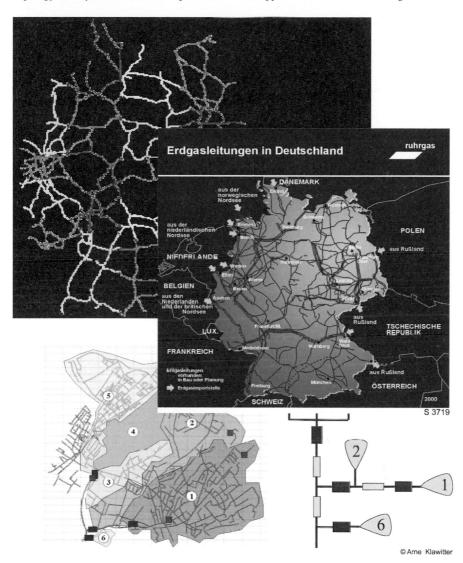

Fig. 2 Infrastructures: transportation networks (traffic, gas, water)vspace*-3mm

other than, say, positivity of energy?' The question points at 'utopia materials' and answers to be given should be capable of explaining both the utopia result and which part of the utopia concept can be actually realized, thus its projection to reality. The current construction of 'meta-materials', those with e.g. negative Poisson ratio, negative refractive index etc., has much to do with this kind of optimization.

In these note we will focus on success stories connected with concrete applications where the process of modeling, simulation and optimization such as described above has proved to be cutting edge.

As a first summary we can state: mathematics gains structural insight by abstraction—stepping back from the immediacy of the application context enables across-the-board visions and opens windows to other disciplines. Technology transfer is thus a genuine mathematical business. But even beyond the qualitative structure-oriented point of view, mathematics via quantitative simulation takes a giant step forward towards concrete applications and thus works as a catalyst for key technologies. Mathematics in its polyvalent function works behind the scene and contributes to the screen-play of new technologies for the future.

3 Success Story: Structure-, Topology- and Material-Optimization

One of the main problems in structural mechanics can be described as follows: *Given displacements and forces along the boundary of a body, to find lightest structure sustaining the loads.*

The optimization of structures is typically realized via variation of the geometry of the structural elements involved (like rods, beams, plates, shells, elasticity). The variation of the geometry, in turn, may be realized via so-called 'sizing' ('variable thickness sheet') or 'shape-optimization' and 'topology-optimization'. These more or less classical approaches can be replaced by a more 'radical' methodology the 'free-material-optimization' (FMO) in which the material tensor is considered as free variable all by itself [3, 4, 13–15].

The first results in that direction date back to the 1980ies. They refer to the theoretical background of shape- and topology optimization via homogenization, a relaxation method of the original design problem [1, 2]. Due to the growing interest also numerical methods for this approach has been devised and implemented [1, 12]. However, after a promising start the interest lost momentum once it became clear that the homogenization method lead to iterated or nested lamination, resulting in a very complex and hard to mechanically realize microstructures. The mathematical reason for this fine-grain material structure is the very nature of the relaxation procedure which replaces the black-white-design (material-no-material paradigm) into a density driven material-mix of two materials, matrix-material and weak-material. While black-white design is represented by so-called characteristic functions, being 1 where there is material and 0 else, a density is continuously distributed between the values 0 and 1. In a sequential optimization procedure a black-white material distribution will in general converge to density driven material mix, the so-called homogenized material. The problem of characterizing all such homogenized materials is up today solved only for special cases (like 2-d problems with isotropic material mix). On the other side, the homogenization approach marks a break-through in material design. It revealed the insight that each material can be viewed locally as a microstructure [4].

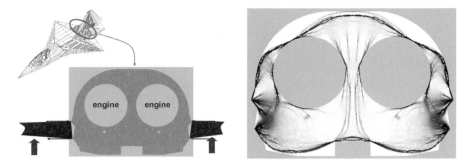

Fig. 3 Aircraft design. 40 000 Finite Elements (source: EADS)

For the optimization of materials the reverse process the so-called 'inverse homogenization' is the crucial concept: while in classical 'forward' homogenization out of the local microstructure in a reference volume element (RVE) is obtained via an averaging procedure (the homogenized limit), in inverse homogenization one tries to construct via optimization a local microstructure within the RVE's such that the homogenized material satisfies certain requirements on the macro-level. It is, however, crucial to obtain such micro-structures that are mechanically implementable. Free material optimization is another very promising approach which, as mentioned above, focuses on the material properties at each material point by optimizing on the full or a material tensor, keeping only the mandatory positive semi-definiteness and bounds on the material tensor's entries. This somehow produces 'utopia-material' with guaranteed optimal properties which, however, has to be realized in a true product. There are various ways to deal with this 'interpretation problem': one may draw stress-lines which can be taken as a basis for what has come to be known as 'tape-laying', a procedure developed and realized in aircraft design at EADS in Munich, or one may use the density information inherently produced by the method to recover holes in the structure (see Figs. 3 and 4). Combined with suitable visualization tools, such realizations have been implemented in certain ribs of the wings for the Airbus 380 (see Figs. 5 and 7). The amazing result was a weight reduction by 33%! (see Fig. 6) (cf. [8]).

Regardless of the grand success in this case, the interpretation of FMO results is still a challenging problem. Of particular interest are post-processing tools that help to convert the FMO results into classes of real materials and black-white designs thereof. The finiteness of the classes available brings in a mixed-integer aspect of stacking sequences. In any case, domain-decomposition techniques are necessary in order to handle very large such problems (cf. [9]).

Nevertheless, the success quickly turned into an international and interdisciplinary research project involving academia and industry which is funded by the EU (PLATO-N: www.plato-n.org) in which particularly FMO-techniques are further developed and implemented on a cutting edge level.

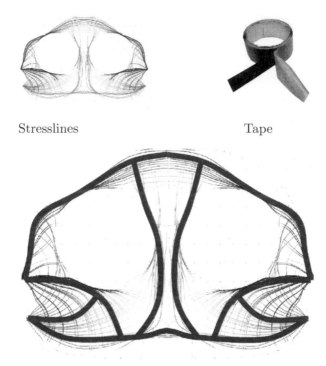

Stresslines   Tape

Fig. 4 Tape-laying technique (source: EADS)

3.1 Resume/Recommendations

Shape- and topology optimization is a grand challenge for industrial applied optimization where continuous as well as discrete paradigms have to be interlinked with high-end simulation techniques and engineering knowledge. The particularly excellent resources of young scientists in this area make a coordinated funding a must.

4 Optimization of Transportation Networks

Networks of pipes serve for the transport of gas from the suppliers to the customers. In contrast to the former situation where only a very small number of providers offered gas at a defined quality, the current gas market sees a large number of providers offering gas-products at various different quality levels. The customers, by now, may choose among the providers and the corresponding gas-prices. This makes the gas distribution problem through pipeline networks a highly dynamic and complex real-time optimization- and control problem for the gas-pipeline management to deal with. One of the subproblems to be handled is connected with the pressure lost due to friction in the pipes. Due to that phenomenon, a few compressors are installed

Fig. 5 Practical realization: leading edge of Airbus 380

within the network that are all by themselves very complex. These compressors consume a considerable amount of gas from the pipeline in order to increase the pressure by the action of pumps. While the pressure has to be increased in some possibly remote pipes the pressure has to released e.g. in some pipes in the vicinity, and, more importantly, flow reversal has to be avoided everywhere. Therefore, together with the operation of costly compressors opening and closure of valves and other steering measures have to be taken in Fig. 8.

The full gas-pipeline network including all of its subparts is way too complex in order to model and automatize its real-time capable control and optimization based on distributed and physically sound laws of gas-flow. Rather one introduces a hierarchy of models of gas-flow in a pipe ranging from simply algebraic nonlinear relations up to nonlinear hyperbolic balance laws. In case of just algebraic equations describing the pressure and flow between two end-points of a pipe (nodes in a graph) and proper transmission conditions at the nodes one can derive a discrete nonlinear network flow problem in which an open valve and an active compressor are represented by 1, a closed valve and an inoperative compressor by 0. Flow-rates as well as the pressure are subject to legal and technical constraints. As a further constraint, admissibility of the flow has to be guaranteed in the sense that the demand can always be satisfied during the optimization procedure. Problems of this kind can be embedded into so-call (dynamic) mixed integer problems (DMIPs or MIPs). Such a MIP can be seen as a coarse grain approximation of the problem as the local flow reality of the gas in the pipelines is only poorly represented and the problem formulation relies on node-to-node interactions. This coarse grain approx-

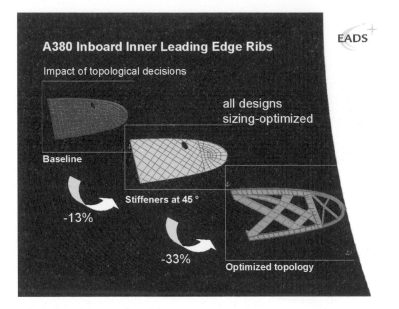

Fig. 6 Practical realization: weight reduction (source: EADS)

Fig. 7 Practical realization: Airbus 380: left prototype, right product show (source: EADS)

imation can be complemented by a small grain modeling which takes into account
local effects via inclusion of partial differential equations (PDEs) in the model. This
small grain modeling, analysis and optimization based on PDEs can however performed only on small local subnetworks (see Fig. 9). The coarse-grain versus small
grain modeling and optimization reflects also an alternation between global optimization in the context of MIPs and local gradient-based optimization with respect
to the PDE-models. One of the challenges is to truly combine discrete and continuous methods. This is also present in the problem of model switching and switching
of controls [7].

Fig. 8 Compressor (source: E.ON Ruhrgas)

The complexity of the problem given its importance prompted the former Ruhrgas AG (now E.ON Gastransport AG & Co. KG) to first contact the Konrad-Zuse Institute in Berlin which after some time led to the initiation of an interdisciplinary research project funded by the German ministry for education and research (BMBF) and more recently be the DFG [10, 11].

Very similar questions arise in the equally important problems of water irrigation, sewer- and freshwater systems, where thousands of open and closed channels (canals) are interconnected. Maintenance of the pressure in freshwater systems, avoidance of pollution and overflow in waste-water and sewer systems, regulation of resources in irrigational networks etc. are issues of great importance. Again simple hydrological models have to combined with free-surface water flow or pressure flow models in networked pipelines. The automatic real-time optimal control of such systems is a grand challenge. In another coordinated interdisciplinary research group, funded by the BMBF, scientist from the Friedrich-Alexander-University of Erlangen-Nuremberg (FAU) and the Technische Universität Darmstadt (TUD) work together with industry e.g. SIEMENS, Hessenwasser, Steinhardt and the Dresden municipal water-works. (www.odysseus.tu-darmstadt.de) [6] (see also the references therein).

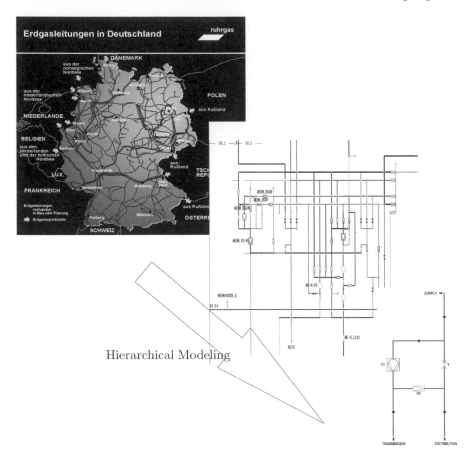

Hierarchical Modeling

Fig. 9 Network hierarchy

4.1 Resume/Recommendations

Optimization, maintenance and planning of infrastructures are problems of extreme relevance in our future society! Billions of dollars will have to be invested all over the world in the near future. The distribution of resources—energy or water—is the most difficult problem of future societies. Most of the underlying mathematical control problems are decision driven, incorporate incomplete information, stochastic variations of parameters and involve a hierarchy of discrete and continuous models. They are multi-scale problems involving hundreds of million variables. Thus, there is a huge demand in robust discrete-continuous and multi-level optimization of transport processes on networks. International as well as national funding agencies should realize (and have in fact realized in part) that the investment in fundamental research is mandatory now, before a more costly and much less effective hands-on parameter-tuning trial-end-error based empirical optimization is triggered on the level of profit-makers.

We have demonstrated that real processes in key technologies can only be dealt with in the framework of multi-scale and multi-level hierarchical models, where forward simulation based parameter optimization has to be replaced with model-based constrained discrete- and or continuous numerical optimization techniques. Optimization appears in the context of process control as well as in topology, material (formation as well as damage) and shape. The latter may be formally included in the framework of 'control-in-the-coefficients'. Some national and international coordinated programs are currently funded by the German DFG and the EU, e.g. the DFG-Priority Program 1253: 'Optimization with PDE-constraints' and a project by the European Science Foundation (ESF) with the same title. Altogether there are more than 60 research groups in Europe involved in PDE-constrained optimization. There is huge chance to link these efforts to optimization groups involved in discrete optimization in order to bring to bear the immense potential of senior and young scientists in the field.

Acknowledgement   The pictures and figures related to the aircraft design have been generously provided by EADS Deutschland GmbH, and those for the gas-network by Ruhrgas AG (now E.ON Ruhrgas).

References

1. Allaire, G.: Shape Optimization by the Homogenization Method. Applied Mathematical Sciences, vol. 146. Springer, Berlin (2002)
2. Allaire, G., Kohn, R.: Optimal design for minimum weight ans compliance in plane stress using extremal microstructures. Eur. J. Mech. A, Solids **12**, 839–878 (1993)
3. Ben-Tal, A., Kocvara, M., Nemirovski, A., Zowe, J.: Free material design via semidefinite programming. The multi-load case with contact conditions. SIAM J. Optim. **9**(4), 813–832 (1999)
4. Bendsøe, M.P., Sigmund, O.: Topology Optimization. Theory, Methods and Applications. Springer, Heidelberg (2004)
5. Durand, M.: Architecture of optimal transport networks. Phys. Rev. E **73**, 016116 (2006)
6. Gugat, M., Leugering, G.: Global boundary controllability of the Saint-Venant system for sloped canals with friction. Ann. Inst. H. Poincare Non Lin. Anal. **26**(1), 257–270 (2009)
7. Hante, F., Leugering, G., Seidman, T.I.: Modeling and analysis of modal switching in networked transport systems. Appl. Math. Optim. **59**(2), 275–292 (2009)
8. Kocvara, M., Stingl, M., Werner, R.: MOPED user's guide. Version 1.02. Research Report 262, Institute of Applied Mathematics, University of Erlangen (2000)
9. Lagnese, J.E., Leugering, G.: Domain Decomposition Methods in Optimal Control of Partial Differential Equations. ISNM—International Series of Numerical Mathematics, vol. 148. Birkhäuser, Basel (2004). xiii, 443 p
10. Martin, A., Möller, M., Moritz, S.: Mixed integer models for the stationary case of gas network optimization. Math. Program. B **105**, 563–582 (2006)
11. Mahlke, D., Martin, A., Moritz, S.: A simulated annealing algorithm for transient optimization in gas networks. Math. Methods Oper. Res. **66**, 99–116 (2007)
12. Rozvany, G.I.N.: Topology optimization of multi-purpose structures. Math. Methods Oper. Res. **47**(2), 265–287 (1998)

13. Stingl, M., Kocvara, M., Leugering, G.: Free material optimization with fundamental eigen-frequency constraints. SIAM J. Optim. **20**(1), 524–547 (2009)
14. Stingl, M., Kocvara, M., Leugering, G.: A sequential convex semidefinite programming algorithm for multiple-load free material optimization. SIAM J. Optim. **20**(1), 130–135 (2009)
15. Zowe, J., Kocvara, M., Bendse, M.: Free material optimization via mathematical programming. Math. Program. B **79**, 445–466 (1997)

Energy and Structural Engineering

Capacity Planning and Scheduling in Electrical Power Systems and in Chemical and Metallurgical Production Plants

Sebastian Engell, Edmund Handschin,
Christian Rehtanz, and Rüdiger Schultz

1 Executive Summary

One of the greatest challenges for humankind in the 21st century is the sustainable, economically efficient and politically fair management of the resources available on earth. It implies a shift of paradigm in all areas of public life. Traditionally, electrical power systems and chemical production plants are fields in which an optimal management of precious and limited resources is of great importance. Mathematical methods have been used for a while already to solve the complex decision problems that result in this area. Yet, by far not all problems that have occurred can be tackled in a sound manner. The growing economic and ecological pressure on the efficiency of the production and supply systems as well as stronger fluctuations of supply and demand lead to new challenges for which new mathematical methods and algorithms have to be developed.

This article reports on the current state of affairs and the significance of mathematics for the capacity planning and scheduling of electrical power systems and

S. Engell (✉)
Technische Universität Dortmund, Fachbereich Bio- und Chemieingenieurwesen, Lehrstuhl
für Systemdynamik und Prozessführung, Emil-Figge-Str. 70, 44221 Dortmund, Germany
e-mail: s.engell@bci.tu-dortmund.de

E. Handschin · C. Rehtanz
Lehrstuhl für Energiesystem und Energiewirtschaft, Technische Universität Dortmund,
Emil-Fligge-Str. 70, 44227 Dortmund, Germany

E. Handschin
e-mail: edmund.handschin@udo.edu

C. Rehtanz
e-mail: christian.rehtanz@tu-dortmund.de

R. Schultz
Universität Duisburg Essen, Fachbereich Mathematik, 47048 Duisburg, Germany
e-mail: schultz@math.uni-duisburg.de

M. Grötschel et al. (eds.), *Production Factor Mathematics*,
DOI 10.1007/978-3-642-11248-5_15, © Springer-Verlag Berlin Heidelberg 2010

chemical and metallurgical production plants, addresses the new challenges and offers suggestions for innovative research projects to respond to these challenges.

Among the examples for an effective use of mathematics in the area of electrical power systems and chemical production plants are the management of power plants to ensure an efficient supply and distribution of electrical power, the control of power grids along with a guarantee of their stability across the continent, the optimized scheduling of the batch production in production processes such as the production of copper or the optimal production planning and scheduling when producing active pharmaceutical ingredients.

What all these examples have in common is that the mathematical modeling and optimization so far have been carried out based on the assumption that the relevant problem data are fully known. In fact, however, we are dealing here with decision problems in dynamic contexts in which the future development is afflicted with uncertainties but nevertheless has to be considered in the decision. Mathematical modeling and optimization have to take into account incomplete information and the arrival of new information in the course of the optimization.

Due to the shift towards liberalized energy markets, the power grid, for example, changes from a merely technical entity to a trading platform with an bidirectional exchange of information between producers, merchants, and customers. When further exogenous influences such as the stochastic wind energy supply arise, the capacity planning and scheduling is faced with highly complex decision problems under incomplete information.

A similar situation can be observed for the optimization of chemical production processes. Here, too, the adequate handling of incomplete information is of utmost significance for capacity planning and scheduling. Such uncertainties can exist with respect to processing times, commodity prices, customer demands, or the availability of plants. In the light of the complexity of the decision problems, not only the incorporation of uncertainties in the decision-making but also a user-friendly communication of solutions with the help of innovative man-machine-communication mechanisms is of major importance.

In close collaboration with other branches of mathematical optimization, stochastic optimization is aiming at developing models and algorithms to improve decision-making under uncertainty. First applications of stochastic optimization are already in use. Nevertheless, further intensive fundamental and application-oriented interdisciplinary studies by mathematicians, engineers and computer scientists are required to respond to the challenges described above. The authors therefore suggest four exemplary subject areas:

- Optimization strategies for the operation, the extension, and the renewal of power grids of the 21st century,
- decision support for optimizing the bidirectional information and energy management in new structures of energy supply,
- problem-specific modeling and efficient handling of uncertainties of resource allocation problems in chemical production processes,
- man-machine-interaction during the optimization of complex production processes.

Based on interdisciplinary networks which already exist or should be created, the BMBF (Federal Ministry of Education and Research), the BMWi (Federal Ministry of Economics and Technology) or the DFG (German Research Council) can help to trigger ground-breaking research projects in these fields.

2 Success Stories

2.1 Supply and Distribution of Electrical Power

Constant compliance with all technical limits as well as the $(n-1)$-principle[1] for managing and optimizing the electrical transmission and distribution networks are a precondition for a safe, economical and ecological supply and distribution of electrical power. In order to determine the current condition of the networks in a safe and reliable manner, extensive network simulations are needed. These simulations are based on the solution of very large systems of nonlinear equations so that even for networks with a large number of network busses and conductors the desired solutions along with a graphical display can be obtained within a few seconds.

Network Control and Reserve Power

The high-voltage supergrid (380/220 kV) is divided into control areas in which the power balance between the generation and the demand has to be kept stable at any point. The transmission system operator (TSO) has to purchase and provide a sufficient amount of balancing power so that a frequency of 50 Hz can always be provided. A primary, a secondary, and a tertiary reserve can be distinguished. The primary reserve is automatically activated peripherally at the power station when needed. It has to be made available immediately and to remain in place for up to 15 minutes. The secondary reserve is activated centrally at the control center and can be delivered by the power stations within or outside of their own control areas. It has to be able to be activated no later than 30 seconds after a power imbalance occurred and to be available for more than 15 minutes. When the tertiary reserve has been demanded by phone, the primary and the secondary reserve can be reduced to their original value again. In view of the size of the European UCTE-system with an installed capacity of 600 GW it becomes clear that this is certainly the most complex control task of a technical system which was ever created by mankind. Figure 1 shows the three reserve types mentioned above and their corresponding time demands.

---

[1] I.e. the functional capability of the grid even in the possible event of the failure of an arbitrary component.

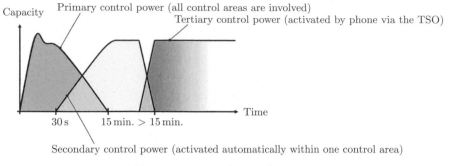

Capacity Primary control power (all control areas are involved)
                          Tertiary control power (activated by phone via the TSO)

                                                        → Time
    30 s   15 min. > 15 min.

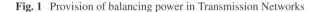

Secondary control power (activated automatically within one control area)

TSO = Transmission System Operator

Fig. 1 Provision of balancing power in Transmission Networks

Similar to the TSO, who is in charge of the high-voltage system and the super-grid, the distribution system operator (DSO), who is responsible for the medium-voltage and low-voltage power grids, has to supply sufficient capacity for distributing the electrical power that customers demand. Due to the liberalization of the energy market and the strong support for renewable energies, of which the wind energy in particular is characterized by strong fluctuations, the demands on the network operation have increased significantly. Due to the disentanglement of production, commerce, and application on the one hand and the grid operation as a natural monopoly on the other hand new challenges evolved which can only be overcome when the parties in charge of the balancing group and the TSO and the DSO cooperate.

Since the physical and the contractual electrical energy trading have been separated, the network operator is required to check whether the requested transport capacity is feasible. He can do so with the help of a comprehensive set of methods provided by mathematical optimization:

- State detection based on interpolation,
- optimal network calculation which requires the numerical solution of a very large system of nonlinear equations to minimize the transmission losses,
- mains network security calculations to test the $(n-1)$-principle and the short-circuit level of all network components,
- diagnosis and elimination of grid bottleneck problems.

The growing European trade with electrical power increased the occurrence of local grid bottlenecks in the European grid significantly. In practice, auction methods have prevailed so far, even though they exhibit considerable disadvantages when closely meshed grids are concerned. First positive results could be obtained from the use of the hierarchical optimization, which takes into account that not all network data of the individual TSOs have to be exchanged [3].

Figure 2 shows the network control center of a large transmission grid. From here, the complete transmission network can be monitored, operated, and optimized.

Fig. 2 Control center of a
large transmission grid
operator (source: RWE)

Stability in Real Time Across the Continent

Due to the feed from strongly fluctuating wind energy plants the stability of the
whole UCTE-transmission network has to be monitored continuously. In the process, it becomes apparent that natural frequencies occur not only in north-south
direction but also in east-west direction which can lead to undesired performance
shifts. Figure 3 shows the results of the analysis of the intrinsic values in the UCTE-
network [33].

Such natural frequencies have to be monitored carefully. To do so, transcontinen-
tal wide area measurements and the respective analytical methods are applied today
since an extensive signal analysis of the obtained data can yield reliable results concerning the type and the location of places of action of damping components from
the area of power electronic controllers (so-called FACTS-devices) [30].

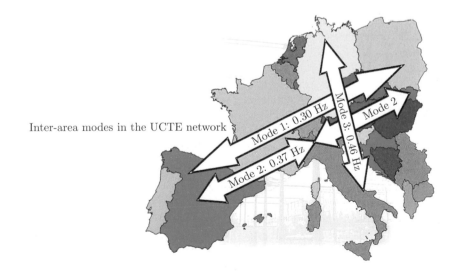

Fig. 3 Natural frequencies in the European UCTE-network

Power Plant Optimization for Efficient Resource Utilization

In terms of power plant optimization, not only a scheduling problem but also the economical load dispatch have to be solved. For the capacity scheduling, short, medium and long term decisions have to be made to determine which generation units should be used to cover the load at what time. Besides dynamic programming, mixed-integer optimization plays a major role here. For mathematical modeling it has to be kept in mind that besides performance requirements additional constraints have to be fulfilled. They have to be incorporated as integral variables in the optimization. Furthermore, minimum operation times and minimum off-periods have to be taken into account. Nowadays, optimization methods are available which can be used to solve even major problems successfully [7, 11].

As for the commercial load distribution, it has to be determined which generating unit to use at which capacity. Here, the load and network side conditions have to be strictly obeyed at any point in time. The uncertainties of the price development as well as of the supply of regenerative producers will lead to the replacement of the established deterministic optimization procedures by stochastic optimization within a foreseeable period of time. Thereby, the uncertainty, which is inherent in the predictions, can be considered within realistic limits [39].

2.2 Capacity Planning and Scheduling in Chemical and Metallurgical Production

Chemical production plants differ from machines used in production engineering in terms of the longer duration of the individuals steps, more complex transition procedures between different operations points and batches, and complex material couplings. Adequate timing and coordination of the individual steps is of crucial importance for an economical and energy-efficient operation of the production plants as well as for reducing the risks that are involved in storing hazardous and toxic substances. During the past few years, the efficient use of production capacities has become increasingly important in chemical production processes. On the one hand, this approach aims at saving energy and resources, and at reducing the amount of waste and of products that are out of spec and may cause high disposal costs, on the other hand the goal is to increase the economic efficiency.

Problems related to planning and scheduling occur in particular when the production is carried out in small batches and when a multitude of different products has to be produced in parallel and sequentially in order to meet the specific demands of the market. Below, two examples will show how optimization techniques can be implemented in this particular area.

Simultaneous Optimization and Scheduling of the Copper Production

No other metal has been used by humankind for such a long time as copper, and it is still of great importance for a multitude of products today, for example in electrical

Fig. 4 Copper production

and structural engineering. One of the greatest advantages of copper is that it can be recycled without any major problems. Impure copper can be cleaned easily and afterwards has the same properties as 'fresh' copper that was extracted from ore. To a large extent, the demand can thus be covered by melting and cleaning scrap metal and waste material from the production process.

During the production process (Fig. 4) the copper passes through several furnaces and converters in which all undesired components are removed until finally a purity of 99.6% is attained, and the copper can be poured into continuous casters. In a typical copper refinery, several furnaces and converters are available in parallel. The metal can be moved between them with the help of large cranes. These cranes share a work area so that the movements of each crane are limited by those of the others. The production in the converters and the anode furnaces has to be arranged in a way that utilization of the shared resource 'cranes' is timed accordingly and that the following continuous caster is always supplied with material. At the same time, waiting times should be avoided as far as possible since they would lead to a loss of energy and because the plants would no longer operate at full capacity. Since slag is returned to the previous levels with a significant copper concentration, a detailed record of the material balance has to kept and considered in production planning and scheduling.

The problem of planning and scheduling the production steps of the individual batches of material is complicated by two factors: the processing times and the achieved cleanliness of each of the production steps depend strongly on the composition of the incoming material, and irregular scheduled maintenance of the equipment has to be taken into account. Apart from the scheduled maintenance operations, unpredicted failures occur when the plants are operating so that the capacity planning might have to be adapted accordingly.

In a solution for the simultaneous recipe optimization and for the scheduling of the copper production that was developed by ABB, the overall problem was formulated as a mixed-integer linear program (MILP) [14]. The optimization problem contains almost 1000 continuous variables, for example the starting times of operations on the respective units, and almost 100 binary variables for the allocation of the resources. With the help of a problem oriented modeling it was possible to solve an integrated optimization problem for a time period of $1 - 1\frac{1}{2}$ days with a commercially available MILP-solver (CPLEX [19]) within a few seconds. Thus, it is possible to react to disturbances or changes in the material composition with hardly any time lag. The movements of the cranes which are necessary for conducting the desired production steps are simulated after the production sequence has been determined to ensure that the calculated production plan can actually be carried out. If it turns out that this is not the case, the times for the crane transports that were estimated in the planning model have to be adapted and the planning has to be repeated. The industrial implementation of this solution has proven to be robust and efficient. In the implemented system, the current status of the production is displayed for all plant components to inform everyone in charge about the current schedule of the next steps and to provide an overview of the status of the plant and the utilization of the units in the overall process [13].

Production Planning and Scheduling in the Production of Pharmaceutical Ingredients

Production planning and scheduling problems as well as control problems in the chemical and pharmaceutical industry differ from the classic 'flow-shop' and 'job-shop' problems of part production by the presence of a number of additional constraints and by very long planning horizons. A typical planning and scheduling problem involves 10 000 different substances (raw materials, intermediate and final products) which are produced in up to 30 or more steps according to specific recipes, several hundred pieces of equipment or packing machines, a number of locations and a planning horizon of several years. Common for chemical production planning problems is a 'diverging tree' where a large number of different intermediate and final products are produced from a few raw materials. In many cases, side products or unused material remain which, if possible, are fed back after purification, or are used elsewhere in the production process. Typically, the transition from one (intermediate) product to another involves a significant cleaning and refitting effort which is why a so called campaign production mode is preferred. In a campaign production mode, only one product is produced on the same part of the plant for a period of time. On the other hand, however, this results in material that has to be stored and thus costs, and both the quantity and the duration of the interim storage can be limited. In addition to the resource scheduling of the plant, the availability of staff is a considerable limiting factor, especially when refitting is required.

Such planning and scheduling problems cannot be solved today or in the near future 'as a whole' by using optimization algorithms. There are commercial tools, however, which provide approximate solutions. They are based on two basic methods of simplifying the solution: Hierarchical decomposition and heuristics and local optimization. For the hierarchical decomposition, usually a simplified problem is solved over the complete planning horizon. In this problem only the more important constraints as well as the main products and main ingredients are considered.

Furthermore, resources which rarely cause bottlenecks as well as relatively short non-productive times are not taken into account. In the optimization, clearly defined time periods of for example one month each are planned; the exact timing within these slots is left to an underlying scheduling algorithm. The optimization problems can be written and solved as mixed-integer linear programs. Planning projects of the dimension described above lead to optimization problems with several 100 000 continuous and approximately 1000 binary variables which can be solved within an hour [8]. Since the solutions of the simplified problem cannot be implemented directly, the production is planned in detail for an interval in the immediate future, for example for three months. In doing so, the results of the long term optimization are considered as orders which have to be carried out. The two levels use models of different accuracy so that their coordination/tuning has to be designed carefully [8].

Whereas in the hierarchical decomposition the problem is mapped with varying accuracy onto planning intervals of different lengths in order to make a solution with mathematical optimization algorithms possible, the heuristic planning approach makes use of a uniform model. Here, all dependencies and restrictions throughout the planning interval are considered in detail. Thereby it can be prevented that a solution which was determined based on simplified conditions later turns out to be impossible to implement. On the other hand, planning the remote future in detail based on assumptions about the plant that might be outdated quickly involves a large effort. The core of successful tools for solving such high-complexity planning tasks over long intervals is the iterative improvement of feasible solutions [29]. At first, a feasible plan is designed on the basis of heuristic principles for example by backward scheduling from the delivery dates with the help of assumed processing times for the individual units and solutions to conflicts following priority rules such as 'most urgent order'. Then this plan is improved by using local optimization methods and heuristics. The user can control this search process by manual rescheduling, securing certain productions and selecting the search criteria and the permissible computing times.

Both approaches are implemented in commercial planning tools [8, 29] and applied in industry.

3 Status Quo

3.1 Power Supply

Change of Paradigm from the Technical Grid to a Trading Platform

In order to accomplish the energy policies demanded by the European Union by 2020,

1. to produce 20% of the electrical energy with regenerative plants,
2. to reduce the CO_2 emissions by 20%,
3. and to increase the efficiency of the production of electrical energy by 20%,

the grids of the future have to meet great challenges. Within the change of paradigm towards liberalized energy markets, the EU-technology-platform 'Smart Grids' offers the key to achieving the goals described above. In the future, the unidirectional energy flow from the power plant to the customer will be complemented by new plants in close proximity to the customers in the kW-region. This development is strongly supported by the development of new, small power plants that generate power and heat [15]. This development is referred to as the Internet of energy since in the future millions of plants will be running simultaneously and exchanging energy over the distribution network. Yet, these plants can only be integrated into the grid in an economically sensible way if there is a bidirectional information exchange between the trading, the network operator and the customer plants.

The installation of digital measuring instruments to determine the power consumption provides the basis for exchanging information in both directions by use of such gateways at the customer's site. Here it has to be kept in mind that only the combined, coordinated operation of many distributed plants produces economical solutions. For this reason, an electronic platform has to be designed which can be accessed easily by everyone involved without distinction. The Internet will play a key role in establishing such a shared trading platform. In order to make a smooth communication between everyone involved possible, shared data logs have to be developed. At the same time, much attention has to be paid to the problem of data integrity by ensuring that only those who have a valid claim receive access to the pertaining data. This can be achieved by means of cryptographic techniques. With the help of a agent-based system, not only the concerns of the trading but also those of the network provider should be considered at any point in time. From an economical point of view such a virtual plant has to be controlled and optimized from one location in 15-minute intervals [25]. In doing so, it is of major importance to consider the uncertainties with regard to future power requests and price developments since this is the only way of approaching real time rates. Multi-stage stochastic optimization procedures have great potential to meet the extremely complex demands of such systems. Moreover, there has to be an option to switch the individual pieces of equipment on and off at the customer's site by using the gateway mentioned above to allow the customer to make optimal use of periods during which the electricity rates are low (e.g. due to a large supply from wind power).

Optimized Energy Management for Virtual Power Plants

In order to operate virtual power plants (see Fig. 5, [25]), innovative energy management systems are needed which are based on stochastic programming. The virtual plant consists of a number of small, peripheral energy conversion plants which are all used and optimized together very close to the customer's site. Since the virtual plant is confronted not only with inaccurate predictions for the power and heat demands but also with the unknown electricity rates of the spot market, a stochastic method has to be applied as it takes uncertainties into account. In close collaboration with the authors of this article it was possible to develop an optimization

Fig. 5 Energy management
for a virtual power plant

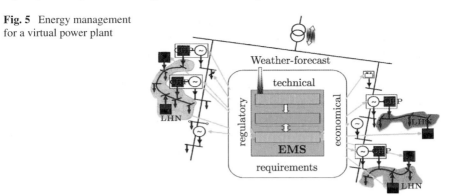

method which is very well suited for practical applications. It calculates and monitors the optimal operation schedule for all plants that are included in the virtual plant in fifteen minute intervals. When all details are taken into account, the result is a very complex optimization problem that for the first time could be solved under realistic constraints. Due to the fact that the stochastic optimization considers two stages, a variety of scenarios with different probabilities can be assumed for the uncertainty of the second stage. The computing time will obviously become longer with an increasing number of scenarios so that here a careful selection is absolutely necessary. The developed solution was applied to an actual virtual plant. It was possible to show that from an operational point of view major technical and economical benefits can be achieved.

Complexity of the Grids and Mathematical Challenges

Together with the development of distributed generation suppliers, the construction of large wind farms on-shore as well as off-shore inevitably leads to the construction of new power lines within the existing transportation grid in order to be able to transport this energy to the respective load centers. Due to the heavily fluctuating supply behavior of wind energy plants stabilizing measures have to be found by which the undesired load oscillations can be eliminated. To do so, an extensive network analysis has to determine where the needed FACTS-equipment has to be installed to achieve the greatest damping effect possible. The resulting dynamic simulation models are outstandingly large and complex because they have to cover a long period of time. The stiff systems of differential equations that have to be solved are very demanding for numerical solution methods since otherwise the capacity of the computers that are available today would be exceeded quickly. With the help of adequate parallelization methods the problems can be divided into manageable sub-problems.

Apart from the grid-specific problems of the wind farms, the balancing of wind energy has to be considered. Since only constant power bands of 15 minutes periods each are traded on the electricity market, the stochastic wind supply has to be converted into a band (wind energy upgrading) [36]. In practice, it used to be customary to bargain and bill the wind supply within each of the four German control areas in individual accounting grids.

Apart from a reliable wind forecast, optimization-based methods have to be developed in order to purchase the required control and balancing energy on time and at a low price. This is the only way to ensure that the costs that are involved in the balancing of wind energy remain reasonable. Since the technical requirements for the wind energy plants have been met already today, it has to be determined in the future to which extent the plants can be relied on to provide grid management services. This is particularly the case for providing the control reserve to be able to limit this very expensive energy despite the additional expansion of the wind energy.

New Challenges of the 21st Century

In summary, the grids of the 21st century have to perform completely new and challenging tasks. Table 1 shows the main differences compared to the traditional grids in a simplified fashion.

Of course the transition to the grids of the 21st century is a long-term and cost-intensive process since the equipment used in power systems technology has a long service life of 40 and more years. Therefore, future investments in the grid have to be made in close connection with a modern asset management of all pieces of equipment [20]. Apart from carefully selecting adequate maintenance and servicing strategies with simultaneous consideration of the risks, renewal strategies have to be defined which help to realize this transition in a technically and economically viable manner. In view of the fact that the planning and the construction of new plants and the peripheral supply have to be carried out independent of the network provider, large efforts have to be made to avoid unproductive investments in the grid.

Table 1 Power grids of the 21st century

Grids of the 20st century	Grids of the 21st century
Unidirectional communication	Bidirectional communication
Few sensors and data acquisition devices	Sensors and data acquisition devices in the complete system
Limited power-flow control	Extensive control options in stations and grids
Suboptimal utilization of the equipment	Modern asset management of all pieces of equipment in combination with a survey of the status quo and dynamic threshold values
Information systems that operate in isolation within an electric power company	Integrated information systems based on the Internet with non-discriminating access for all market actors
Small, individual peripheral supply	Millionfold peripheral supply and active involvement of the customers due to active load management
Limited rate details for customers, fixed rates	Complete price transparency, real time rates, active customer involvement

3.2 Chemical and Metallurgical Production Processes

Supply-Chain Optimization

In every large company that is involved in the metal production, in specialty chemistry, in food production, projects for improving the manufacturing logistics have been carried out. Usually they are referred to by the catchphrase 'supply-chain-optimization'. This means an overall view of the production and distribution process ranging from the purchase and storage of the raw materials over production to the distribution by peripheral storage facilities or wholesale dealers [40]. However, an integrated optimization of these overall processes is hardly possible at present.

Currently, the main instrument for the investigation and the improvement of the production logistics is discrete-event modeling and simulation [37]. Discrete-event simulation means that unlike continuous simulation not the temporal changes of process variables are determined but sequences of discrete events, for example the start and the end of production, cleaning and transportation steps. In contrast to the production of parts and pieces, however, substances properties such as the concentration of the different substances in a batch or a tank have to be tracked because of the changes that occur during the production process or when additional amounts are filled into the tanks. With the help of the discrete-event simulation, different scenarios (for example different numbers and capacities of production and supply units or different allocations of the individual productions to the plants or sites) can be simulated. For the capacity and production scheduling decisions, the results of the simulation of different scenarios are taken into account. An optimization in the strict sense is still uncommon.

The critical point of such an approach is evidently the selection of the scenarios of the demand and, if applicable, the evolution of the prices that is considered. They involve a significant uncertainty which the decision makers deal with in an intuitive fashion. It can be noted that the handling of this uncertainty is currently the weakest point in the application of logistic optimization. The marketing departments usually forward the predictions to the 'production people' as constraints for their planning decisions. The considerable uncertainty of these data is rarely quantified. As a consequence, great efforts may be put into implementing an optimization which is based on assumptions that will most likely never become true.

Medium-Term Production Planning

Medium-term production planning refers to the planning of production processes over periods ranging from a few months up to several years. The aim is to allocate orders to production units and to sequence them. To do so, the campaign production mode is of major importance, as already mentioned above. By producing as many batches of the same kind in a row, refitting and cleaning expenses can be avoided. Very long campaigns, however, require much storage space to meet the demands, which leads to costs and involves the risk that products that are not in demand are produced needlessly and thus have to be sold at loss or disposed of at high cost.

As described in the previous section, powerful tools are now available for production scheduling. From a mathematical point of view these are large optimization problems with integer or binary variables. Normally, modeling as a mixed integer linear program (MILP) is sufficient. Nonlinear phenomena are usually approximated by linear correlations which causes only small additional errors compared to the uncertainties due to the long planning horizons. Models for medium-term planning essentially contain mass balances for the stored or transported substances and limitations on the use of the resources and the sequencing. There are manifold constraints. Apart from the availability of the pieces of equipment, the staff capacity and the allocation of the material are of great importance. Constraints on the production sequences define the order of the production steps, delays due to refitting and cleaning operations, required off-times or maximum lifetimes of intermediate products, times for laboratory analysis, etc. In the case of simplified modeling the resulting problems are such that they can be solved by state-of-the-art MILP-solvers with computing times in the range of hours. These computing times are acceptable for planning tasks, yet hardly sufficient for evaluating a large number of alternatives based on optimization when planning the supply-chain. For these cases, semi-heuristic methods which can produce feasible solutions in a relatively short period of time are more practical.

The availability of the required data is a decisive problem in medium-term production planning. On the order of 10 000 recipes have to be made available in which the required resources for every step have to be defined precisely. It is necessary to automatically generate these data from a recipe data base as provided by today's integrated production planning and control systems (ERP-systems).

Online-Production Scheduling

The results of the medium-term production planning have to be transformed into a detailed schedule to carry out the production. In doing so, not only more detailed restrictions than for the medium-term planning but also the unavoidable deviations from the plan and from the assumed availability of capacities in the planning process have to be taken in to account and, as far as possible, to be compensated. This requires detailed models with a high temporal resolution and short computing times in order to be able to respond to disruptions and changes, such as rush-orders, quickly. The scheduling horizon is naturally shorter than the planning horizon since a detailed forecast for the remote future would not be efficient. Although optimization-based tools are available for this task as well, in practice manual control on the basis of a visualization of the process along with a forecast of further developments is usually preferred. This seems to be caused primarily by the general problem of missing acceptance of non-transparent solutions that were obtained by mathematical optimization. Additional factors are the large effort which is involved in a detailed modeling of all the constraints that the plant operator is aware of, and the update of this data. After all, the algorithmic problems increase compared to the medium-term planning when complex constraints and nonlinear correlations are involved whereas the time that is needed to solve the problem decreases.

3.3 Mathematical Methods

Mathematical methods, already for a while, are well established for formulating and solving problems in capacity planning and scheduling of electrical power systems and chemical production plants. Electrical power systems and chemical productions today are no longer imaginable without computer-based decision support as highlighted in the previous sections. The mathematical core of such decision support consists of models and algorithms from different areas of optimization. Most prominently, linear mixed-integer and nonlinear optimization are to be mentioned here. With stochastic optimization, a discipline is gaining ground that makes proposals for optimization under incomplete information, a feature which is becoming more and more indispensable.

Mixed Integer Linear Optimization

Capacity planning and scheduling typically involve both continuous variables such as consumption and production rates and variables that model indivisibles. Indivisibles can be in the nature of things as is the case of unsplittable goods or production batches. It is even more significant, however, with regard to yes/no-decisions which normally account for vital aspects in capacity planning and scheduling: Should a generating unit be started or not, on which of the plants should a batch be produced, should A or B be produced first? In capacity planning and scheduling it is of paramount importance how the ingredient 'time' is dealt with. A commonly used method is to divide the time horizon into suitable decision intervals (e.g. hourly, quarter hourly, or even shorter intervals). With many of such intervals and lots of possible alternatives in each interval, this often leads to decision problems with several thousands, sometimes several hundred thousands or even millions of unknowns. Mixed-integer linear optimization offers a flexible spectrum of methods for the mathematical modeling and solution of such decision problems. For example for operation and investment planning in power supply systems or for production planning and scheduling in the chemical and pharmaceutical industry, models and algorithms of mixed-integer linear optimization have become relatively well established by now. The basis of this development was provided by the availability of powerful software, in which the latest results of mathematical research were implemented in a user-friendly way. There are, however, problematical cases such as the so-called job-shop-scheduling problems. These concern the allocation of orders to a number of possible resources, which mixed-integer programming cannot solve efficiently. Here constraint programming methods have shown to be more effective.

The combination of both approaches to problems featuring aspects of both pure sequential planning and linear optimization is a promising field of work for the future. In current mathematical research, mixed-integer linear optimization also contributes to topics such as approximation of nonlinearities and decomposition of stochastic optimization problems.

Nonlinear Optimization

Load flows in power grids and yields of chemical production processes are examples of nonlinear effects which can have a decisive influence on the results of capacity planning and scheduling. If linear approximations of such phenomena are not accurate enough, the direct treatment of nonlinear models becomes inevitable. As long as the inclusion of combinatorial and/or stochastic effects is avoided, large-scale, nonlinear optimization problems arise from the mathematical modeling. These often can be solved within acceptable computing times by available algorithms and software. 'Solve', however, normally refers to the computation of local optimal points or merely points which fulfilling necessary optimality conditions. This practically means that there is a chance that better solutions exist (global optima). Yet, many practical optimization problems include both nonlinear equations and discrete decision variables, so called mixed-integer nonlinear problems (MINLP). The development of solution methods for this problem class is still in the early stages.

Stochastic Optimization

As mentioned above, on closer inspection many tasks of capacity planning and scheduling in electrical power systems and chemical production plants turn out to be optimization problems under incomplete information: At the time of the decision-making essential model data such as e.g. the inflow and outflow of energy, availabilities of plants, or purchase and sales prices are not fully known. In the past, a commonly adopted method was to replace unknown parameters by estimates (expected values). To some extent this approach works relatively well, for example for the load prediction in electrical power supply systems in the period of regulated energy markets. As a result of the opening and the decentralization of the market, however, capacity planning and scheduling now are subject to such a diversity of uncertainties that simple estimates are no longer sufficient and that the interplay between decision-making and obtaining information has to be incorporated explicitly into the optimization models. The supply of wind energy for power grids may serve as an example. If only one estimate is used in the model, larger deviations can either result in serious operational disruptions (e.g. deficits in the load coverage or overload of the transmission lines) or conservative actions have to be taken a forehand to guard against all eventualities, which can be very expensive and can weaken the competitive position. If instead a probability distribution is assumed for the wind supply and if it is taken into account which decisions have to be made before and which have to be made after the observation of the actual wind supply, the possibility arises to include this gain of information in the model and to find solutions which are less conservative but nevertheless feasible and thus competitive. Stochastic optimization offers flexible tools for taking into account uncertainties in optimization models. Figures 6 and 8 show risk neutral and risk averse two-stage stochastic optimization models which are described mathematically in more detail. Stochastic optimization approaches are becoming increasingly popular in the operation of electrical power systems and chemical production plants. Some examples are:

- earlier work dealing with power plant resource planning and scheduling under incomplete information [26, 27, 41, 42],
- work dealing with batch scheduling under uncertainty in chemical multi- product plants [2, 4, 43],
- work dealing with the tapping of oil and natural gas fields via oil rigs and drill ships [5, 45],
- first work dealing with stochastic optimization in power supply systems with distributed generation [12, 25].

4 Perspectives and Challenges

4.1 Power Supply

The change of paradigm in the power supply and distribution that was caused by the deregulation leads to new challenges for mathematics not only with respect to the planning of the extension of the grid but also to its operation. Due to the fact that in the future all considerations will be centered around the customers, their consumption profiles will be available as new optimization variables in addition to the previously used variables. Their efforts will be focussed on procuring low-priced power.

Optimized Efficient Power Procurement

In close collaboration with the trading, optimized procurement strategies have to be developed which can adequately factor in the uncertainties with regard to future prices and loads. Multi-stage stochastic optimization methods take into account the fact that the near future can be predicted with greater accuracy than events in the more distant future. By formulating realistic scenarios of the future evolution, these situations can be taken into account which greatly impact the short-term decisions. Due to the large number of parties involved the development of decentralized or hierarchical optimization methods will gain importance in order to be able to prevent the potential establishment of sub-networks with effective counter measures.

Production planning and scheduling under uncertainty. The ultimate goal is to fulfil the demand at minimum cost. At the time of planning, the demand, costs, or future disturbances are subject to uncertainties. After planning and observation of certain previously uncertain variables, further decisions can be made, e.g. those which compensate the deviations between demand and output of the previously planned production. These decisions have to meet constraints and create costs. They thus usually result from a second optimization.

The planning decision variables are called x and are to be selected from a set X. They create direct costs $c^\top x$. The compensations are called y. They are restricted to the set Y. Furthermore, there are constraints that link x and y. These constraints as well as the costs caused by y are random. We formulate the constraints as $T(\omega)x + W(\omega)y = z(\omega)$ and the costs as $q(\omega)^\top y$ with randomness marked by ω. The variables x are called first-stage decisions. They have to be made 'here and now', i.e. in-

dependently of ω, and thus are not allowed to anticipate stochastic entities which cannot be observed until later. The compensations y are called second-stage decisions ('recourse'). They can be selected adaptively depending on the planning x and the observation (or scenario) ω (Fig. 7). The classical risk neutral two-stage stochastic optimization problem can now be formalized as follows

$$\min\{c^\top x + \mathbb{E}[\Phi(x,\omega)] : x \in X\}$$

with

$$\Phi(x,\omega) := \min\{q(\omega)^\top y : T(\omega)x + W(\omega)y$$
$$= z(\omega),\ y \in Y\}.$$

For all feasible plans $x \in X$, the sum of the direct costs $c^\top x$ and the expected value of future costs that are caused by x is minimized. For the calculation of the latter it is assumed that the best possible adaptive decisions $y = y(x,\omega)$ are made.

Fig. 6 Risk neutral two-stage stochastic optimization model

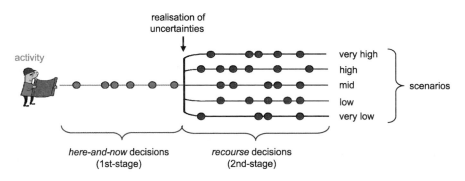

Fig. 7 Two-stage stochastic optimization

Risk Optimization

For the area of asset management new risk-based methods have to be developed. In order to implement these, however, a reliable and comprehensive data base has to be made available. The respective data have to be obtained in a complex but nec-

Consider the random optimization problem

$$\text{“min”}\{c^\top x + q(\omega)^\top y : T(\omega)x + W(\omega)y = z(\omega),\ x \in X,\ y \in Y\}$$

together with the stipulation that (in the first stage) x has to be selected, then the realization of the random variable $\xi(\omega) := (q, T, W, z)(\omega)$ becomes known, and finally a decision about $y = y(x, \omega)$ can be made (for the second stage). It is assumed that the probability distribution of ξ is known and—for the situation considered here—independent of x.

Since at the time when x is chosen the constraints of the problem are not fully known, the above 'minimization' is not yet well-defined. We are going to look at the problem from the following angle:

$$\text{“min”}_x \{ c^\top x + \underbrace{\min_y \{q(\omega)^\top y : W(\omega)y = z(\omega) - T(\omega)x,\ y \in Y\}}_{\Phi(x,\omega)} : x \in X \}$$

$$= \text{“min”}_x \{ \underbrace{c^\top x + \Phi(x, \omega)}_{f(x,\omega)} : x \in X \}$$

$$= \text{“min”}_{x \subset X}\ f(x, \omega)$$

The choice of an optimal x therefore corresponds to picking a member from the family of random variables $(f(x, \omega))_{x \in X}$ where x is considered an index. Statistical parameters can be used to evaluate the random variables with the aim of choosing the 'smallest' one. For the risk neutral case this is the expected value $\mathbb{E}$, when taking into account the risk aversion it is a weighted sum $\mathbb{E} + \rho \cdot \mathcal{R}$ consisting of the expected value and a risk measure $\mathcal{R}$ with which different concepts of 'risk' can be formalized (see Fig. 9).

Thus we can obtain (well-defined!) mean-risk optimization problems of the type

$$\min\{\underbrace{(\mathbb{E} + \rho \cdot \mathcal{R})[f(x, \omega)]}_{Q(x)} : x \in X\}.$$

A non-anticipative first-stage decision x is determined which minimizes the weighted sum of the expected value and risk of the overall costs $f(x, \omega)$. Here it has to be kept in mind that x has to be chosen without knowledge of $\xi(\omega)$ and that after the selection of x and the observation of $\xi(\omega)$ the remaining decisions y are chosen best in compliance with the now fully known constraints.

Fig. 8 Abstract approach to two-stage stochastic optimization with risk aversion

essary process in order to be able to use new optimization methods which factor in limitations of the budget. These methods have to be developed in close collaboration with the grid regulator since the grid operator's frame of action is becoming more and more restricted due to the regulation by incentives. Especially the correct formulation of a quality component for the regulation by incentives is going to play an important role when formulating the target function for optimal asset management. As bounds of the uncertainties can be stated, methods for 'unknown but bounded' uncertainties will have to be refined in order to be able to derive adequate solutions [38].

Optimized Operation Under Uncertainty

From an operational point of view the grid operator is required to load his grid to its technical limits. For this reason, more attention will have to be paid to the stability

For a random variable $X(\omega)$ and a cost level η the *expected excess* $EE_\eta(X) = \mathbb{E}[\max\{X(\omega) - \eta, 0\}]$ shows by how much the value of $X(\omega)$ exceeds the level η on average:

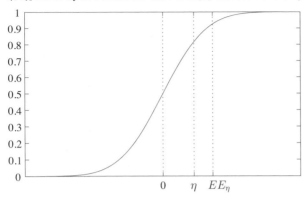

For a random variable $X(\omega)$ and a confidence level α the *Value-at-Risk* $\mathrm{VaR}_\alpha(X) = \min\{t : \mathbb{P}[X(\omega) \leq t] \geq \alpha\}$ is the most favorable of the $(1 - \alpha) \cdot 100\%$ most unfavorable realizations of $X(\omega)$. The *Conditional Value-at-Risk* $\mathrm{CVaR}_\alpha(X) = \mathbb{E}[X(\omega)|X(\omega) \geq \mathrm{VaR}_\alpha(X)]$[a] is the mean value of the $(1 - \alpha) \cdot 100\%$ most unfavorable realizations of $X(\omega)$:

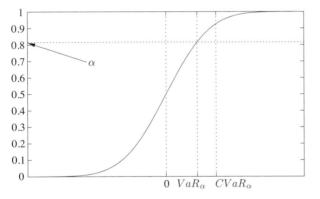

[a]For discretely distributed $X(\omega)$ this expression has to be modified.

Fig. 9 Examples for risk measures

analysis in the future. In doing so, both the frequency and the voltage response have to be simulated and analyzed in detail with the help of dynamic models. By using new control methods based on optimization as well as up-to-date power electronics the TSO and the DSO have to ensure that the best possible capacity of transporting is available in his grid at any time. New data acquisition systems such as wide range monitoring provide additional information which can be analyzed using signal analysis in order to determine the current state of the grid in an optimal and reliable fashion. In coordination with the market and the customers, it has to be checked frequently when and where grid-bottlenecks can occur so that appropriate, optimization-based defense measures can be initiated.

4.2 Chemical Production Processes

Problem-Specific Modeling

The mathematical modeling and the optimization-based solution of capacity planning and scheduling problems in the chemical production have made significant progress during the past 10 years. For many subproblems there are now standard model formulations and showcase solutions and also commercial products which offer solutions for the application in industrial practice. The secret of the success in applications is the problem-specific modeling of the underlying problem [32]. Many questions related to the modeling, for example which description of time (intervals, continuous, time-slots) is optimal for which problem, are still open [23]. A recent development are disjunctive formulations instead of the traditional big-M-methods which lead to more narrow bounds in the mixed-integer linear programs [21].

Dealing with Uncertainty

All planning and control problems that were described above are characterized by significant uncertainties in the underlying data. If the data are interpreted as mean values or as most probable values for the respective variables, the result is not the solution that is optimal for the expected value of the occurring cases. From a practical point of view it is furthermore essential to take into account the effects of the unavoidable deviations between model and reality (different demand profiles, failures of pieces of equipment, the varying yield, etc.) on the profitability and especially on the compliance with constraints. In a tight plan, many disruptions can cause the plan to fail, for example if a needed resource is not available for a batch that has to be handled without a waiting period. In principle, the aim should be to implement the optimization for stochastic data. Here modeling on the basis of different scenarios with associated probabilities seems to be the most realistic approach. The systematic and well-formulated generation of sets of scenarios which cover the occurring cases with as few elements as possible is currently being studied.

Man-Machine-Interaction

This problem primarily occurs where with the help of the optimization decisions have to be made about what should be produced where next and with which resources. Plant operators, dispatchers, etc. deeply distrust 'mechanical' solutions for which they do not understand how they were reached, which they wrongly or rightly consider disadvantageous, and into which their experience with regard to what will work is not incorporated. Improving the optimization as such in terms of factoring in uncertainties is going to be a major contribution towards gaining wider acceptance but will not suffice. In the future, research in planning and scheduling should aim at combining mathematical optimization with human overview and experience.

This applies not only the communication between algorithm and operator, for example by a good depiction of alternatives, but also the algorithm itself. If the user can affect the course of the optimization for example by providing preferences or by fixing selected important decisions, he or she is more likely to accept the results in terms of the many individual decisions and can thus be relieved from extensive detailed planning without having to fear to lose his or her task and status.

4.3 Mathematical Methods

The appropriate and efficient handling of incomplete information is becoming more and more the central challenge for capacity planning and scheduling of electrical power systems and chemical production plants. Mathematics is making first important contributions to respond to these challenges. Yet, the potential of the existing mathematical insight has by far not been exhausted. At the same time practical needs demand further mathematical analysis which in part has been initiated already.

Adequate Modeling of the Random Supply

Often the initially lacking model information is not unveiled all at once as in the two-stage model shown in Fig. 8 but step by step. This leads to multi-stage extensions of the mentioned model which are far more complex and, if integer variables are involved, are understood theoretically only to a very small extent. The gain in information that underlies the model and that mirrors the stochastic data process is also more complex than in the two-stage case. It is usually incorporated into the model in discretized form, as a scenario tree. The optimal sizing of these trees (not too big since this would cause algorithmic problems; not too small since then the randomness would not be included sufficiently) is often done empirically. Thanks to latest findings of mathematical research, first theory-based and numerically efficient methods are now available, e.g. [16].

Risk Management

The needs of financial markets have led to developments in risk management that optimization under uncertainty can benefit from. In current research, integration of one- and multi-periodic risk measures into stochastic optimization models is particularly significant, [28]. Special attention is paid to those risk measures which can be treated numerically, e.g. by decomposition, in large-scale decision models that are relevant in practice.

Moving Horizons

A conceptually attractive approach to capacity planning and scheduling with sequentially arriving information is stochastic optimization on moving horizons. As result of the optimization one obtains implementable immediate decisions, whose computation took into account some forecast horizon. After the arrival of new information the forecast horizon is shifted and updated, and the optimization initiated again. The implementation of this approach for problems of realistic size is a major challenge. It requires fundamental analysis of the structure and the numerics of the resulting optimization models. In addition, the coordination with upper planning levels has to be kept in mind.

Human-Algorithm-Interaction

Similar to the embedding of newly received model information, the consideration of user expertise in the course of the optimization is a critical factor for the applicability and acceptance of mathematically determined solution proposals. The mathematical foundation of such human-algorithm-interactions and their implementation offer great challenges and interesting perspectives for highly innovative interdisciplinary research projects.

Linear vs. Nonlinear Modeling

Often, nonlinear effects of resource management have to be taken into account as, for instance, electrical power flows or process dynamics in chemical engineering. The direct treatment of such nonlinearities in the typically large-scale mixed-integer optimization models for capacity planning and scheduling is usually not possible anymore. Present research therefore aims at developing mixed-integer linear approximations for such nonlinearities that offer algorithmic advantages. Specially structured large-scale optimization problems with nonlinearity, stochastics and integer requirements are getting within the reach of practical computation this way, [22, 24].

Algorithms and Software for Practical Usage

For discrete probability distributions two- or multi-stage stochastic optimization problems can be written as large-scale mixed-integer optimization problems with block structure. Even the most advanced commercial solvers for this class of problems, however, have to surrender to the sheer size of models that are derived from stochastic optimization problems. Decomposition methods, a core issue of algorithmic research, provide remedies. The initial problem is divided into subproblems which can be solved more easily. From their solutions, a solution to the overall problem can be generated, possibly only approximately, [31, 34, 44]. There are no fixed recipes for decomposition methods, however. Depending on the features of the particular problem (two- or multi-stage, risk neutral or risk averse, integer or continuous variables) different mathematical approaches have to be taken. In particular, for multi-stage and integer models lots of questions have yet to be answered. Stochastic optimization software is currently available only as research code. This applies to both decomposition methods and model reduction via customized scenario trees.

Decision dependent distributions

Those stochastic optimization models, which, so far, are understood best, almost exclusively assume that the random influences are purely exogenous. This means that the underlying probability distributions do not depend on the decisions that are to be optimized. For a number of phenomena, such as, e.g., the wind energy supply to power grids or the heat demand of a service area, this assumption is true. It is not longer sustainable, if e.g. a participant with market power acts on an electricity market, or if chemical processing times are afflicted with considerable uncertainties, or if the yield of oil fields is determined by the first drilling and thus the decision variable. Research dealing with these matters is still in its initial stages although the topic is highly relevant to practice and some existing results, e.g. in the area of stochastic scheduling [9] could serve as starting points, [1, 5, 6, 45].

Market design—Equilibrium design

The formulation of transparent rules for the emerging liberalized energy and commodity markets is of great importance for electrical power systems and chemical and metallurgical production plants. Mathematically, equilibrium concepts from game theory and their implementation in optimization models play an important role. In professional circles this is called EPECs (Equilibrium Problems with Equilibrium Constraints) and MPECs (Mathematical Programs with Equilibrium Constraints). Examples for more recent publications that are relevant for electricity markets are [17, 18]. Here, too, many questions, in particular algorithmic ones, have yet to be answered, especially when the incorporation of stochastics or integer decisions is an issue.

5 Visions and Recommended Courses of Action

Our vision is to strengthen the role of mathematics as a production factor in electrical power systems and chemical and metallurgical production plants so that the new economical and technological challenges in these areas can be met. Liberalization and globalization of energy and commodity markets as well as the necessity of sustainable and ecological management set the stage. As presented above, mathematics is making important contributions today, yet has not come even close to exhausting its potential. Realistically, we should aim at an interdisciplinary research mix which is not limited to short- and medium-term activities which are needed for immediate applications. It should also include as core elements long-term projects in application-driven basic research. Fundamental research in electrical power systems and chemical production plants today is dependent on basic mathematical research to a much larger extent than before. The following list indicates possible areas for research projects which will be necessary in the medium term:

- Optimization strategies for the operation, the extension, and the renewal of the power grids of the 21st century,
- decision support for optimizing the bi-directional flow of information and energy in innovative structures of power supply,
- problem specific modeling and efficient handling of uncertainty in capacity allocation problems in chemical and metallurgical production processes,
- man-algorithm-interaction during the optimization of production processes.

From the mathematics point of view, priority should be given to the development of methods for optimal capacity planning and scheduling which not only can include complex technical and economical constraints but are also capable of integrating incomplete information efficiently. In particular, this concerns multi-stage decision problems. For these, considerable progress in mathematics is needed, which in turn can only be achieved in close cooperation with the engineering sciences, industrial experts, and IT-developers.

This cooperation does not have to start from scratch. Based on existing networks, and aiming at their extension, the BMBF/BMWi or the DFG, jointly with industry, should initiate such activities in the short or medium term.

The DFG-Priority Program 469, Online Optimization of Large Scale Systems [10], and the successful implementation of mathematical research in the course of the Networks of Basic Research in Renewable Energy and Rational Energy Utilization [35] may serve as successful examples from the past. In both programs, the promotion of collaboration among engineers and mathematicians has pushed ahead both application driven research and research on mathematical foundations. Furthermore, lasting cooperations have been established this way. A DFG Priority Program 'Stochastic Optimization' and extended funding of related applications-oriented research initiatives by the BMBF could set the frame for a broad range of highly relevant and innovative activities.

References

1. Anderson, E.J., Philpott, A.B.: Optimal offer construction in electricity markets. Math. Oper. Res. **27**, 82–100 (2002)
2. Balasubramanian, J., Grossmann, I.: Approximation to multistage stochastic optimization in multiperiod batch plant scheduling under demand uncertainty. Ind. Eng. Chem. Res. **43**, 3695–3713 (2004)

3. Brosda, J.: Hierarchische Optimierung für ein zonenübergreifendes Netzengpass-Management. Dissertation, Fakultät für Elektrotechnik und Informationstechnik, Universität Dortmund (2004)
4. Engell, S., Märkert, A., Sand, G., Schultz, R.: Aggregated scheduling of a multiproduct batch plant by two-stage stochastic integer programming. Optim. Eng. **5**, 335–359 (2004)
5. Goel, V., Grossmann, I.E.: A stochastic programming approach to planning of offshore gas field developments under uncertainty in reserves. Comput. Chem. Eng. **28**, 1409–1429 (2004)
6. Goel, V., Grossmann, I.: A class of stochastic programs with decision dependent uncertainty. Math. Program. **108**, 355–394 (2006)
7. Gollmer, R., Nowak, M.P., Römisch, W., Schultz, R.: Unit commitment in power generation—a basic model and some extensions. Ann. Oper. Res. **96**, 167–189 (2000)
8. Göbelt, M., Kasper, T., Sürie, C.: Integrated short and midterm scheduling of chemical production processes—a case study. In: Engell, S. (ed.) Logistic Optimization of Chemical Production Processes, pp. 239–261. Wiley-VCH, Weinheim (2008)
9. Grigoriev, A., Sviridenko, M., Uetz, M.: Machine scheduling with resource dependent processing times. Math. Program. **110**, 209–228 (2007)
10. Grötschel, M., Krumke, S.O., Rambau, J. (eds.): Online Optimization of Large Scale Systems. Springer, Berlin (2001)
11. Handke, J.: Koordinierte lang- und kurzfristige Kraftwerkseinsatzplanung in thermischen Systemen mit Pumpspeicherkraftwerken. Dissertation, Universität Dortmund (1994)
12. Handschin, E., Neise, F., Neumann, H., Schultz, R.: Optimal operation of dispersed generation under uncertainty using mathematical programming. Int. J. Electr. Power Energy Syst. **28**, 618–626 (2006)
13. Harjunkoski, I., Beykirch, G., Zuber, M., Weidemann, H.J.: The process "copper": copper plant scheduling and optimization. ABB Rev. **4**, 51–54 (2005)
14. Harjunkoski, I., Fahl, M., Borchers, H.-W.: Scheduling and optimization of a copper production process. In: Engell, S. (ed.) Logistic Optimization of Chemical Production Processes, pp. 93–109. Wiley-VCH, Weinheim (2008)
15. Hauptmeier, E.: KWK-Erzeugungsanlagen in zukünftigen Verteilungsnetzen—Potential und Analysen. Dissertation, Fakultät für Elektrotechnik und Informationstechnik, Universität Dortmund (2007)
16. Heitsch, H., Römisch, W.: Scenario tree modeling for multistage stochastic programs. Math. Program. doi:10.1007/s10107-007-0197-2
17. Hobbs, B.F., Pang, J.S.: Nash-Cournot equilbria in electric power markets with piecewise linear demand functions and joint constraints. Oper. Res. **55**, 113–127 (2007)
18. Hu, X., Ralph, D.: Using EPECs to model bilevel games in restructured electricity markets with locational prices. Oper. Res. **55**, 809–827 (2007)
19. ILOG Inc., Mountain View, CA, USA
20. Jürgens, I.: Langfristoptimierung für das risikoorientierte Asset Management von elektrischen Energieversorgungssystemen. Dissertation, Fakultät für Elektrotechnik und Informationstechnik, Universität Dortmund (2007)
21. Lee, S., Grossmann, I.E.: New algorithms for generalized disjunctive programming. Comput. Chem. Eng. **24**, 2125–2141 (2000)
22. Martin, A., Möller, M., Moritz, S.: Mixed integer models for the stationary case of gas network optimization. Math. Program. **105**, 563–582 (2006)
23. Mendez, C.A., Grossmann, I.E., Harjunkoski, I.: MILP optimization models for short-term scheduling of batch processes. In: Engell, S. (ed.) Logistic Optimization of Chemical Production Processes, pp. 163–184. Wilry-VHC, Weinheim (2008)
24. Moritz, S.: A mixed integer approach for the transient case of gas network optimization. Dissertation, Fachbereich Mathematik, TU Darmstadt (2006)
25. Neumann, H.: Zweistufige stochastische Betriebsoptimierung eines virtuellen Kraftwerks. Dissertation, Fakultät für Elektrotechnik und Informationstechnik, Universität Dortmund, Shaker Verlag, Aachen (2007)
26. Nowak, M.P., Römisch, W.: Stochastic Lagrangian relaxation applied to power scheduling in a hydro-thermal system under uncertainty. Ann. Oper. Res. **100**, 251–272 (2000)

27. Pereira, M.V.F., Pinto, L.M.V.G.: Multi-stage stochastic optimization applied to energy planning. Math. Program. **52**, 359–375 (1991)
28. Pflug, G.Ch., Römisch, W.: Modeling. Measuring and Managing Risk. World Scientific, Singapore (2007)
29. Plapp, C., Surholt, D., Syring, D.: Planning large supply chain scenarios with "quant-based combinatorial optimization". In: Engell, S. (ed.) Logistic Optimization of Chemical Production Processes, pp. 59–91. Wiley-VCH, Weinheim (2008)
30. Rehtanz, C.: Autonomous Systems and Intelligent Agents in Power Systems and Control and Operation. Springer, Berlin (2003)
31. Ruszczyński, A., Shapiro, A. (eds.): Handbooks in Operations Research and Management Science. Stochastic Programming, vol. 10. Elsevier, Amsterdam (2003)
32. Sand, G.: Engineered Mixed-integer programming in chemical batch scheduling. In: Engell, S. (ed.) Logistic Optimization of Chemical Production Processes, pp. 137–161. Wiley-VHC, Weinheim (2008)
33. Schnurr, N.: Potential multifunktionaler FACTS-Geräte zur Erhöhung von Übertragungskapazität und Mittelzeitstabilität im Europäischen Verbundnetz. Dissertation, Universität Dortmund, VDE Verlag (2004)
34. Schultz, R.: Stochastic programming with integer variables. Math. Program. **97**, 285–309 (2003)
35. Schultz, R., Wagner, H.-J. (eds.): Innovative Modellierung und Optimierung von Energiesystemen. LIT Verlag, Münster (2008)
36. Schulz, W.: Strategien zur effizienten Integration der Windenergie in den deutschen Elektrizitätsmarkt. Dissertation, Fakultät für Elektrotechnik und Informationstechnik, Universität Dortmund (2007)
37. Schulz, M., Spieckermann, S.: Logistics simulation in the chemical industry. In: Engell, S. (ed.) Logistic Optimization of Chemical Production Processes, pp. 21–36. Wiley-VHC, Weinheim (2008)
38. Schweppe, F.C.: Uncertain Dynamic Systems. Prentice-Hall, New York (1973)
39. Spangardt, G.: Mittelfristige risikoorientierte Optimierung von Strombezugsportfolios. Dissertation, Universität Dortmund, Fraunhofer IRB Verlag (2003)
40. Stobbe, M.: Supply chain and supply chain management. In: Engell, S. (ed.) Logistic Optimization of Chemical Production Processes, pp. 3–18. Wiley-VHC, Weinheim (2008)
41. Takriti, S., Birge, J.R., Long, E.: A stochastic model for the unit commitment problem. IEEE Trans. Power Syst. **11**, 1497–1508 (1996)
42. Takriti, S., Krasenbrink, B., Wu, L.S.-Y.: Incorporating fuel constraints and electricity spot prices into the stochastic unit commitment problem. Oper. Res. **48**, 268–280 (2000)
43. Till, J.: New hybrid evolutionary algorithms for chemical batch scheduling under uncertainty. Dissertation, Fachbereich Bio- und Chemieingenieurwesen, Universität Dortmund, Shaker Verlag, Aachen (2007)
44. Till, J., Sand, G., Urselmann, M., Engell, S.: A hybrid evolutionary algorithm for solving two-stage stochastic integer programs in chemical batch scheduling. Comput. Chem. Eng. **31**, 630–647 (2007)
45. Van den Heever, S.A., Grossmann, I.E.: An iterative aggregation/disaggregation approach for the solution of a mixed integer nonlinear oilfield infrastructure planning model. Ind. Eng. Chem. Res. **39**, 1955–1971 (2000)

Simulation-Based Optimization in Structural Engineering—New Concepts from Computer Science

Dietrich Hartmann, Matthias Baitsch,
and Van Vinh Nguyen

1 Executive Summary

Engineers have always been keen to develop technical systems in an "optimal" and efficient manner, considering the amount of attention they give to thorough planning, sound design and meticulous production. In particular, the processes themselves by which these structures are designed and built are also expected to be reliable, highly efficient, well-planned and, therefore, optimal.

By using powerful computer models and applicable software as a communication basis, researchers from the fields of mathematics, computer science and engineering have collaborated time and time again to not only successfully solve current engineering problems, but also to tackle completely new, cutting-edge research problems.

Recently, the introduction of holistic aspects into the optimization process has become an important research goal. A holistic approach within optimization is desirable for various reasons, in particular because it can incorporate economic and environmental factors such as reduced energy consumption or minimal construction costs. Ultimately, optimal technical solutions will contribute to the standard of welfare or the prosperity of all people. This applies in particular to the technical systems

D. Hartmann (✉)
Ruhr-Universität Bochum, Lehrstuhl für Ingenieurinformatik im Bauwesen,
44780 Bochum, Germany
e-mail: hartus@inf.bi.rub.de

M. Baitsch
Ruhr-Universität Bochum, Lehrstuhl für Informatik im Bauwesen, Gebäude IA 6/151,
Universitätsstraße 150, 44780 Bochum, Germany
e-mail: matthias.baitsch@rub.de

V.V. Nguyen
Lehrstuhl für Informatik im Bauwesen, Ruhr-Universität Bochum, IA 6/44, Universitätstr. 150,
44780 Bochum, Germany
e-mail: vinh.nguyen@rub.de

M. Grötschel et al. (eds.), *Production Factor Mathematics*,
DOI 10.1007/978-3-642-11248-5_16, © Springer-Verlag Berlin Heidelberg 2010

and processes of structural engineering and urban development, encompassing the fields of civil, architectural and environmental engineering.

A holistic approach encompassing a large portion of relevant aspects of a complex system was only possible since the development of high performance computers that provide the needed amount of large computational resources. Also, integrating such computers into local or wide area networks to create a platform of distributed or grid computing has increased the computational resources available for "grand challenge" projects to an even greater extent. Also, continued miniaturization and sophisticated manufacturing techniques of computer hardware in the past have also contributed to the computational resources available today that allow such grand challenge problems to be tackled.

Besides the approach of theory and experiment, numerical simulation methods have become a third paradigm in the solution of realistic computer models. Numerical methods based on sound mathematical models can solve existing engineering problems otherwise not tractable and, in fact, can validate solutions or test their applicability in the design stage and thus possibly eliminate the need for costly experiments.

In particular, simulation based engineering allows a designer to evaluate consequences, impacts, hazards or other possible disadvantages when a system or process is being modified, both quantitatively and qualitatively. Thus, numerical simulation is the driving force behind the successful optimization of engineering systems and engineering processes. Although simulation based optimization is a promising technique, it must also be regarded as a technological challenge, because the underlying numerical simulation components must often be executed hundreds or even thousands of times to find a (hopefully global) optimum value. To solve real world problems, therefore, requires a close interaction between the optimization methods and the simulation tasks.

The computational difficulties are further increased because simulation-based optimization often leads to so-called non-standard optimization problems that are characterized by unexpected non-linear behavior, a large number of dimensions, many local optima and non-smooth feasible regions. Often, the given "objective criterions" and "constraints" required to accurately describe an engineering optimization problem are not closed functions in the classical mathematical sense, rather, they are given as general algorithms or software components. If, additionally, some optimization parameters are time-dependant or of stochastic nature, the demand for computational resources is increased even more. The only viable solution to this dilemma is to develop new types of mathematical or computer science approaches which are capable of handling such complex problems. This article will show some possible approaches from the field of structural engineering based on the experiences gained by our own research and give some proposals for similar tasks in other areas.

2 Success Stories

Several research projects that have been successfully completed in recent years, including projects promoted by the German Research Foundation (DFG), DFG graduate schools, DFG research groups and DFG priority programs and collaborative research centers. They are evidence that cooperation between researchers in mathematics and engineering can be very fruitful. Because realistic problems can't be solved today without the use of computer systems or computing techniques, computer scientists along with mathematicians and engineers also contribute to the success of research problems.

Structural Optimization of Large Structures

The optimal design of large structural systems (structural optimization) with respect to properties such as weight, costs, physical behavior or reliability, or defining constraints with respect to allowable stresses, deformations or other design requirements usually leads to development of correspondingly large software models that require high-performance, iterative optimization methods. Because of the large number of dimensions and non-linearity of the such optimization problems, the optimization process places a high demand on the amount of computing resources needed, especially when the computer hardware is "only" a sequentially operating von Neumann architecture as found in today's typical workstations and PCs. This disadvantage must be overcome, since computer-based optimization promises invaluable advantages: The labor-intensive dimensioning of structural systems, which is of course done today by using computer based methods such as the finite element methods, and the design of structures using CAD software can carried out in a more cost effective manner using optimization strategies because the degree of automation can be increased. For this reason, it is important to reduce the demand on computing and storage resources as much as possible in the optimization of large structural systems. This can be achieved by a consistent parallelization of structural optimization problems and by decomposing and partitioning structures into a number of smaller (and less complex) substructures that can be distributed to solvers on a parallel computer system.

Initial pioneering approaches have been developed in the DFG Research Group "Optimization of Large Systems" [12]: The decomposition problem that arises, namely if and how to create overlapping and non-overlapping subproblems, and also the implementation of the corresponding parallel algorithms could only be solved by the close cooperation of researchers from mathematics, computer science and structural engineering. As an example, a stadium roof was decomposed into four non-overlapping substructures (see Fig. 1), each of which could be solved on individual processors for various load cases.

Based on decomposition, partitioning and parallelization structural designs were discovered that could not have been identified without the use of parallel optimization techniques. Figure 2 shows the result of a cooperative structural optimization for a quarter of the stadium roof, including a plot of the objective function (weight) vs. the number of iteration steps.

Discrete and Discrete-Continuous Structural Optimization

Discrete and discrete-continuous structural optimization is characterized by the inclusion of structural parameters (optimization variables) that can only have discrete values, in contrast to optimization variables that are usually continuous in

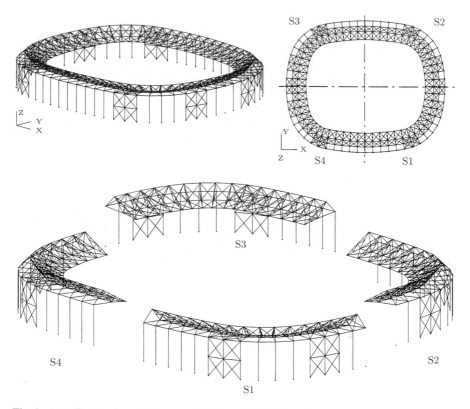

Fig. 1 A stadium roof modeled as a spatial truss with 1660 members

nature. A typical example are the geometric dimensions of standard cross-section profiles (HE-A, HEB, HE-IPB, etc.), which can only have fixed values. For these categories of discrete optimization problems numerous research results have been accomplished that also have great practical value. This again shows that the close integration of mathematics and engineering sciences can solve challenging problems in structural engineering that would otherwise not be possible.

Many of the popular optimization methods used in conventional numerical optimization, in particular those based on gradients or Newton methods, are unable to solve design problems containing discrete optimization variables. Such conventional methods are either only capable of providing an initial rough approximation or they require a substantial reworking, possibly including a remodelling of the underlying optimization problem.

Backed by mathematical theory regarding aspects of uniqueness, stability and divergence behavior, so-called evolutionary algorithms (EA [1]) and other specifically tailored algorithms have been successfully applied for some time in the investigation of discrete-continuous structural optimization problems. In our own research, especially evolutionary strategies (ES [5]) have proved to be highly successful for structural optimization problems.

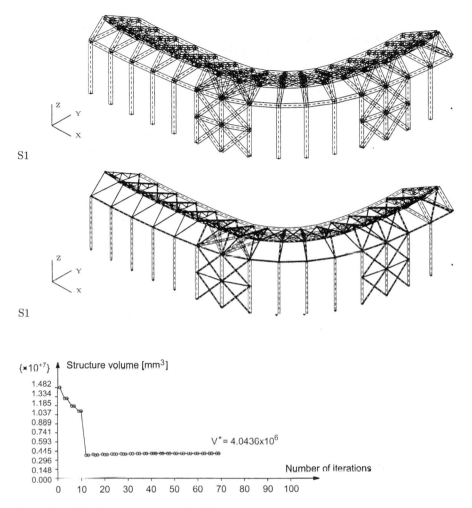

Fig. 2 Spatial substructure model with 415 members

Evolution Strategies (ES) are robust and well manageable stochastic search methods based on the concept of biological evolution, i.e. the collective learning process of a set of interacting individuals (the population). Biological populations show evidence of "evolving" from generation to generation based on the concepts of mutation, recombination and selection, each of which can be considered to be efficient mechanisms within an optimization process. In fact, biological evolution can be considered as a form of self-optimization within the population. The consequent transfer of such biological optimization principles (mutation, recombination, selection, lethality, etc.) to mathematical and engineering research fields opens up entirely new approaches to problems that could not be solved previously by classical mathematical methods alone. The principles of mutation and recombination provide the large spectrum of variety necessary for finding possible solutions in engineering design problems. Then, the principle of selection can guide the search process towards a (global) optimum, without the need for first or second order derivatives. This guided search can be viewed as a "diffusion" of candidate individuals towards an optimum value.

Another approach to structural optimization, besides coupling structural analysis models with numerical optimization methods using ES, is the concept of object-oriented modelling, analysis and design, which has its historical roots in artificial intelligence and is further proof of the synergetic effects that emerge when research in mathematics and engineering combines. Founded on a concise and precise descriptive calculus (Unified Modeling Language, UML) with globally recognized symbols and notation, highly complex numerical and non-numerical problems from engineering as well as object oriented principles (principles of modularity and inheritance, information hiding, etc.) can be modelled rigorously and understandably. The semantics of objects and classes, as defined by properties, behavior, associations and relationships, reflect the terms and concepts as they appear in the real world. Thus, the semantic loss of information from the initial modelling of a "real world problem" to its final implementation as computer software code can be kept to a minimum, which is often much less than that of other software development methods.

Figure 3 depicts some results of discrete and discrete-continuous structural optimization that was carried out at our Institute (see, for example, [10]). It must be noted that the results from this example are completely different from previous, similar examples where discrete optimization variables were either approximated by continuous variables or completely ignored.

Lifetime-Oriented Design Concepts

Anyone who drives his car over a bridge or, for example, walks through a factory building with cranes moving overhead obviously assumes that "nothing will happen". Even in old buildings we presume that the stability, durability and reliability of the structure is completely adequate because structural designers and construction engineers somehow allowed for the aging process. Computing the statics and dynamics, the service strength and service life as well as the reliability and stochastic effects of structural systems has undoubtedly reached a high level of technological and scientific maturity within the last few decades. However, computer-based design and management systems are needed to accurately simulate the entire service life of a structural system, especially if unexpected damages can occur during its life time. In particular, such a simulation can then form the basis for an optimization system which can be used to find an optimal design for large, complex structural systems.

Finding appropriate lifetime-oriented designs allowing for randomly induced damages (deteriorations) proves to be a very difficult problem. From a mathematical point of view, these design problem can be transformed into equivalent stochastic optimization problems with a range of scales of time and space. This results in a

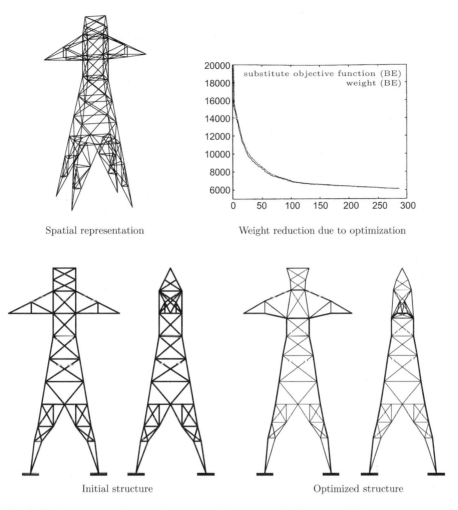

Fig. 3 Optimization of a free standing transmission tower with discrete variables

type of a new multi-level, multi-scale problem that is highly non-linear, both in a mechanical as well as in a mathematical sense. Such complex design problems can only be solved by using appropriate simulation-based methods carried out on high performance computers.

Within the scope of the Bochumer Collaborative Research Center SFB 398 "Lifetime-oriented design concepts" [18] new approaches from mathematics, computer science and structural mechanics have been applied to these special type of problems. In particular, design concepts for industrial buildings (see Fig. 4) and steel tied arch bridges (see Fig. 5) have been developed in the last 12 years, the duration of the DFG research project.

Hanger connection plates, the connecting element between the hanger and the main suspension beam of a tied arch bridge, are highly prone to deterioration. In particular, lateral oscillations induced by wind lead to a so-called lock-in effect, resulting in cracks to the

$$\min \left\{ A(X_1, X_2) \;\middle|\; \begin{array}{c} 0,1 \le X_1 \le 1,0 \\ 0,1 \le X_2 \le 0,3 \\ g\big(\mathbf{X}, \mathbf{Y}, \mathbf{Z}(t)\big) = P_f\big(\mathbf{X}, \mathbf{Y}, \mathbf{Z}(t)\big) - P_{zul} \le 0 \end{array} \right\}$$

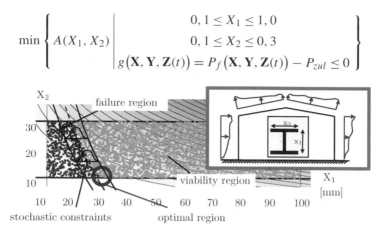

Fig. 4 Lifetime-oriented design optimization of a structural frame

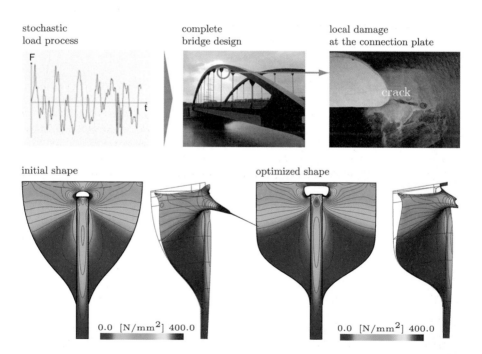

Fig. 5 Lifetime-oriented design optimization of a hanger connection plate of a tied arch bridge: *Above*: Cause of damage to the hanger connection plate: a stochastic load process (i.e., wind) induces high frequency oscillations to the round hanger bar, resulting in cracks in the hanger connection plate. *Below*: Optimization of the hanger connection: Shown are the initial and final shape of the hanger connection plate in a top view and the 3-dimensional graph of the stress state. The reduction of the stress peak compared to the initial shape can be clearly seen

plates. A major result of the Bochumer SFB 398 research project was to model such tied arch bridges by a multi-level simulation model and to find an optimal shape of the hanger connection plates that can considerably increase their lifetime expectancy (see Fig. 5). The calculation of the stresses in the hanger connection plates, which is the basis for estimating lifetime expectancy, was carried out using the finite element method (FEM). The optimization changed the shape of the hanger connection plates so that the calculated increase in lifetime expectancy increased by an order of magnitude [8]).

Shape Optimization Using the *p*-Version of FEM

In any kind of simulation based optimization there is an underlying computer model which allow one to determine the effects caused by modifications of the optimization variables as prescribed by the optimization method. For most simulation problems this computer model is based on the finite element method (FEM), which calculates the values of physical quantities such as deformations or a temperature distribution by approximating functions defined piecewise on discrete regions, the so-called finite elements. In particular, in the shape optimization of continuous structures, the so-called *p*-version of FEM has proved to be very successful. In the *p*-version, the entire structure is split into relatively large elements and the partitioning does not need to be modified during the optimization process. This is in contrast to other versions of FEM, which are based on very small elements and require frequent remeshing by optimization (see Fig. 6).

An example from [3] is shown in Fig. 7: The shape of a dam is to be optimized with respect to the amount of material used, where the calculated stresses in the reinforced concrete must not exceed the allowable limit. The optimized result requires about 30% less material than the initial design. The model of the structural system is based on the *p*-version of the FEM and includes various techniques from Computer Aided Geometric Design (CAGD). In order to reduce computing time, the optimization was carried out on a cluster computer system consisting of 56 dual-core compute nodes connected with a high capacity network. The coupling of the FEM software with CAGD as well as the subsequent parallelization was greatly eased

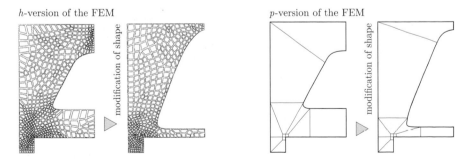

Fig. 6 Optimization of continuous structural systems. Comparison of the *h*-version and the *p*-version of the FEM for a shape optimization problem: The *h*-version requires the generation of a completely new mesh after each modification of the geometry. In contrast, the mesh in the *p*-version does not need to be changed at all

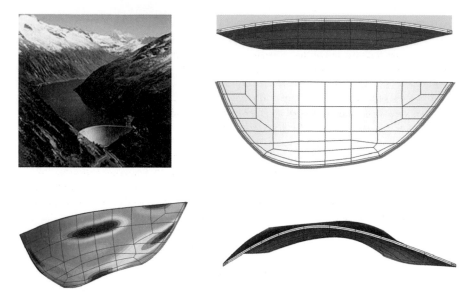

Fig. 7 Optimization of an arch dam: *Above*: Arch dam of the Zillergründl reservoir. Shown to the right is a finite element model of a comparable structure. *Below*: Optimization of the arch dam: Shown left is the simulated material stress as the basis of a shape optimization and shown right is the optimized shape of the structural system

by the use of object-oriented software methods. In fact, only by using object-oriented design was it possible to develop the software within the alloted time span.

Robust Optimization of Slender Structures Subjected to Compressive Stress

In structural optimization, the attributes of a physical structural system are mapped to an artificial mathematical (computer) model. Therefore, the optimization process can only find "optimal" designs with respect to the underlying mathematical model. The solutions themselves must be considered as "special" solutions produced by the optimization strategy. If the structural system that is later erected at the site does not completely adhere to the exact specifications (due to imperfections caused by faulty assembly, material errors, etc.), then the actual load bearing behavior will depart from the calculated results significantly [19]. Because the optimum design is usually located at the boundary of the feasible region, implying that the design is just about allowable, any deviation from the "optimal" value, however small, will violate the safety of the structural system. This is particularly true for slender steel structures subjected to compressive stresses, where even small deviations from the projected geometry can lead to large changes in the behavioral response. For the design of such structures that are sensitive to imperfections a systematic approach is required that incorporates this sensitivity to imperfections directly into the design goal.

In research carried out at our department [2] we therefore included possible geometric imperfections as part of the structural optimization model, which is itself composed of four interacting sub-models: (1) In the geometry model the overall design is given by basic geometric elements using NURBS to efficiently describe the shape of an element using only a few key nodes. (2) The probabilistic attributes of the geometric imperfections are realistically described by a corresponding imperfection model based on so-called random fields. The covariances of random deviations defined by the random field are evaluated at some given discrete points and assembled into a covariance matrix. After performing a spectral decomposition of the covariance matrix, the imperfections can be written as a linear combination of deterministic basis vectors (the eigenvectors of the covariance matrix) with random amplitudes. The random amplitudes themselves are independent Gaussian variables. (3) The structural analysis carried out within the optimization process is done with the help of the finite element method, where stability analysis requires a geometrically non-linear approach. (4) An uncertainty model is needed to capture the inherent uncertainty induced by the geometric imperfections. Because it is not possible to rigorously state how geometric imperfections influence the probabilistic properties of complex structural systems, a substitute convex model of uncertainty is used. More specifically, a convex hull containing the all possible imperfect designs is defined such that the norm of the vector of imperfection lies within a given quantile value (95%, for instance). With this requirement, it is possible to find the most unfavorable (worst case) vector of imperfection with an additional optimization task ("anti-optimization"). The entire optimization process therefore consists of two nested loops: An outer optimization loop to optimize the structural system as a whole and an inner optimization loop to find a worst case imperfection vector needed to calculate behavioral response. The example that follows shows how a arch frame consisting of individual members can be optimized with this model (see Fig. 8).

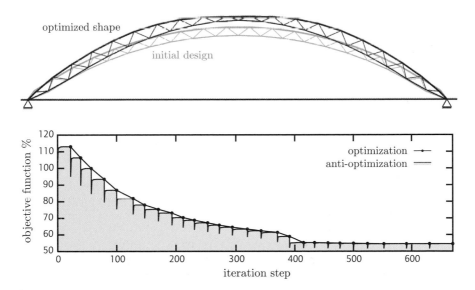

Fig. 8 Robust optimization of a slender structure subjected to compressive stress: *Above*: Comparison of the initial and final shape of the arched girder. *Below*: Optimization of the system stiffness: The most unfavorable imperfection shape for each optimization step is colored *blue*

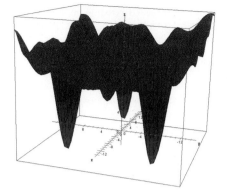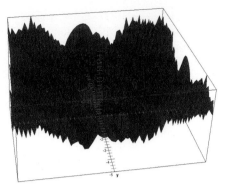

Fig. 9 An optimization landscape with multiple maxima

3 Optimization of Non-Standard Problems with Multi-Agent Systems (MAS): The Status Quo

Finding a good solution in a non-standard optimization problem based on a numerical simulation can be a very challenging task for an expert because, as mentioned above, the complex and irregular topology of the search space implies that properties such as non-linearity, multi-modality or discontinuities are to be expected. If the optimization problem is also multi-leveled and multi-scaled with respect to time and space, then the expert must rely on optimization support based on knowledge and previous experience, even if generally applicable optimization methods, such as evolutionary strategies (EA), are used. Further, results of numerical simulation are often confounded with random noise that must also be considered. Consequently, the search region defined by the constraints and the objective function has to be viewed as a jagged landscape containing many local maxima and minima. This can be seen in Fig. 9, which depicts the landscape of a typical two-dimensional problem as found in realistic problems. How is it possible to find the global optimum in such a craggy region? What happens when the dimension of the search space is increased from two to, say, ten as is often the case in many engineering problems? The answer is: high-performance, highly efficient simulation-based optimization strategies must be developed. In fact, not just one single good optimization strategy is sufficient, but a suite of competing and interacting strategies is needed, each of which is individually tailored to solve a specified type of problem. The development of such high innovative optimization strategies is a worthwhile task for mathematicians, engineers and computer scientists alike.

3.1 Agent Systems

The paradigm of multi-agent systems (MAS), only recently being applied to engineering research fields, can provide the theoretical framework needed to solve such

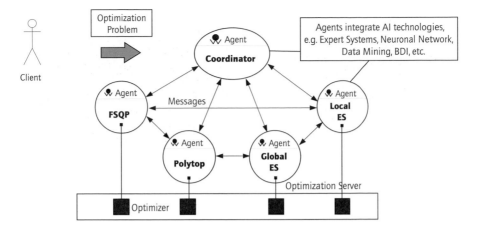

Fig. 10 The network of relationships within a multi-agent system

non-standard optimization problems discussed above [4, 17]. Research carried out using multi-agent systems have shown impressively the advantages of this technology, in particular if unforeseen or unexpected events arise during the solution processes that require a very flexible response. Also, if a problem can only be solved as a team effort, then multi-agent systems can coordinate the communication between the collaborating members. The collaboration itself can be heterogeneous and time variant, requiring only a very loose organizational structure for the team members. This type of situation principally applies also to the solution of non-standard design optimization problems based on numerical simulation. In this case, the "team members" are now the individual optimization strategies, competing and cooperating with one another to find optimal solutions (see Fig. 10). The cognitively motivated approach found in multi-agent systems can therefore define a workable basis to overcome the many hurdles that arise in the search for the global optimum value.

3.2 Strategy Networks

It has been demonstrated that the coupling of stochastic, globally operating optimization strategies (such as evolution strategies [5], genetic algorithms [9], particle swarm methods [6] or ant colony concepts [14]) with deterministic, locally operating optimization strategies can solve multi-modal, non-linear optimization problems much more efficiently than any individual method alone, especially if the objective function or constraints are defined algorithmically (for example, in terms of a finite element code). It remains, however, an open question to decide when to use which optimization strategy which is sufficiently efficient and robust in the current optimization context. Optimization methods which perform satisfactorily in a given situation can become completely inefficient or even instable if some parameters of the design model are slightly different. Because of the large amount of computational resources needed in the evaluation of simulation based optimization problems (remember: these problems use non-linear finite element methods to analyze large structural systems), it is completely infeasible to compare various optimization strategies without the help of computer software.

Applying an agent-based network of optimization strategies, however, a systematic evaluation of optimization strategies becomes possible and is preferable to an assortment of strategies based on an intuitive selection. Because the autonomous and emergent behavior that characterizes multi-agent systems can be defined in a formal manner, such knowledge of a selection process can be coded into the components of a multi-agent system, replacing the traditional tedious trial-and-error method. This is the added value of multi-agent systems for the design optimization process, especially if multi-level, multi-scale problems with exponential complexity are involved.

Intelligent, distributed optimization components, coded as agents and containing either knowledge and rule bases or so-called artificial neural networks (ANN [16]), can greatly assist optimization experts in the solution complex non-standard design optimization problems. Because of the inherent emergence of agent systems, one can assume (or at least hope) that a cooperating network of optimization strategies will provide much better results than purely the sum of each individual strategy.

The Four Pillar Concept

In a research project supported by the German Research Foundation (DFG), the classical "three pillar" concept of design optimization consisting of simulation, optimization models and strategies [7] is being extended with a fourth pillar, the multi-agent system pillar (see Fig. 11). This fourth pillar represents the "artificial intelligence" component of the optimization concept and couples the structural optimization model with the available optimization strategies. In particular, it is possible to code a generally valid solution methodology for simulation-based design optimization, independent of a particular application.

Analysis Tools, Visualization and Steering

In the conventional approach to simulation-based optimization without the use of agent-based systems, typical subtasks such as pre- and post-processing, numerical simulation, etc. are usually carried out with the help of individual computer programs and tools not designed as integrated components. Thus, for example, the exchange of data between such components is often file based, meaning that the output results generated by one component must be specifically formatted to the specifications of the following component. In the worst case, this conversion must be done by hand, requiring a lot of time and practically eliminating the possibility of an automatic, iterative computing cycle.

One alternative in the elimination of these disadvantages in the traditional approach is the use of "computational steering" which supervises a program during

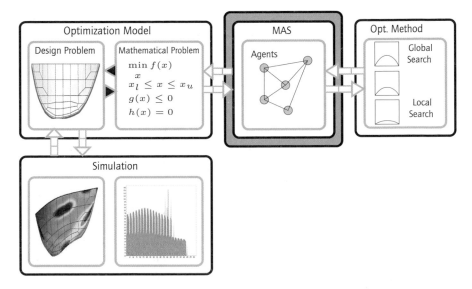

Fig. 11 The four pillar concept with multi-agent systems (MAS). The multi-agent system (MAS) represents the "artificial intelligence" component of the simulation-based optimization concept. With the help of this concept, engineering problems based on computer simulation, such as traffic simulation, explosive demolition, system identification or optimal sensor placement can be solved in a general and intelligent manner

its execution and supports the interaction between the user and the program. Figure 12 shows the software architecture of JASOE, the Java Agent Simulation and Optimization Environment developed as an experimental simulation platform at our institute. This "computational steering" application provides many control and interaction options within the optimization process, allowing the user to "experiment" with different types of optimization strategies and software agents to try and find a suitable global optimum design.

A glance at the graphical user interface of the computational steering platform (Fig. 13) shows that many of the operational and steering components, tailored to be used by optimization experts, have been implemented. In particular, the emphasis was placed on graphically visualizing the underlying structural system and the details of the optimization process. As it is well known, graphical displays have always had a strong appeal to users, because they allow a clear and concise description of complex systems and processes. This is also valid for the optimization with MAS, where many different components and subtasks need to be coordinated and fine tuned. Real time information on the progression of an optimization process in the form of diagrams, plots, message and event reports, etc. can give an optimization expert detailed insight into the workings of a MAS. Ingo Rechenberg, one of the fathers of the evolution strategy, also pleads in his book "Evolutionsstrategie'94" [15] for a systematically driven interaction between human experts and optimization processes.

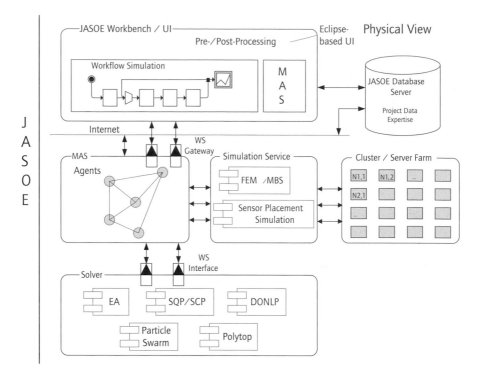

Fig. 12 Architecture of the agent-based optimization system JASOE

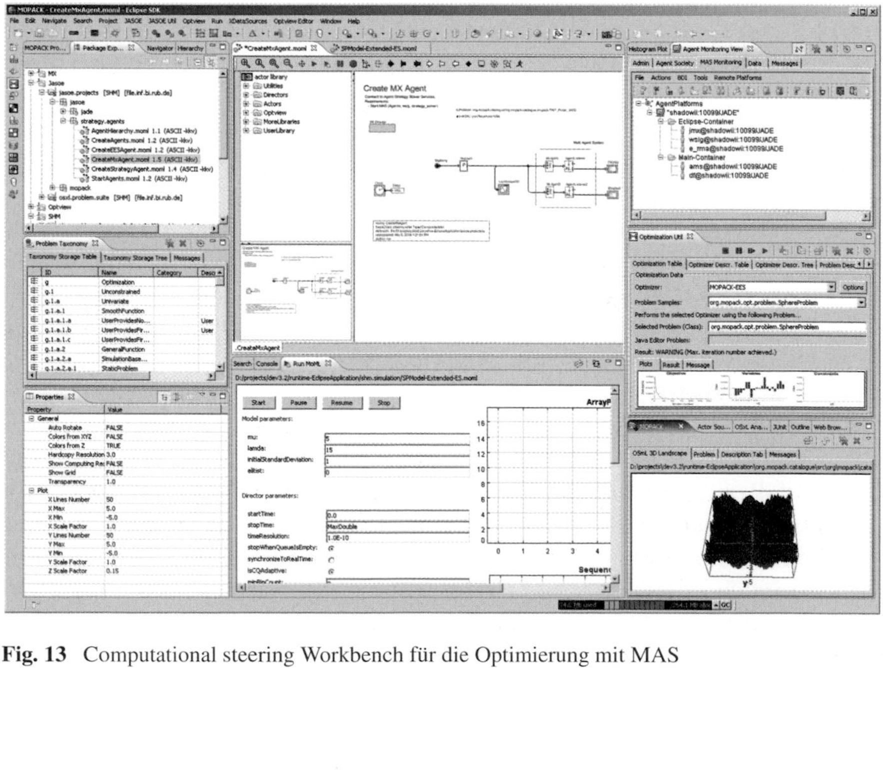

Fig. 13 Computational steering Workbench für die Optimierung mit MAS

4 Strong Points, Weak Points and Challenges

Even though the previous chapter has indicated how the coupling of engineering know-how with mathematics and computer science allows the solution of complex engineering problems previously thought to be intractable, there is still no reason to be euphoric. A solid and unemotional analysis of the strengths resulting from the synergetic effects of collaboration also shows the weak points that arise. The elimination of these weaknesses still present major challenges in complex structural optimization.

4.1 Practical Aspects

Large optimization problems typically found in engineering practice have a type of complexity that mathematicians can only grasp with the help of optimization experts having an engineering background. This is due in part to the type of training mathematicians receive and correspondingly to their way of thinking about abstract optimization problems. This is not a judgement for or against any type of science, rather it expresses the dire need for team work and mutual understanding.

The complexity of problems found in structural and environmental engineering, which are the focus in this article, is mainly due to the fact that the solution of real world problems requires that many different participants, often with conflicting interests and competing approaches, must work together for various periods of time. For example, a structural engineer or a design engineer usually has a different idea what structural optimization should be compared to the concepts of a mechanical engineer or mathematician with a numerical optimization background. In turn, an architect or an engineer concerned with building physics or foundation construction will probably have a completely different focus on relevant optimization objectives.

For mechanical engineers and mathematicians it usually suffices to map the objective function (as defined in the mathematical optimization model) to a simple computable quantity such as total mass or total volume of a structure. Also, the constraints are usually given as closed, analytical expressions involving well-defined and readily computed quantities such as stresses and displacements. By contrast, the non-linear structural behavior should be simulated as realistically as possible by using advanced non-linear finite element programs, such as those based on the p-version discussed above.

Design engineers, on the other hand, want to code their objectives and constraints not as classical mathematical functions, but rather as general algorithms, where real world concepts such as low costs can be defined as an objective function. The inclusion of technical specifications, detailed design aspects or construction site logistics as optimization constraints all require algorithms that are not easily coded as simple mathematical functions. This also applies to objective functions and constraints that can only be computed by using numerical simulation. Examples from research carried out at our own institute include lifetime-oriented structural optimization, both in time and space (so-called semi-infinite stochastic optimization), inverse problem solving (system identification for the localizing damages in structural systems using back analysis), optimization of the collapse process in the precision blasting of complex structural systems (optimization in the temporal-spatial micro, meso and macro ranges) or the optimal placement of sensors in structural health monitoring (mixed discrete optimization problem with event driven, time variant optimization variables and constraints).

4.2 New Optimization Strategies/Algorithms

To be able to solve the type of optimization problems discussed above, the cooperation between engineers, mathematicians and computer scientists must increase still further. Current optimization methods are often geared towards solving complex mathematically oriented problems and not the type of real world problems design engineers are usually confronted with. Optimization problems typically found in engineering have objective functions and constraints that are highly non-linear and most often not even differentiable. The search space is irregular and contains multiple (local) optimum values. If "time" dependent parameters are included, then the optimization problem is of semi-infinite type. Increasingly, optimization problems can only be solved by numerical simulation based on discrete approximation methods such as finite element methods, boundary element methods or particle methods, or by using models based on multi-body physics. Thus, one aim of cooperation between engineers, mathematicians and computer scientists is to improve and refine the efficiency of existing optimization methods and to develop completely new optimization strategies. On the one hand, mathematician and computer scientist must understand the basics concepts of structural engineering, on the other hand engineers must observe the demands mathematicians place on the quality of solutions, such as proofs of uniqueness, computational efficiency, convergence behavior, etc.

4.3 Extension of Mathematical Fundamentals for Engineers

Concepts of graph theory [11], often underrated in current practice, can be used successfully in many types of engineering problems. As an example, the collapse process mentioned above that occurs due to precise demolitions of structural systems can be accurately defined by an object-oriented model. In particular, it is important to identify possible open or closed kinematic chains that are decisive for the course of the collapse process and are therefore needed to implement (optimal) countermeasures and to offset unwanted effects. Other examples on the use of graph theory in engineering problems include the cooperation of federative agents, the placement of sensors in a structural system and the monitoring of work flows with the help of colored and noisy Petri nets.

The mathematical modeling of fuzziness, uncertainty and incompleteness of available data and engineering models also requires a radial rethinking. The significance of fuzziness in advanced engineering problems (such as fuzzy control and fuzzy-randomness in the design of structural systems [13]) as well as the acceptance of such solution approaches by engineers requires the continued intense effort of mathematicians for better methods. In particular, the tight meshing of fuzzy randomness with design optimization, which leads to entirely new types of problems, is still in a nascent stage, although developments are long overdue. The safety and reliability acquired by such solutions can be increased significantly and such scientifically backed solutions would have a higher degree of acceptance from practicing engineers.

5 Visions and Courses of Action

Research results available up to now, but also research being carried out currently, suggest that highly complex, simulation-based structural optimization problems (and other problems from related fields) can soon be solved efficiently and to a high degree of precision. This will open a new dimension for the engineering sciences. In fact, it can be considered as the beginning of a new era, where engineers need not be satisfied with "just" finding local optimization values that depend entirely on good or bad initial values. Although it is not disputed that the approach up to now has been very beneficial, there is still much room for improvement.

New developments in design optimization are due to the application of standard concepts from computer science to engineering problems, including agent technology as well as object-oriented modeling and programming (OOMP). This allows the development of a new type of "computational steering" for design optimization, where engineers can significantly improve the quality of detailed design and dimensioning without completely relinquishing control of the design process. The global solution of the "optimization problem" will then be found by a set of competing software agents, each representing a specific optimization strategy.

Irrespective of the advances in the development of optimization methods, the underlying computational models need to use CPU and memory resources very efficiently to be able to solve complex structural engineering problems. This, again, can only be solved by the close cooperation of researchers from the fields of mathematics, computer science and engineering.

The help of mathematicians is particularly needed when, for example, the globality of optimal solutions is to be discussed. Good upper and lower bounds for each optimization variable are important; they can be computed approximately from the information obtained during the evaluation of the objective function and the constraints by response surfaces, for example. But also the precise formulation of the manner in which optimization strategies will cooperate with each other can be implemented by software agents. Also, defining how certain strategies are to be initialized or how termination criteria are to be checked are still open questions. The mathematical treatment of fuzziness and time variance with regards to the uniqueness, stability and convergence of the solution method also requires further research activities. Only when researchers in the fields of mathematics, computer science and engineering continue to work together as closely as they started to do in the past will it be possible to close the remaining gaps in both theory and practise.

References

1. Bäck, T.: Evolutionary Algorithms in Theory and Practice. Oxford University Press, Oxford (1996)
2. Baitsch, M., Hartmann, D.: Optimization of slender structures considering geometrical imperfections. Inverse Probl. Sci. Eng. **14**(6), 623–637 (2006)
3. Baitsch, M., Hu, Y., Hartmann, D.: Parallel shape optimization of three-dimensional continua with high-order finite elements. In: Herskovits, J. (ed.) Proceedings of the International Conference on Engineering Optimization, Rio de Janeiro, Brazil, June 2008
4. Bellifemine, F., Caire, G., Greenwood, D.: Developing Multi-Agent Systems with JADE. Wiley, New York (2007)
5. Beyer, H.-G.: The Theory of Evolution Strategies. Springer, Berlin (2001)
6. Clerc, M.: Particle Swarm Optimization. ISTE Ltd (2006)
7. Eschenauer, H.A., Koski, J., Osyczka, A.: Multicriteria Design Optimization. Springer, Heidelberg (1990)
8. Gálffy, M., Baitsch, M., Wellmann-Jelic, A., Hartmann, D.: Lifetime estimation of vertical bridge tie rods exposed to wind-induced vibrations. In: Mota Soares, C.A., Martins, J.A.C., Rodrigues, H.C., Ambrósio, J.A.C. (eds.) Proceedings of the IIIrd European Conference on Computational Mechanics Solids, Structures and Coupled Problems in Engineering, Lisbon, 2006. Springer, Berlin (2006)
9. Goldberg, D.E.: Genetic Algorithms in Search, Optimization, and Machine Learning. Addison-Wesley, Reading (1989)
10. Grill, H.: Ein objektorientiertes Programmsystem zur gemischt-diskreten Strukturoptimierung mit verteilten Evolutionsstrategien. Dissertation, Institut für konstruktiven Ingenieurbau, Ruhr-Universität Bochum (1997)
11. Gross, J.L., Yellen, J.: Graph Theory and Its Applications. Chapman & Hall/CRC, London (2006)
12. Hartmann, D., Leimbach, K.-R.: Analyse von Strukturoptimierungsmodellen im Hinblick auf ihre parallele Realisierung und Umsetzung, 1996. Abschlussbericht, DFG-Forschergruppe Optimierung in der Strukturmechanik
13. Möller, B., Beer, M.: Fuzzy Randomness. Springer, Berlin (2004)
14. Onwubolu, G.C., Babu, B.V.: New Optimization Techniques in Engineering. Springer, Berlin (2004)
15. Rechenberg, I.: Evolutionsstrategie'94. Frommann-Holzboog, Stuttgart (1994)
16. Rey, G.D., Wender, K.F.: Neuronale Netze: Eine Einführung in die Grundlagen, Anwendungen und Datenauswertung. Huber, Bern (2008)
17. Russell, S., Norvig, P.: Künstliche Intelligenz: Ein moderner Ansatz. Pearson Studium, Upper Saddle River (2003)
18. SFB-398: Life time oriented design concepts. Monographie zum DFG-Sonderforschungsbereich SFB 398 "Lebenesdauerorientierte Entwurfskonzepte", ca. 600 Seiten, in Druck, erscheint 2008
19. Thompson, J.M.T.: Optimization as a generator of structural instability. Int. J. Mech. Sci. **14**, 627–629 (1972)

Object-Oriented Modelling for Simulation and Control of Energy Transformation Processes

Dirk Abel

1 Introduction and Overview

Mathematical modelling of technical systems and processes holds a well—established position in all engineering disciplines—not least driven by powerful digital computers that allow for simulation, analysis and design. This is stimulated by new developments in the domain of modelling languages and computer tools that possess modern object-oriented concepts and algebraic/symbolic calculating capacities. With two current investigation projects on energy conversion—in the domains of power plant engineering and combustion engines—the object-oriented modelling and its potential in comparison with conventional approaches will be presented from the engineering perspective, especially from that of automatic control.

The importance of mathematical models for the automatic control is obvious. (Almost) Each control unit design is based on a model of the process to be controlled. The functional diagram on the basis of the principle of cause and effect is an important graphical tool to describe complex dynamic systems. Therefore, the functional diagram reappears in the user interface of many popular block-oriented simulation programs (such as, for example, SIMULINK). Since the late nineties MODELICA provides a so-called object-oriented description language for physical models, for which various graphical development and simulation programs, like DYMOLA, exist. MODELICA allows its application in a wide range of scientific domains: mechanics, electrical engineering and electronics, thermodynamics, hydraulics and pneumatics, automatic control and process technology.

Object-oriented modelling and simulation tools like DYMOLA are perfectly suitable for the simulation of complex systems consisting of multiple interacting components, due to their model structure and thus the possibility of modularising. The modelling of subsystems is based on algebraic equations, where all the existing signals are categorised into potential and flow variables. The subsystems can then be

D. Abel (✉)
RWTH Aachen, Institut für Regelungstechnik, 52056 Aachen, Germany
e-mail: d.abel@irt.rwth-aachen.de

M. Grötschel et al. (eds.), *Production Factor Mathematics*,
DOI 10.1007/978-3-642-11248-5_17, © Springer-Verlag Berlin Heidelberg 2010

connected to each other via vectorial interfaces, the so-called *Connectors*. The essential characteristic of such connectors is that subsystems of a simulated system, such as dynamic models of components of a process to be controlled, are connected to each other and begin to interact via interfaces that present an analogy to reality. Moreover, due to the algebraic/symbolic calculating functionality of the tool, the user is relieved of the duty to decide on the direction of cause and effect that is unavoidable with conventional modelling and simulation tools.

2 Application in Power Plant Domain: Project OXYCOAL-AC

2.1 Goals and Requirements

The project OXYCOAL-AC is one project of the COORETEC Programms (see [5]) of the Federal Ministry of Economics and Technology that emerged from the obligation of the United Nations according to the Kyoto Protocol to reduce drastically the CO_2 emissions. The treaty sets target values for the emission of the greenhouse gas as the main cause of the global warming.

The OXYCOAL-AC project investigates a process of CO_2-emission-free coal combustion where the emitted carbon dioxide is separated from the process to be further stored in geological formations. The important requirement of an economic separation of the CO_2 from the flue gas during the power plant process is realised with combustion of the fuel with a flue gas/oxygen gas mixture. So, the combustion gas basically consists of carbon dioxide and water. A combustion process sketch of the OXYCOAL-AC process is presented in Fig. 1. The oxygen supply proceeds through a high temperature membrane system. Compressed air, preheated in a heat exchanger, is blown on the one side (air side) of the ceramic membrane, and hot flue gas is blown on the other side (flue gas side). Due to the high oxygen partial pressure ratio between the air and the flue gas side, oxygen diffuses from the air side through the membrane to the flue gas side. As can be seen from the figure, one part of the flue gas is circulated by the exhaust gas blower, oxygenated and used for the combustion, whereas the other part is excluded from the process.

The above described structure of the OXYCOAL-AC process requires new control concepts for the combustion gas supply (see also [7]). Whereas at conventional power plants combustion air mass flow can quickly be adjusted by varying the speed of the air blower when fuel mass flow changes, the new process has a more complex character. It should be awaited that changes of the diffusion mass flow over the membrane will be comparatively slow due to heat transfer in the heat exchanger and the membrane, as well as delay time in tubes. It should also be taken into account that the ceramic materials of the membrane are highly sensitive to spatial and temporal temperature changes, and thus only a moderate temperature rate of change is allowed to prevent material damage. A dynamic process model is needed to develop a suitable control concept for the non-linear, highly coupled multivariable system with delay time. The model must give a detailed representation of the membrane with regard to the thermal gradients in the control concept.

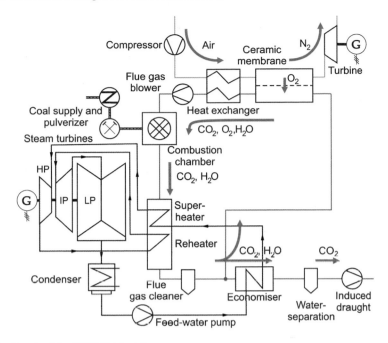

Fig. 1 Sketch of the OXYCOAL-AC process

2.2 Object-Oriented Modelling of Power Plant Processes

An increasing demand concerning security, reliability and efficiency in power plants makes a detailed mathematical and physical analysis unavoidable during the development phase. Due to the complexity of the systems of equations, such analyses can not be accomplished without appropriate numerical modelling and simulation tools. Whereas the efficiency of power plant processes increases through optimisation of the process structure based on static simulations, a dynamic modelling is necessary for the optimal design of the control loops to ensure further safe and reliable operation of the plant. For dynamic modelling, it is beneficial to use a component-based approach with standardised interfaces in order to obtain a multilateral and universal transferability of the individual power plant components' models onto processes with different structures and dimensions. Thus, a high re-application rate of the developed models is achieved, and the number of components to be modelled is reduced.

The implementation of a component-based modelling and the application of standardised interfaces for information exchange between the sub-models is the key idea of the object-oriented modelling language Modelica. Hence, it provides a flexible approach to describe models of power plant components. The acausal modelling concept of Modelica allows the formulation of equations, also implicit equations, in the manner they are known from physical relations, with no need to define the signal flow direction and to rearrange the system of equations according to input, output and state variables. Thus, one and the same model with corresponding parameterisation can be flexibly applied in various configurations and contexts with variable boundary conditions. Moreover, it is easier for the user to understand such models and modify them according to one's own needs.

The object-oriented principles of inheritance and polymorphy additionally enhance modelling flexibility. The polymorphy mechanism enables describing a component model with various selectable systems of equations through redefinition. Thus, given the same model, it is possible, for instance, to vary between accuracy and simulation duration by selecting systems of equations of different levels of detail for various simulations.

The developed models can then be structurally integrated into libraries. Due to the above mentioned advantages in the object-oriented approach, there already exist freely available component libraries in Modelica which are especially suitable for modelling power plant processes. Based on the Media library (see [4]) owned by the Modelica language standard, the open ThermoPower library (see [2]) offers a good basis for power plant modelling. The modelling procedure of the OXYCOAL-AC process will be elucidated in the following sections.

2.3 Object-Oriented Modelling of the Oxycoal Power Plant

To construct a model of the process, described in Sect. 2.1, a library with all of the process components has been built. As shown in Fig. 2, the OXYCOAL-AC library is based on the ThermoPower library, which, in its turn, is based on the Modelica standard library. It consists of three main parts: pckBasis, pckComponent and pckSimulation. The package pckBasis defines units, constants, connectors and base classes; the package pckComponent provides the compilation of component models; and the package pckSimulation contains models of the entire process for simulation, which are composed of the individual components. The OXYCOAL-AC library with all the components, necessary for the design of the entire process, is built on the basis of standardised interfaces, available from the ThermoPower library, for standardised data exchange. Standardised interfaces to describe the interchange of gases and for the mechanical rotatory links are applied. The application of the interfaces of the ThermoPower library for the interchange of gases allows the development of a component model independent of the used fluid. The state of a fluid at the interconnection of two components is defined by the mass flow w, the pressure p, the specific enthalpy h and the vector of mass fractions ξ. The gas fluid, defined for the design of the OXYCOAL-AC process, consists of seven components: nitrogen (N_2), carbon dioxide (CO_2), oxygen (O_2), carbon monoxide (CO), water (H_2O), sulphur dioxide (SO_2) and nitric oxide (NO). For the description of the fuel flow, a connector has been defined which provides the mass flow w, the temperature T and the composition ξ of the coal. The designed coal consists of the following components: carbon (C), hydrogen (H), water (H_2O), oxygen (O), sulphur (S) and nitrogen (N). The connection of the mechanical rotatory components is carried out via the rotation angle φ and the torque M.

Fig. 2 Structure of the
OXYCOAL-AC library

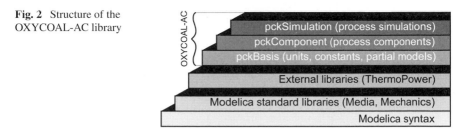

Based on the developed interfaces, dynamic models of the most important components of the OXYCOAL-AC process have successively been constructed. Models of the particular components have consequently been described on the basis of physical mass, energy and moment balances. Below, the ceramic membrane model and the model of the heat exchanger of the process presented in Sect. 2.1 will be used as an example to demonstrate the advantages of the object-oriented modelling and simulation.

Ceramic High-Temperature Membrane

The ceramic membrane for air separation is an essential element of the OXYCOAL-AC process. Operation of the sensitive ceramic component is a particular challenge for control design. On the one hand, the values of pressure, temperature and mass flow at the membrane inlet, on both the air and the flue gas side, must be set for effective operation; and, on the other hand, safe operation of the membrane must be ensured. Thereby, an important task is to avoid high temperature gradients, for they lead to thermal stress that can destroy the membrane. These gradients can develop either through temporal changes in the ambient temperatures of the flue gas or air flow, or spatially, for instance, along the flow direction. The difficulty of avoiding critical temperature gradients makes the membrane model necessary for representing these conditions close to reality.

Generally, in modelling of physical systems lumped and distributed models can be singled out. In lumped models the dynamic behaviour of the system is described by variables which are location-independent, and thus, allow a description with ordinary differential equations. In contrast, the dynamics of location-dependent variables are described by partial differential equations. Modelling a distributed system with partial differential equations is usually complicated, and an analytical solution of these equations is in general impossible to obtain. An approach to make a model of a distributed system without partial differential equations is provided by a finite element method (cf. [3]). Thereby, a distributed system, described with partial differential equations, is spatially discretised, and each finite element is then described through a system of ordinary differential equations.

A finite volume approach has been chosen for modelling the temperature gradients lengthwise of the membrane. With the help of spatial discretisation along the flow direction of the membrane, temporal temperature changes, as well as the spatial

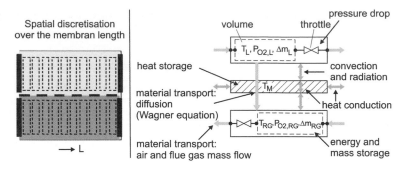

Fig. 3 Model of the membrane module

temperature distribution along the membrane can approximately be considered and investigated. The left part of Fig. 3 schematically demonstrates the spatial discretisation of the membrane module. According to this concept, the air flow (above) and flue gas flow (below) are separated from each other by the permeable membrane material. The model of the entire membrane module can be subdivided along the flow direction into a parameterizable number of N discrete volumes. The conservation equations for mass and energy are formulated for each of these volumes. The entire membrane model is obtained through an automatic connection of the separate finite volumes.

A physically-oriented approach has been used for the whole process of membrane modelling in order to make a transfer of the model to other geometries or materials possible. The number of discrete volumes is parameterizable. Likewise, the system of equations used for each individual volume is adjustable. The mechanism of polymorphy, supported by the Modelica language, offers the possibility to redeclare the model of a finite volume, defined in the membrane module. Due to the possibility of choosing between different systems of equations of various complexities for the finite volumes, a high degree of flexibility is achieved concerning accuracy and computation time.

The right part of Fig. 3 shows the model of a membrane volume's element. The models of the finite membrane volumes are built up with the help of subcomponents. They contain two volume elements that function as energy and mass storage, one for the flue gas side and one for the air side. A throttle on both sides accounts for the pressure losses. The membrane substance between the both volumes acts as a heat storage. To describe the heat transfer in the membrane module different systems of equations can be chosen. The most complex selectable system of equations comprises lengthwise and crosswise heat conduction in the membrane material and convective heat transfer between the membrane material and the gas flows on both sides. Additionally, heat transfer by radiation is implemented. The heat transfer coefficients are continuously adjusted to the particular flow conditions. Less complicated modellings forego, for instance, the description of the effect of the heat conduction lengthwise the membrane material, or consider solely the convective heat transfer and neglect radiation.

The mass transfer through the membrane is described by the so-called Wagner equation

$$j'' = \frac{C_1}{d} T e^{\left(\frac{C_2}{T}\right)} \ln\left(\frac{p_{O_2,L}}{p_{O_2,RG}}\right)$$

for the diffusion process [12]. Here, j'' stands for the area-related oxygen mass flow through the membrane, d for the membrane thickness, T for the membrane temperature and $p_{O_2,L}$ and $p_{O_2,RG}$ for the oxygen partial pressures on the air (L) and flue gas (RG) sides. C_1 and C_2 are constants to describe the diffusion properties of the membrane material.

Gas-Gas Heat Exchanger

The function of the gas-gas heat exchanger is additional heating of the compressed air with hot flue gas. As in case with the membrane, the heat exchanger is a distributed system, which can be described with a finite volume approach. Since the only feature that basically differentiates the heat exchanger from the membrane is the absence of mass transport, the subcomponents that have been used for the model of the membrane will also be applied for the description of flow processes and heat exchange modelling of the heat exchanger. Figure 4 shows the Dymola model of the heat exchanger. The storage effect of the heat exchanger, on both the flue gas and

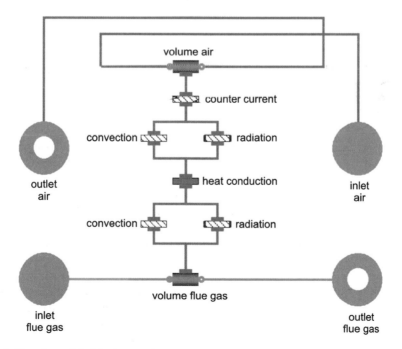

Fig. 4 Dymola model of the heat exchanger

the air side, is designed with the subcomponent volume. The both volumes are entities of the same class but parameterised differently. The both volumes are separated from each other by a tube wall. The heat transfer over the tube wall from the flue gas side to the air side of the heat exchanger is defined by three entities: convection, radiation and heat conduction. The flue gas transfers a heat flow to the flue gas side of the tube wall through convection and radiation depending on the temperature and flow conditions of the flue gas side volume. Through heat conduction in the tube the heat flow is transmitted to the air side of the tube wall and then transferred in the air side volume again due to convection and radiation.

2.4 Results

The afore-described oxyfuel power plant process is part of a current investigation project and is presently neither completely realised, nor in all its components. To design the entire process the development tool *Epsilon* is used for steady state process calculations (cf. [8]). Due to the lack of data, for dynamic simulations of the future OXYCOAL-AC power plant the models of the components are first adapted to the results obtained from the steady state design. The unknown model parameters are determined in such a way that at the operating point the dynamic model exhibits the same steady state values for the steam parameters and other mass flows, pressures and temperatures. For this purpose the *Calibration-Toolbox* of the commercial development and simulation tool Dymola is applied. This tool provides an identification method which adjusts selectable model parameters in a way that certain process variables follow the pre-set trajectories as well as possible. In case of the membrane module, for instance, the coefficients of the valves are adjusted to the pressure drops of the steady state calculations.

After identifying the component models a so-called sensitivity analysis is carried out on the model of the entire process. Therein, different input variables, as for example the speed frequencies of compressors and fans, are raised one after another stepwise or rampwise and the reaction of different output variables, like mass flows, temperatures and pressures, etc., are recorded. Sensitivity tests can give important information on linkage of input and output variables and time constants of the oxyfuel power plant. They thus present the basis for the development and design of an automatic control concept.

Figure 5 displays the membrane temperatures in the finite volumes as a reaction on a rampwise temperature reduction of the flue gas flow at the membrane inlet. The membrane module of the presented simulation has an area of one square meter at a length of one meter and is divided for the modelling in ten discrete volumes (see Fig. 5, left). The temperature of the flue gas flow is reduced by 50 Kelvin within 100 seconds starting at time $t = 2000$ s. As the figure shows, the membrane temperature decreases with a temporal delay because of the modelled heat exchange and heat storage between flue gas and membrane and a temperature drop happens along the flow direction of the flue gas. Thus, with the selected modelling temporal

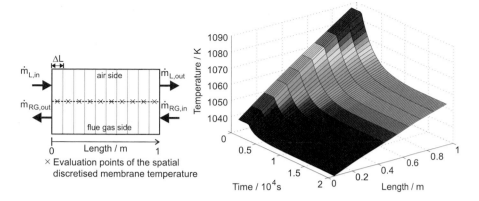

Fig. 5 Spatial and temporal characteristics of the membrane temperature

as well as local phenomena can be shown. Information about temporal temperature characteristics of the discrete membrane volumes after a change of air and/or flue gas temperature can be considered developing a control system to avoid high temperature gradients.

2.5 Conclusions

The object-oriented modelling method with Modelica offers a flexible approach to create a model of a power plant process. The component-based approach and the application of the already available models in the free libraries proved to be especially suitable for the modelling of the OXYCOAL-AC process. The application of standardised interfaces is advantageous because it enables a flexible combination of component models in differently structured models of the entire process. The interfaces for fluids defined in the ThermoPower library allow the formulation of a system of equations of a component independent of the used fluid, and thus, make the re-application of single components possible in other contexts. Thus, in the OXYCOAL-AC process the air compressor as well as the flue gas blower can be modelled with the same model of a compressor, but differently parameterised, although they convey different fluids. Hence, the library with the developed models for the flue gas circulation is restricted not only to the OXYCOAL-AC process and can be further applied

The component-based modelling method helps modelling distributed systems. Once a model of a single finite volume is developed, it is possible to combine an optional number of finite volumes to make a model of the whole component. Thus, and further on with the possibility of redefining the system of equations of a component, simulations with different levels of detail can be realised with the same models.

In addition, flexible configuration of components enables analysis of different process configurations or building of sub-process models. This is of special interest for the development of automatic control concepts because various control scenarios, such as, for example, the start-up process of the OXYCOAL-AC power plant, can be examined.

3 Application in the Field of Combustion Engines

3.1 Model-Based Control of Combustion Engines

Steady decreasing emission limits of combustion engines highly require improvement of combustion processes, as well as related control methods. For diesel engine, that is taken here as an example, a special focus is on nitrogen oxide (NO_x) and soot emissions. For reduction of NO_x emissions the exhaust gas recirculation (EGR) has proved reliable, which recirculates exhaust gas from the turbine after the compressor back to the incoming air mass flow. Controlling this system is unavoidable, otherwise the combustion process can not be optimally effective and emissions can even increase. A highly appropriate control method is a model-based predictive controller that internally applies a simple model of the controlled system to provide the best possible actuating variable. The standard software for the development of controllers is Matlab/Simulink, as the engineers find the signal-oriented development suitable.

Simulation of the control path is extremely important for a cost-effective development of a controller in general and is, thus, an essential element of the development of each controller. Although Simulink can also be applied to simulation of systems, the signal orientation is not always profitable. Complex systems can quickly become confusing, and error rate increases. For that reason, the object-oriented modelling language Modelica gains more popularity, because it offers ideal conditions for simulations of complex systems in a wide range of disciplines.

The following section elucidates two essential differences between the two modelling methods on the basis of the following examples: the development of a controller-internal model for the model-driven predictive controller in Simulink and the development of a simulation model for a diesel engine with the modelling language Modelica using the simulation tool Dymola. The models developed for a controller with Simulink possess a low level of detail as compared to the models applied for simulation of the control path. The main difference is still obvious and becomes greater as complexity grows. This does not affect the application of Simulink for the development of a controller.

The schematic design of a diesel engine, as well as the control and simulation operation made with the two modelling methods are further explained in Sect. 3.2. Subsequently, Sect. 3.3 dwells on the examples of the different modelling methods. Section 3.4 gives a conclusion of the experience.

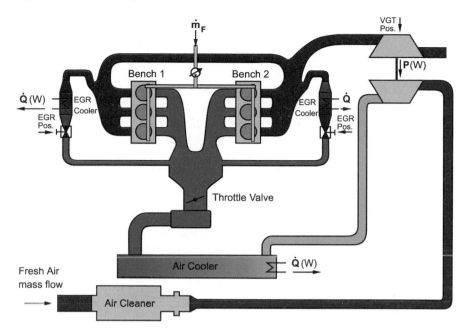

Fig. 6 Schematic model of the engine

3.2 Structure of the Control Path

Responding to increasing requirements to combustion engines, control and simulation tasks during the development process become more important. The example of the development of a controller for the air path of a diesel engine helps to show which models emerge and which tools are used for it.

Figure 6 represents a schematic construction of a diesel engine to be designed. The engine in question is a six-cylinder diesel V-engine[1] with common rail direct fuel injection. The air path consists of a high-pressure exhaust gas recirculation (EGR) to minimise nitrogen oxide emission and of a turbocharger with variable geometry turbine (VGT) for boost pressure generation. Moreover, the engine has a throttle valve to generate enough pressure drop within the EGR path at low engine operation points as well.

Due to EGR, the combusted gas recirculates back to the incoming air mass flow. This helps to lower combustion chamber temperature, and thus the nitrogen oxide emissions. The turbocharger is installed to use the exhaust gas energy for the compression of the incoming air. This, in turn, enhances the power density of the engine and can also be used for consumption reduction.

---

[1]The engine described here is applied in the ZAMOMO Project, supported by BMWI, which investigates and promotes interlocking between modelling of embedded software and control engineering on the example of an engine controller (see [6, 11]).

From the point of view of control engineering this system is a coupled two-variable system, because the actuating variables of VGT-position and EGR-setting interchangeably influence the control variables of boost pressure and incoming air mass flow. This coupling of the multivariable system can well be decoupled with a model predictive controller (see [9, 10]). This kind of a controller has an internal model of a control path to calculate the future behaviour and to provide ideal actuating variables in an optimizing procedure.

In order to test the developed controller at an early stage, a detailed simulation model of the control path is simultaneously built. This model allows testing the controller functions already before the application in a real engine. To be able to test a real control unit with a Hardware-in-the-Loop-Test (HIL), these models must additionally be capable of a real time use. For that reason, mean value models for air path simulation are mostly employed which calculate an average operation of the engine per cycle. Now Modelica is used as an object-oriented modelling language for the construction of this model.

Due to the fact that the same control path is designed for the controller and the simulation model, the model development with Simulink can quite well be compared with the development of Modelica models. The both models include only the dynamic elements[2] of the entire system that are relevant for the control of the air path. Various levels of detail of the models must be taken into account while comparing. For it is true that the calculating capacity of a control unit is not comparable with a real time simulation platform.

3.3 Comparison of Object-Oriented and Signal-Oriented Modelling on the Example of Air Path Components

This section shows two important differences between signal-oriented and object-oriented modelling methods on the examples of volume and turbocharger models. These both elements are chosen, amongst others, because they represent the both essential elements that are responsible for dynamic operation of the air path.

Volume

A volume has normally a zero-dimensional design in the values of pressure and temperature for average-value models. Dynamic wave progressions in the tubes and volumes are neglected. Volume is a most frequently applied element in simulation of an air path. Therefore, the modelling must be re-applicable and easy to parameterise.

---

[2]Further on, only these elements are understood under the term of dynamics. High frequency elements, as, for instance, resonance in the intake tube, are neglected in the control of the air path.

The simplified basis equations for matter and energy conservation are as followed for an adiabatic volume:

$$\frac{dm_V}{dt} = \dot{m}_{In} - \dot{m}_{Out},$$

$$\frac{dH_V}{dt} = \dot{H}_{In} - \dot{H}_{Out} + \overbrace{\dot{W} + \dot{Q}}^{=0}.$$

These equations, which are basically true for various flow directions are common for the both Simulink and Modelica models, and they must be realised respectively.

That makes yet no problem in Simulink for a specified flow direction, as is shown in Fig. 7. Because of signal orientation, already three signal paths (in- and output mass flow and input temperature) as input variables and two signal paths (output temperature and volume pressure) as output variables must be connected with the surrounding components. Altogether, five signal paths in this simple volume model must be linked to each individual component.

Moreover, it is also clear in the realisation of a Simulink model that a flow inversion is not possible here. Besides, the temperature of the mass flow entering the volume depending on the flow direction must be taken into account. This means that for this model the output temperature as an input variable must also be available, followed by setting the flow direction. The originally simple equation becomes confusing because of these additional components and signal paths.

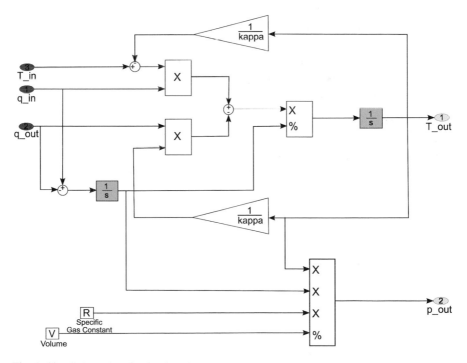

Fig. 7 Simulink-model of a simple volume

Basic equations	Conditional query	
```M = gas.d * V "Gas mass";E = M*gas.u "Gas internal energy";der(M) = inlet.w + outlet.w "Mass balance";der(E) = inlet.w* i + outlet.w*ho "Energy balance";```	```if inlet.w >= 0then hi = inlet.hBA; Xi i = in-let.XBA;else hi = gas.h; Xi_i = gas.Xi;end if;```	```if outlet.w >= 0then ho = outlet.hAB; Xi_0 = out-let.XA;else ho = gas.h; Xi_o = gas.Xi;end if;```

Fig. 8 Example of a volume model of the ThermoPower [2]

Generally, four flow directions[3] can be singled out in a volume. It turns more complicated, if, apart from pure substance systems, also multi-component and multi-phase systems must be considered. In these cases, the comparatively simple governing equations are not apparent any more due to signal distribution of a volume model. The model file is no longer sufficient to document the model. It is necessary to maintain a separate file which stores the related equations and assumptions. This is still possible, with a reasonable complexity, for relatively simple equations of an internal controller model.

Modelica allows nicely solving the problem of traceability and reusability through application of *connectors* and acausal modelling. The connectors unite all the variables that are exchanged between components over the system boundary. Distinction should be made between flow and potential variables. In addition, various flow directions can be taken into account straight here. The connector from the ThermoPower-Library can serve here as an example (see [2]), which defines the variables of *mass flow*, *pressure*, *specific enthalpy* and *material composition* for thermal systems. The latter two variables must additionally be divided into input and output values.

For this volume the inlet and outlet connectors are determined with this connector's definition. Due to the mass flow definition as a flow variable the equation for mass conservation is very easy to accomplish (see the third line of the governing equations on Fig. 8). Different flow directions of the both connectors (inlet and outlet) are automatically entered in the equation with the correct sign. For the energy equations, in each case, a variable for the inlet and outlet specific enthalpy is defined that is assigned against the flow direction. The process is apparent if carried out with the help of two conditional queries (see Fig. 8 right) that assign the respective variable of the connector to the variables *hi* and *ho*. In consideration of multi-material components material composition is needed as well. The latter can be implemented within the existent conditional queries. Clarity and traceability of this solution is a substantial advantage as compared to the Simulink solution.

Turbocharger

The second important element of the air path dynamics is a turbocharger. It extracts energy from the hot flue gas with a turbine and delivers it into the compressor which

---

[3]Combination of, in each case, two flow directions for the input and output.

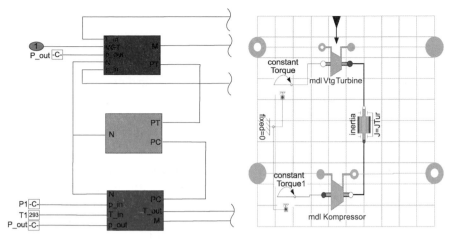

Fig. 9 Comparison of modelling a turbocharger with Simulink (*left*) and Dymola (*right*)

compresses the incoming air. In diesel engines a turbocharger with a variable geometry turbine (VGT) is often installed, with the help of which the turbine can better be adjusted to the related engine operating point. This process leads to a higher efficiency, and the boost pressure can, to a certain degree, be taken under control with the VGT setting

For the simulation the VGT turbocharger is designed with the help of characteristic diagrams that describe characteristics of the turbine and the compressor. Figure 9 presents the comparison of the turbocharger modelling with Simulink and Modelica's simulation tool Dymola. First of all, on this level many signal paths of the Simulink model are again conspicuous. In contrast, the structure of the Dymola model is much clearer and is close to physical reality.

In addition to the mentioned differences in the display format of physical systems, there is a further substantial difference between signal-oriented and object-oriented modelling, which consists in the absence of the necessity to make manual transformation of equations. This task is implemented by the simulation tool in the object-oriented modelling. The following example illustrates the advantage of this procedure.

To develop a model of a compressor, conventional conservation equations are applied for material and energy, as well as some parameters for turbo-machines. The description of the compressor characteristics requires two characteristic diagrams. The first one describes compressor efficiency against the reduced turbocharger speed and the reduced incoming air mass flow. The second diagram determines, according to the modelling method, compressor pressure ratio against the reduced mass flow and the reduced turbocharger speed, or the reduced mass flow against the reduced turbocharger speed and pressure ratio. Hence, the only difference of the two modelling methods lies in input and output variables. The governing equations are identical.

For a Simulink compressor model this means that for each case a separate model must be built. Re-usage of the existing components, especially those of governing equations, can not be provided within signal-oriented modelling. The operator must compute all the equations new, depending on the input and output variable choice.

In contrast, the entire equations can be re-applied within the object-oriented modelling. The both modelling systems only differ in characteristic diagrams. Transformation of the equations according to the respective in- and output variables is implemented internally by the simulation tool. Simple algebraic equations can be solved with the simulation tool in the same way too. This automatic transformation of equations is, however, not unlimited. It can be absolutely effective if the operator himself solves certain formulae or uses an alternative calculating rule. To illustrate it, an example of a simple electrical circuit with non-linear resistance is taken from [1]. The resistance is described with the equation

$$i = a \cdot \tanh(b \cdot u)$$

In this model the current i is a state variable, thus, this equation must be solved for u. Dymola is not capable of solving these equations analytically and uses a numerical algorithm. An alternative method is to choose another defining equation for the hyperbolic tangent,

$$\tanh(x) = 1 - \frac{2}{e^{2x} + 1}$$

and set up the equation directly against the state variable current intensity i.

$$u = \frac{1}{2b} \cdot \log\left(\frac{1 + i/a}{1 - i/a}\right)$$

A comparison of the two implementation strategies shows that the second solution, given the similar results, is considerably quicker, because the numerical solution of the equation is not applied.

With this example it is argued that automatic methods for solving equations do not always give better results, but rather that the operator can influence the performance either positively or negatively.

3.4 Conclusions

The comparison of signal- and object-oriented modelling has shown that the object-oriented modelling has many advantages against the signal-oriented modelling. Closeness of the structure of the Modelica model to the physical reality helps the operator to better understand the models. The error-prone transformations of mathematical equations typical of the signal-oriented modelling are not present, and that gives the operator more freedom of action. Furthermore, it is necessary for constructing such models to make an insight into the world of the object-oriented modelling, which, as it has been shown, substantially differs from the signal-oriented modelling.

For simulation of complex systems the object-oriented modelling has proved to be effective, to some extent also due to many free access libraries that presently exist for the most application domains. Every operator can freely use the libraries and apply them according to his own needs.

In the development of controllers the signal-oriented modelling, as applied in Matlab/Simulink, can show its strengths against the object-oriented modelling. The object-oriented modelling is not yet competitive in this area. Another advantage of Simulink is that it has meanwhile become a quasi-standard program for co-simulations. Almost each commercial simulation program has an interface to Simulink. In using both tools and their data coupling—which is not the topic of the present paper—advantages of both approaches can nicely be combined.

4 Summary

After insights taken into power plant and combustion engine engineering, the project-specific characteristics and advantages of the object-oriented modelling and simulation approaches should once again be briefly summarized. An essential feature is undoubtedly the abandonment of the cause-and-effect principles.

Besides from other aspects, essential advantages of the object-oriented models are their universality and re-application (among others, in view of various important output variables), as well as clarity of interfaces, which helps to support the construction of modularised model libraries to a great extent. All this is possible due to the advanced, namely algebraic/symbolic computational capacity of executive calculation tools which substantially reduce the operator's workload previously implemented in model description with intellectual power. Especially remarkable is the distinctive support of the interdisciplinary communication with the help of mathematics: representatives of various disciplines can interact with each other much better concerning modelling, because the descriptive terms are free from the computational burden.

Acknowledgements I would like to thank those who supported the investigation projects described at length in this article. I am especially thankful to my research assistants, particularly to Sarah Dolls, Sebastian Hölemann and Frank Heßeler who provided material on their ongoing research projects.

References

1. Bachmann, B.: Mathematical aspects of object-oriented modeling and simulation. Tutorial auf der 6. Internationalen Modelica Konferenz, Bielefeld 2008

2. Casella, F., Leva, A.: Modelica open library for power plant simulation: design and valida-tion. In: Fritzson, P. (eds.) Proceedings of the 3rd International Modelica Conference 2003, Linköpings, Schweden
3. Casella, F., Schiavo, F.: Modelling and simulation of heat exchangers in Modelica with finite element methods. In: Fritzson, P. (ed.) Proceedings of the 3rd International Modelica Confer-ence 2003, Linköping, Schweden, pp. 343–352
4. Casella, F., Otter, M., Proelss, K., Richter, C., Tummescheit, H.: The Modelica fluid and me-dia library for modeling of incompressible and compressible thermo-fluid pipe networks. In: Proceedings of the 5th International Modelica Conference 2006, Wien, Österreich, pp. 631–640
5. COORETEC: Forschungs- und Entwicklungskonzept für emissionsarme fossil befeuerte Kraftwerke. Bericht der COORETEC-Arbeitsgruppe (2003)
6. Drews, P., Hesseler, F., Hoffmann, K., Abel, D., Schmitz, D., Polzer, A., Kowalewski, S.: En-twicklung einer Luftpfadregelung am Dieselmotor unter Berücksichtigung nichtfunktionaler Anforderungen. In: AUTOREG 2008 – Steuerung und Regelung von Fahrzeugen und Mo-toren. 4. Fachtagung Baden-Baden, 12. und 13. Februar 2008. VDI-Berichte Nr. 2009. ISBN 978-3-18-092009-2
7. Hölemann, S., Nötges, T., Abel, D.: Modellgestützte Prädiktive Regelung zur Sauerstoffver-sorgung eines oxyfuel Prozesses. In: GMA-Kongress 2007, 12.–13. 6. 2007, Baden-Baden. VDI-Berichte, pp. 729–738. VDI-Verlag, Düsseldorf (1980)
8. Kneer, R., Abel, D., Niehuis, R., Maier, H.R., Modigell, M., Peters, N.: Entwicklung eines CO_2-emissionsfreien Kohleverbrennungsprozesses zur Stromerzeugung. VDI-Berichte Nr. 1888 PL03 (2006)
9. Rückert, J.: Modellgestützte Regelung von Ladedruck- und Abgasrückführrate beim Diesel-motor. Dissertation, RWTH Aachen (2004)
10. Richert, F.: Objektorientierte Modellbildung und Nichtlineare Prädiktive Regelung von Dieselmotoren. Dissertation, RWTH Aachen (2005)
11. Schmitz, D., Drews, P., Heßeler, F., Jarke, M., Kowalewski, S., Palczynski, J., Polzer, A., Reke, M., Rose, T.: Modellbasierte Anforderungserfassung für softwarebasierte Regelungen. In: Herrmann, K., Brügge, B. (eds.) Software Engineering 2008. Fachtagung des Gl-Fachbereichs Softwaretechnik, 18.–22. 3. 2008 in München. LNI 121 Gl (2008). ISBN 978-3-88579-215-4
12. Wagner, C.: Equations for transport in solid oxides and sulfides of transition metals. Progr. Solid State Chem. **10**(1), 3–16 (1975)

Design Tools for Energy Efficient Architecture

Dirk Müller

1 Executive Summary

In the Federal Republic of Germany, the heat supply of buildings has a 30% share in the total energy demand of the Republic [1]. The major part of the final energy is consumed by the heating of residential buildings, whereas gas and fuel constitute the main sources of energy.

The building envelope and the building services are especially interesting targets for a possible reduction of the required energy and pollutant emission. Often, even with simple techniques one can achieve significant improvements. Mathematics substantially contributes to the process of building planning and design. Different computational tools can assist during the three phases of energetic optimization, the reduction of thermal load by means of beneficial design, alignment and form of the building envelope, integration of renewable energy and optimizing operations.

The building envelope can provide many functions, which are essential for energetic efficiency and user's comfort. Sufficient insulation reduces the heat loss and ensures a comfortable internal wall temperature. Additionally, it reduces the probability of occurrence of hygienic problems such as mold build-up. Using finite elements methods, two- and three-dimensional computations of heat conduction processes in wall corners and geometrically complex structural elements, such as the window frame, provide precise data for the expected heat losses.

Sun protection of transparent elements may avoid overheating of the interior in summer, whereas daylight is still conducted into deeper spatial depths in a nondazzling way. Simulation of daylight can predict the light conditions in deeper regions of the interior, which can be adjusted to comply with the requirements. Different methods of beam tracing are employed to account for light diffusion on surfaces and multiple reflexions.

D. Müller (✉)
RWTH Aachen University, E.ON Energy Research Center, Institute for Energy Efficient Buildings and Indoor Climate, Jägerstr. 17–19, 52066 Aachen, Germany
e-mail: dirk.mueller@eonerc.rwth-aachen.de

M. Grötschel et al. (eds.), *Production Factor Mathematics*,
DOI 10.1007/978-3-642-11248-5_18, © Springer-Verlag Berlin Heidelberg 2010

Fig. 1 Components of a
dynamic building simulation

Heat can directly be supplied to or dissipated from all wall units, whereas external air supply is achieved through windows or other front openings. These transport processes of energy and humid air can be modeled and, using numerical methods for coupled differential equations, one obtains instationary solutions. Thus, the temperature distribution in a certain region can be predicted and a heating, ventilation or air-conditioning system for it can be designed.

Modern buildings are usually powered not only by one single (primary) energy source. Instead, several heat and cold production processes are responsible for a comfortable internal climate. Increasingly, sources of renewable energy are employed as additional means or in special cases, to achieve low carbon dioxide emission, as an exclusive energy source. The complex interplay as indicated in Fig. 1 of all components can only be simulated using elaborate computational tools based on numerical solution of systems of coupled differential equations.

And, at the end of the day, we—the users—wish to feel comfortable in a building. Whether we perceive our surrounding as comfortable and enjoyable is directly related to the micro climate of the surrounding area. The connection between our micro climate and the building envelope or the building services can be drawn from the calculation of the interior air flux using mathematical methods. Typically, finite volume methods are used for the solution of Navier-Stokes equations. Thus, already during the design phase, one can predict the comfortableness of users in a virtual building.

An integral design in architecture is nowadays no longer imaginable without the application of mathematical tools. Every building is original and, unlike in other technical development processes, it is not possible to build a prototype that can be used for measurements. This fact by itself leads to the high relevance of the utilization of computational tools during a virtual design process in architecture.

2 Development of a Novel Cooling System by Means of a Thermohydraulic Building Simulation

Even in spring or in fall, higher temperatures are increasingly reached in the interior of office buildings, which narrows the comfort and the productivity of the users. Ventilation or air conditioning with a conventional compression cooling machine cause a high energy consumption and peaks in power demand, which in several countries have already led to a breakdown of the power supply system.

In the scope of a research project, supported by BMWi [2], and in cooperation with partners from industry, it was possible to develop a novel cooling system that can reach a perceptible room temperature reduction without the employment of a chiller machine. The idea is to use the cool night air, that can be stored using a latent heat storage, for air cooling during the day. The system consists of a latent heat-storage unit that is integrated into the fassade and whose phase change temperature is just below the desired room temperature. Other components are a water based ceiling cooling system within the room and a heat exchanger on the outside of the cladding. All components are connected by a water circuit. During the day the latent heat-storage unit melts absorbing the warm room temperature, and during the night it solidifies releasing the stored heat. For the achieved space cooling one only needs to ensure the required pump performance to run the water circuit.

2.1 Application of the Object-Oriented Programming Language Modelica

For the development of the novel cooling system the object-oriented programming language Modelica was used, in which all components of the planned construction can be replicated. Modelica is a physical description language that is different from the common programming languages in several aspects. As a description language it contains constructs for data flow control (such as conditional statements, loops) but, for instance, scarcely communication functions with the user. This is due to the exactly specified application area of the language, which is not the interactive utilization of programs but simulation of predefined physical systems. The advantage of the language is, therefore, the handling of mathematical equations and systems of equations.

While conventional high level languages are not capable of processing mathematical equations but only of handling one-directional assignments, in Modelica it is possible to formulate equations that can be manipulated and solved during program execution. Since the sets of known and unknown variables can change during simulation, the equations can be manipulated in different ways. This allows for a very clearly arranged description of physical systems.

2.2 Extension of a Complex Simulation Model

For the computation of the cooling system with latent heat-storage units, it is indispensable to construct a complex computation model that incorporates all fundamental physical effects: structural-physical behavior of the building envelope, influence of weather aspects, systems and control engineering. The user takes center stage of the system by defining the inner load, which he sets by just using the space, and also by representing the control benchmark, since his comfort demands have to be met. Since the cooling system works with small difference temperatures, smaller than ten Kelvin, the simulation has to meet high requirements with respect to the accuracy.

For the description of the thermic behavior of a room, different physical processes have to be considered. Firstly, inner and outer heat sources have to be distinguished. Inner sources are heat sources (people, illumination, machines, etc.) that

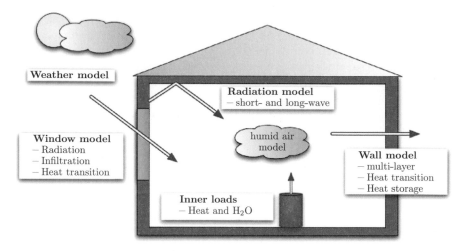

Fig. 2 Physical processes in a room

are located on the inside of a building envelope. Outer heat sources are transported into the interior via the enclosing walls. These include the direct and diffuse radiation, heat transport via the enclosing walls and mass transport (air, materials). In the interior of the room, one has to take into account not only convective transport processes but also long-wave radiation exchange between the enclosing walls, and usually also sources of humidity. A summary of the main processes is given in Fig. 2.

For the simulation of the installation, models of piping, pumps, valves and heat exchangers are available that replicate the hydraulics of the cooling system. In addition, virtual sensors and controllers are employed for the optimization of the cooling system.

To obtain a model of the latent heat-storage unit, it was necessary to develop a model for the phase change of the latent heat-storage material. All other components and physical processes could be taken into account using conventional approaches from building and installation simulation. Many models of the phase change, that are described in the literature, are usually based on either an analytic solution of the phase change of pure materials or, when dealing with mixtures of materials with a phase change region, various grades of simplifications of the phase change process are used.

There are many approaches that use the mean enthalpy of the material for the temperature region of the phase change [3] or linear functions, as described in [4]. Only a few approaches found in the literature use a detailed model of the physical behavior of the phase change of mixtures of materials that would include a possible supercooling or hysteresis of the material.

Suggestions for the description of the phase change using enthalpy as a temperature-independent function were published within this research project [5] and also described in [6]. For a more realistic replication of the phase change for

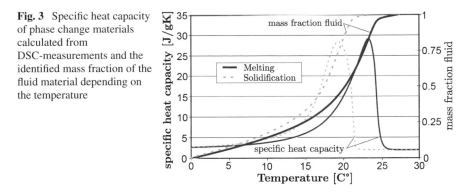

Fig. 3 Specific heat capacity of phase change materials calculated from DSC-measurements and the identified mass fraction of the fluid material depending on the temperature

the design of latent heat-storage systems and the estimation of their performance, it is indispensable to model the melting and solidification processes.

Especially when using natural heat sinks for the cooling of the storage, an exact model of the solidification process becomes essential. A difference in temperature of only a few Kelvin can significantly derogate the overall performance of the system. In Modelica, the model for phase change was developed using temperature-dependent enthalpy of the material, calculated from DSC-measurements (Differential Scanning Calorimetry), see Fig. 3.

For the calculation, the storage capacity is discretized using the finite volumes method. The model replicates not only the state-dependency of the material values density, heat conductivity and heat capacity but also the temperature-dependency of the specific heat capacity in the region of phase change.

A particular difficulty arises from the hysteresis of the specific heat capacity measured during melting and solidification: A unique mapping between temperature and specific heat capacity is only possible knowing its previous history.

2.3 Results of the Calculations

By means of simulations of the novel cooling system, many issues can be addressed that can not be investigated experimentally or at least not without enormous efforts.

In a simulation, unlike in an experimental investigation using a test building, a weather period can be repeatedly simulated using different system parameters. The results provide a direct comparison between different phase change materials and can be used for the optimization of operating control. From year-round simulations, one can determine the mean cooling performance, the cumulative frequency curves for the room temperature and the energy demand for the pumps. Prior to extensive measurements, based on calculations, an optimized cooling system is already available.

Figure 4 shows the simulation results for the operation of a cooling system based on a latent heat-storage during a midsummer period. The room temperatures of the cooled room, the uncooled reference room and the temperature of the ambient air

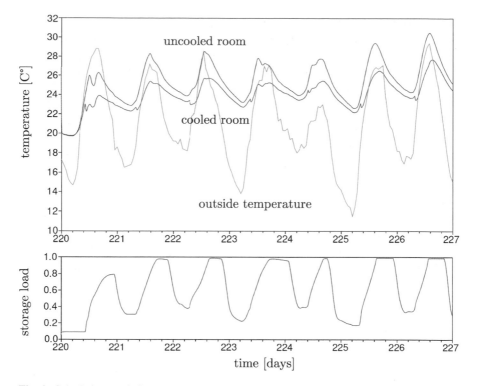

Fig. 4 Calculation result for a room with a cooling system with latent heat-storage and one room without it (*upper diagram*), illustration of storage load (*lower diagram*)

are depicted in the upper part of the diagram. The lower part shows the state of the storage load. Clearly, one can see here that the capacity of the latent heat-storage is not always sufficient.

The obtained information "storage capacity is not sufficient in midsummer" can be used in different ways. If the system has to be sufficiently effective for all simulated cases, then the simulation results can be drawn on to optimize the storage size. Is a further extension of the storage capacity not possible, for instance, if the size of heat exchangers on the fassade is structurally limited and, thus, the regeneration of higher storage capacities is impossible, then the simulation results can be used for an optimized storage operation. To this end, a control unit could be designed that efficiently uses the limited cooling performance. For instance, this can be achieved by a coupling to a usage scheme of the room. In this case, the control system would safe storage capacity by turning off the cooling system when the room unused. For office space, this strategy would imply that on week-ends even at high room temperatures the cooling is turned off, thus, making the full storage capacity available at the beginning of the following week. However, also an optimized control during the day can improve the operation of the latent heat-storage unit. Using the data of the weather forecast, on the days when the storage capacity will presumably not be sufficient for keeping up the usual temperature upper bounds, it can be achieved that from the beginning of the day higher temperatures will be tolerated. In this case, the cooling system would intervene later then usual but, on the other hand, the storage capacity would last until the end of the work day. Using this strategy, a slightly higher but stable temperature level can be achieved, whereas with a "non-predictive" control, the temperature would be downregulated to the usual upper bound for the first part of the day, which then, however, would lead to a complete breakdown of the cooling system. A significant increase in temperature in the afternoon would be the consequence.

All mentioned approaches are based on mathematical models. The results from measurements of a prototype of the cooling system without a chiller machine have shown that the numerical results are in accordance with the measured data. Due to inexact boundary conditions (building circulation, structural tolerances, higher dimensional heat conductance processes), it can not be expected that the calculated temperature distributions exactly replicate the measured data. Differences up to one Kelvin are common and do not delimit the main advantages of a simulation: the comparative estimate of different system variants.

3 Strengths and Weaknesses Analysis, Challenges

Many important functions of the building envelope and the installation can be modeled and evaluated using mathematical methods. The energy efficiency of the building is mainly determined within the first stages of the design process and can be enhanced by a systematic application of different computational tools.

Although nowadays, many tools for virtual design of buildings are available, only rarely the more sophisticated mathematical tools are used in praxis. In most projects, the cooperation between the architectural planners and the specialist consultants is not geared to a virtual development process. Usually, an overall model of the building, that all involved planners could work on simultaneously, does not exist. The planning tasks are accomplished on partial models, the final space geometry and wall constructions are added only at a late stage of the design process.

In larger building projects, thermic building simulations for some particular rooms are carried out already in the first phases of the design process. In addition, day light analysis and in special cases also flow simulations are performed. Unfortunately, the application of these tools is usually limited to very early versions of the designing work, all changes and modifications during the design process are not monitored "online" and evaluated. However, good numerical models could provide many valuable insights for the usually involved building services engineering. The numerical models are furthermore suitable for an optimization and monitoring of the central building control systems.

In addition to a continuous enhancement of the computational methods, special effort needs to be put into the service interface of the assembly section. For the virtual design process, as it is nowadays widely realized in the automotive industry, the construction process is lacking an integral virtual model and the necessary

In the future, considerably less primary energy sources will be available for operating our buildings. For the reduction of the primary energy demand, a reduction of the energy flow through the building envelope as well as the quality of the used energy could be addressed. Many technical processes produce waste heat in large amounts that often remains unused due to the low temperature level. The extension of district heating in urban areas as well as the lowering of the necessary difference temperatures in the building services, may lead to significant savings for heating and cooling.

Therefore, new computational methods do not only have to take into account one single room or one single building. The system boundary is shifted to quarters, districts or whole supply areas. Specifications for the load profiles and the necessary temperature levels have, therefore, to be deduced from the whole system and implemented in the individual buildings.

4 Visions and Recommended Actions

Each new building is individual and, unlike in many other development processes, it is not possible to build prototypes. Therefore, further development of the virtual design process is essential for a continuous enhancement of energy efficiency of buildings. Mathematics is already largely contributing to the development of new components and to the improvement of energetic concepts. In the future, the importance of mathematical methods will grow and increasingly many aspects of the design process will be incorporated into a virtual scheme. This leads to highly complex computational models, which require a suitable data management and evaluation as well as adequate methods for the estimation of numerical accuracy.

One of the most important points for the further development of design tools is surely the interface between the mathematical methods and the users. The creative collaboration between the architectural planner and the specialist consultants may not be limited by the specifications of a virtual development environment. In fact, the team work on the design of a building should be supported by an appropriate visualization and control. Through a simplification of the man-machine interaction, also the client and the eventual user may actively become part of the design process if perceptible or even palpable space emerges from a three-dimensional visualization process.

References

1. Umweltbundesamt (ed.): Nachhaltige Wärmeversorgung – Sachstandsbericht. Dessau (2007). ISSN 1862-4359
2. Verbundforschungsprojekt "Niedrigenergiesysteme für die Heiz- und Raumlufttechnik": Finanzierung durch das Bundesministerium für Wirtschaft und Technologie (BMWi), Betreuung durch den Projektträger Jülich (PTJ). Förderkennzeichen 0327370A

3. Lamberg, P.: Approximate analytical model for two-phase solidification problem in a finned phase change material storage. Appl. Energy **77**, 131–152 (2004)
4. Glück, B.: Einheitliches Näherungsverfahren zur Simulation von Latentwärmespeichern. In: HLH, Lüftung, Klima, Heizung, Sanitär, Gebäudetechnik, vol. 57, pp. 25–30. Springer, Berlin (2006)
5. Tschirner, T., Haase, T., Hoh, A., Matthes, P., Müller, D.: Simulation of phase change procedure in heat storage devices using the object oriented modelling language Modelica. In: Association francaise du froid (eds.) Proceedings of the 7th Conference on Phase Change Materials and Slurries for Refrigeration and Air Conditioning 2006. Dinan, Frankreich (2006)
6. Günther, E., Mehling, H., Hiebler, S.: Modeling of subcooling and solidification of phase change materials. Model. Simul. Mater. Sci. Eng. **15**, 879–892 (2007)

Medicine

More Mathematics into Medicine!

Peter Deuflhard, Olaf Dössel, Alfred K. Louis,
and Stefan Zachow

1 Executive Summary

This article presents three success stories that show how the coaction of mathematics
and medicine has pushed a development towards patient specific models on the basis
of modern medical imaging and "virtual labs", which, in the near future, will play
an increasingly important role. Thereby the interests of medicine and mathematics
seem to be consonant: either discipline wants the results fast and reliably. As for
the medical side, this means that the necessary computations must run in shortest
possible times on a local PC in the clinics and that their results must be accurate and
resilient enough so that they can serve as a basis for medical decisions. As for the
mathematical side, this means that highest level requirements for the efficiency of
the applied algorithms and the numerical and visualization software have to be met.
Yet there is still a long way to go, until anatomically correct and medically useful
individual functional models for the essential body parts and for the most frequent

P. Deuflhard (✉) · S. Zachow
Zuse Institute Berlin (ZIB), Takustr. 7, 14195 Berlin-Dahlem, Germany
e-mail: deuflhard@zib.de

S. Zachow
e-mail: zachow@zib.de

P. Deuflhard · S. Zachow
Freie Universität Berlin, Dept. Mathematics, Arnimallee 6, 14195 Berlin, Germany

O. Dössel
Institut für Biomedizinische Technik, Universität Karlsruhe (TH), Geb. 30.33, Raum 515,
Kaiserstraße 12, 76131 Karlsruhe, Germany
e-mail: Olaf.Doessel@ibt.uni-karlsruhe.de

A.K. Louis
Universität des Saarlandes, Fachrichtung Mathematik, Postfach 15 11 50, Geb. E 1.14.OG,
66041 Saarbrücken, Germany
e-mail: louis@num.uni-sb.de

M. Grötschel et al. (eds.), *Production Factor Mathematics*,
DOI 10.1007/978-3-642-11248-5_19, © Springer-Verlag Berlin Heidelberg 2010

diseases will be at hand. This will only be possible, if more mathematics enters into medicine.

2 Mathematics in Medical Imaging

2.1 History of a Success

When Konrad Röntgen discovered the "X-rays" in 1895 in Würzburg, Germany, he opened a window to non-invasive insight into the human body. The attenuation of X-rays is strongly dependent on the tissue through which they travel, for example, bone attenuates more than fat. The thus produced shadow images deliver visual information about the interior of the body. The disadvantage is that dense tissue like bones masks the tissue along the same path of the X-rays, whence the application was restricted to those parts of the body with few bones. Examinations of the brain were, due to the surrounding skull, virtually impossible without the dangerous addition of contrast agents.

The conceptual and technical breakthrough was achieved in the sixties: Nobel laureates Allan Cormack and Godfrey Hounsfield suggested an imaging system, where shadow images of an imaginary cross section through the body and for many directions are registered simultaneously. Each of these shadow images contributes only little information. The exploitation of the fact that all of these images cover the same part of the body allowed for images of the interior of the body with previously unknown resolution. This was the starting point of a long success story.

Computed Tomography (CT)

Computed tomography, also known as computer assisted tomography (CAT), is the name of an imaging system, where the role of mathematics is evident, see Fig. 1. With the large number of collected data the calculation of the desired image information can be achieved only by means of powerful computers. Fundamental for the use of computers is the development of efficient algorithms, based on a precise mathematical model of the complex connection between measured data and image information to be determined. The measured data, i.e. the attenuation of the X-ray intensity after travelling through the body, is related to the X-ray attenuation coefficient, interpreted mostly as the density of the tissue, along the path of the rays. In mathematical terms, the observed attenuation is related to the line integral of the X-ray attenuation coefficient along the ray path. The arising integral equation is, in the 2D case, named as Radon transform, after the Austrian mathematician Johann Radon. In the first commercial scanners, Godfrey Hounsfield solved this integral equation by standard discretization methods. He projected the solution on a pixel basis, resulting in large, unstructured systems of linear equations that he solved iteratively. Only in the mid-seventies the integral equation was actually recognized as

Fig. 1 Reconstruction of a trans axial cut through the head. Source: Siemens Healthcare, SOMATOM Definition AS+

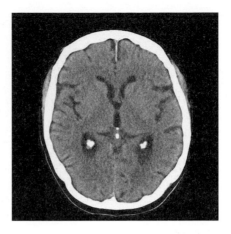

the Radon transform, for which Radon had derived an analytical inversion formula already in 1917.

However, it has been a long way to go from his mathematical formula to working algorithms, questions to be answered first concerned image resolution, non-uniqueness for finitely many data, and optimization of parameters and step sizes. In addition, a phenomenon typical for all imaging methods showed up: These problems are so-called *inverse problems*, where unavoidable input data errors, due to photon scattering, beam hardening, or miscounts in the detector, are extremely amplified in the solution. In order to avoid this unwanted effect, the problem has to be regularized, such that a balance between best possible resolution in the image and maximal damping of the noise is obtained. In the case of CT this is achieved by elimination of the high frequency components in the data. Nowadays, the method developed by the mathematicians Shepp and Logan [17] is well established. The resulting algorithms consist of two steps, a filtering of the data, via discrete convolution in the two-dimensional application or via Fourier techniques in higher dimensions, and a backprojection of the filtered data onto the reconstruction region. Both steps can be performed in parallel during the measuring process, resulting in a dramatic gain in computer time. It is worth mentioning that the acceleration of computing time due to the progress of mathematical algorithms is much higher than the one due to the progress in computer hardware.

Of course, the advancement of engineering performance should be mentioned, which allowed for an essential speed-up in time and accuracy of the measurement process. This led to completely new scanning geometries and thus to new challenges in mathematics. The originally introduced parallel geometry, where one path was measured at a time instance, was replaced by fan-beam geometries, where, at one time instance, the complete plane under consideration was covered by a set of divergent beams. The helical scan, where the patient is moved through the gantry, first realized with a few detector lines, was established in the early nineties. Today real 3D scanning with a detector array is the object of intensive research, already widely used in non-destructive material testing. The mathematical model remains to be line integrals over the paths of the rays. The thus computed reconstructions are

Fig. 2 Reconstruction und
visualization based on
MRI-data. Source: Siemens
Healthcare, Sektor
Magnetresonanztomographie,
MAGNETOM Verio 3T

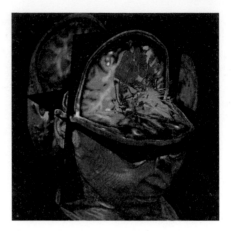

often processed in order to enhance the diagnosis. Methods, wherein parts of the image process steps are integrated parallel to the reconstruction method are presently under development. Among the pioneers from the mathematical side are Gabor Herman, Kennan Smith, or Frank Natterer. An extensive presentation can be found in Natterer's most recent monograph [15].

Magnetic Resonance Imaging (MRI)

Magnetic resonance spectroscopy, 1946 independently developed by Felix Bloch and Edward Purcell, gives information on the chemical surrounding in a molecule, by exciting it to resonance in a strong magnetic field. This was the starting point for magnetic resonance imaging (MRI, formerly called nuclear magnetic resonance (NMR)). Paul Lauterbur, Nobel laureate for medicine in 2003, achieved a spatial resolution by modulating the primary homogeneous magnetic field by so-called gradient fields in such a way that the regions of constant resonance frequency became planes through the body. In that way, plane integrals over the proton distribution inside the body were measured. The corresponding mathematical model is the 3D Radon transform [14]. In the early eighties there were no desktop computers available, allowing for handling these huge data sets within tolerable time. Peter Mansfield, Nobel laureate as well in 2003, further developed gradient fields and excitations such that the Fourier transform could be used to invert the data. The regularization, already discussed above in the context of X-ray CT, was achieved by eliminating the high frequency components. In that way, with high technical complexity, the mathematical problem had been simplified to be solvable with the computers of that time.

Ultrasound

This imaging technique has been used in medicine for quite a time. Well established are B-scan devices, acting as emitter and receiver at the same time sending ultrasound waves into the body. Measured are travel times of the echoes, which are produced at interfaces of tissues with different acoustic impedance and scattering properties. Upon assuming that the speed of sound is independent of the tissue an image can be computed. Even though this is only approximately correct, nevertheless the images contain sufficient diagnostic information. If, in addition, the frequency shift is recorded, then using the Doppler effect, the speed of blood can be calculated.

2.2 Mathematics as Innovation Factor

For many years, medical industry worldwide has played an internationally leading economical role. The following numbers may document this statement for Germany. In 2002, commodities in X-rays & CT for over 2 Billion Euro have been produced. The German trade surplus with respect to X-rays and CT figures around 1 Billion Euro, around 600 Mio. Euro for MRI systems.[1] Leading companies in Germany are Siemens Medical Solutions, but also Phillips Health Care and General Electric. A main location factor seems to be well-trained applied mathematicians and engineers with a sound understanding of mathematics.

2.3 Perspective: New Imaging Methods

The perspective in medical imaging is dominated by the development of newer measuring technologies.

3D Tomography

In non-destructive material testing, *3D X-ray CT* is widely used in connection with a circular scanning geometry; i.e., the X-ray tube is moved on a circle in a plane around the examined object with the detector plane positioned at the opposite side. Helical geometry is favoured in medical applications, but the therein necessary variable shift of the patient has not been solved yet satisfactorily in existing algorithms. In principle, X-ray tube and detector can be moved along arbitrary trajectories around the patient. The determination of trajectories that are optimal with respect to resolution and stability remains a mathematical challenge. Higher hardware capacities will allow for new approaches different from the classical filtered backprojection type.

---

[1] Foreign trade statistics of OECD and German Ministry of Education and Research.

Fig. 3 Three-dimensional
visualization of
two-dimensional CT scans
through a human heart.
Source: Siemens Healthcare,
Sektor,
Computertomographie,
SOMATOM Definition AS+

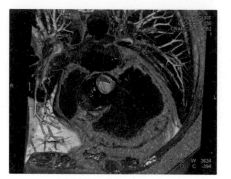

4D Tomography

Since the early eighties, *3D spatial presentations* have been computed from a series of 2D images. If one combines these volumetric images for different instances,
clearly *time* enters as a fourth dimension. The resulting movies, for example from
a beating heart, are impressive: Fig. 3 is a snapshot from such a movie. Certainly,
it will take some time to develop this kind of technique for routine application. At
present, each image frame is treated separately, a regularization with respect to time,
as would be necessary, e.g., in current intensity reconstruction from MEG/EEG data,
is not yet implemented.

Electron Paramagnetic Resonance Imaging

In this technique a decoupling of the fourth dimension is not possible, since there,
besides the three spatial dimensions, a *spectral* dimension shows up additionally.
The corresponding mathematical model is the Radon transform in four dimensions.
This technology is presently studied in the stages of pharmaceutical research and
animal experiments. However, due to the limitations in field strength, the data in
Radon space cannot be sampled completely with the consequence that a limited
angle problem has to be solved. Theoretically, the desired distribution would be
uniquely determined, if all data in the restricted range were available, but instabilities and strong artefacts complicate the reconstruction problem.

Ultrasound CT

In this approach sender and receiver are spatially separated, the corresponding mathematical model is an inverse scattering problem for the determination of the spatially
varying sound impedance and the scattering properties. The difficulty here is that,
in contrast to CT, the paths of the waves depend upon the variable to be computed,
which makes the problem highly nonlinear. A linearization of the problem via Born
or Rytov approximation neglects the effects of multiple scattering and is therefore
not sufficiently accurate. That is why ultrasound tomography today is still a challenge to mathematics and algorithm development.

Transmission Electron Microscopy (TEM)

For the visualization of biomolecules by TEM, various approaches are pursued. If one does not aim at averaging over many probes of the same kind, again a limited angle problem arises. In addition, wave phenomena enter for small sized objects, leading, as in ultrasound tomography, to nonlinear inverse scattering problems. Fortunately, linearization is feasible here, which facilitates the development of algorithms significantly.

Phase Contrast Tomography

In this technology, where complex-valued sizes have to be reconstructed, linearizations are applicable, too. The phase supplies information even when the density differences within the object are extremely small. Using data from a synchrotron, impressive results have been achieved, see Fig. 4. Due to different scanning geometries, medical application still generates challenging problems for the development of algorithms.

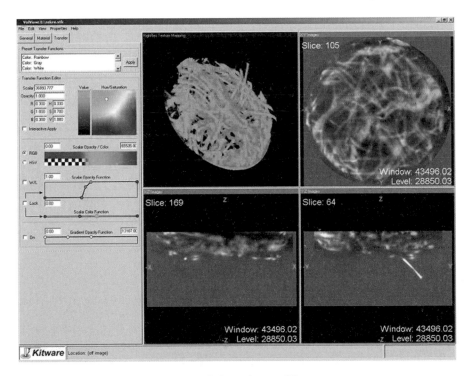

Fig. 4 Phase contrast tomography: phase information, see [9]

Diffusion Tensor MRI

This method provides a tensor at each reconstruction point, thus bearing information concerning the diffusivity of water molecules in tissue, see Fig. 2. Here reconstruction and regularization are performed separately. Only after having computed the tensors, properties of the tensor like symmetry or positive definiteness are produced point by point.

The above list of imaging techniques under present development is by no means complete. Methods like *impedance tomography* are studied as well as those applying *light* to detect objects close to the skin. In all of the mentioned measuring techniques, the technological development is so advanced that the solution of the associated mathematical problems—like modelling, determination of achievable resolution and development of efficient algorithms—will yield a considerable innovation thrust.

Image Fusion

Different imaging systems supply different kinds of information, the fusion of which in a single image can give a lot more insight to the physician. Prerequisite of image fusion is an imaging of (almost) identical cross sections through the body or of volumetric presentations. Different physical formats have to be mapped to each other. This requires methods of image recognition, for example segmentation (see Sect. 4), to determine the geometric mappings. First methods exist, partially applied in connection with the rapidly developing *intensity modulated radio therapy* (IMRT).

Feature Reconstruction

Supporting medical doctors in the interpretation of the ever increasing amount of information is an extremely important task. In non-destructive testing in materials science, where tomographic methods are applied as well, tools for an *automatic* recognition of precisely defined properties are in use, such as the detection of blow holes inside metal. In medical applications, this surely is not the aim at the moment. Nevertheless, a preprocessing of images is often useful. At present, the two

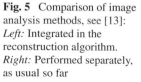

Fig. 5 Comparison of image analysis methods, see [13]: *Left:* Integrated in the reconstruction algorithm. *Right:* Performed separately, as usual so far

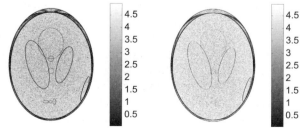

steps, reconstruction and pre-processing, are performed separately. In [13], a novel technique has been presented, wherein the pre-processing step is optimally integrated into the reconstruction algorithm. In this way, significantly refined results can be realized, as shown in Fig. 5. Nearly without any additional effort both the classical reconstruction and the enhanced images can be computed. This approach of combining image reconstruction and image analysis paves the way for *feature reconstruction*.

3 Mathematics in Cardiology and Cardial Surgery

3.1 Success Stories: ECG and Biosignal Processing

Everyday some million Electrocardiograms (ECGs) are recorded, many of them Holter ECGs (long term ECGs). A single 24 h ECG contains about 100 000 heart beats. It is practically impossible that a medical doctor can look through all these data to filter out events of diagnostic relevance. Not only that precious time would be lost, in addition there is the danger that alertness drops quite fast and important events are overlooked.

Telemedicine is an important future market—currently many companies position themselves in the market. Sportsmen and elderly people can buy a continuous monitoring of their state of health—24 hours per day and 7 days per week. In addition to the ECG often also other vital health parameters are measured, like weight, body temperature, blood oxygen saturation etc. Also in this case it is neither practical nor would it be affordable that physicians monitor all these data.

A solution is offered by software packages that support the medical doctor in ECG data analysis. With their aid the essential characteristics of the ECG are revealed and the most important parts of the data stream are highlighted. During monitoring in the intensive care unit of the hospital or in Telemedicine, a life threatening state of the patient must be detected immediately and with high reliability and an alarm must be given. But which algorithm is able to perform this data analysis with the highest possible reliability? These algorithms must be applicable to all (!) ECGs, even if the ECG of an individual patient is extremely seldom and strange. And they must be very robust, since often artefacts are much larger than the biosignals of interest.

Today mathematical methods together with the implemented software are indispensable tools for good biosignal analysis. Companies of medical systems have recognized that the added value in the production chain for ECG and monitoring systems is only to one half due to hardware (e.g. in electronics). The other half comes from intelligent signal processing and analysis. Suppression of artefacts is carried out using filters in the frequency domain, but also wavelet transformation is applied, since it allows for best possible frequency separation simultaneously with good temporal resolution. Detection of data intervals of diagnostic relevance is realized by first extracting characteristic features using mathematical methods followed by a

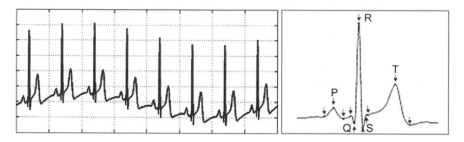

Fig. 6 Measured ECG together with the result of a computer assisted annotation of the most important time instants [11]

mathematical classification (nearest neighbor, Bayes-maximum-likelihood, neural networks etc.). This way, e.g., a QRS complex can be separated from an extrasystole [18]. Also model based techniques like Kalman filters or Markov models are employed. Figure 6 shows a typical ECG, as can be found in long term ECG acquisitions, together with the annotation of the most important time instants like the R-peak, the P-wave, or the T-wave.

For an early detection of trends in the ECG, new mathematical methods of time series analysis are employed. Using these methods, stochastic fluctuations can be separated from deterministic ones and significant changes can be uncovered long before the eye of the medical doctor can find it in the ECG. Currently, one is trying to find mathematical rules (linear and non-linear ones) that are able to predict the ECG of the next beat from the preceding ones. Once these rules will have been found, the detection of any deviation is an indication to an early stage of a disease. This research is right in the centre of the mathematical field of "complex dynamical systems" that will enable to detect many characteristics of diagnostic relevance in the ECG. For example, most often stable trajectories can be found in the state space of the heart, but sometimes they pass through bifurcations into chaos (like fibrillation [5]).

3.2 Mathematics as Innovation Factor

In medical engineering in general and in cardiology and heart surgery in particular it is of increasing importance to recognize the essentials out of a huge amount of information about a very complex system, and to draw the right conclusions. Algorithms and computers can help in this task. By no means it is planned to replace the medical doctor by a computer. But the computer with its implemented algorithms can already today give valuable advice to the physician and this trend will continue and gain relevance.

In the first stage, algorithms can highlight or zoom into information contained in the data that cannot be recognized with the naked eye: small changes of the ECG, unusual objects in the images, characteristic deviations of hematological data etc.

In the second stage, values of diagnostic importance are determined quantitatively. Often the decision to choose one or another line of therapy depends on whether a specific value is above or below a given threshold. Using mathematical methods these values can be found as accurately as possible even in disturbed and noisy data. Finally, mathematical methods can be employed to optimize therapy by trying out various variants and evaluate the outcome using objective and traceable criteria.

In future, it will be decisive for manufacturers of medical devices to integrate the huge amount of patient data into a comprehensive view in an intelligent way and to support the medical doctor in his/her diagnosis and therapy decision. Only companies that can offer these options will play a major role in the future world market.

3.3 Perspective: The Virtual Heart

Integration of Imaging, Electrophysiology, and In-Vitro Diagnostics

The ECG described above is an important source of diagnostic information for the physician, but by far not the only one: Medical images and in-vitro diagnostics like, e.g., hematological data are indispensable additional sources of information. Mathematical methods will play a major role in integrating all these data into a comprehensive picture about the state of the patient. Purely geometrical information of the medical images like projection X-ray, CT, ultrasound or MRI is complemented by functional data. Some of these functional data can be gained from medical imaging devices: metabolism, e.g., from PET, flow, e.g., from Doppler ultrasound, or perfusion, e.g., from contrast enhanced X-ray. In the next step, other data have to be integrated like, e.g., ECG, blood pressure, blood oxygen saturation, hematological data, and enzyme activities and, on the long run, also genetic profiles that show up a predisposition for a disease. The objective is to integrate all these data into a complete patient model so that finally important functional characteristics can be determined. (For readers with medical and physiological knowledge: such functional characteristics are, e.g., elasticity, contractility, propagation of electrophysiological depolarization, perfusion, enzyme activity, etc.)

Such an integrated heart model, a virtual heart, can depict how the real heart of the patient deviates from the physiological average. This opens many new options both in diagnosis and therapy planning.

Therapy Planning

Mathematical models of heart and circulation have many interesting applications in cardiology and heart surgery.

Upon using *electrophysiological models*, it will be possible to

- optimize RF ablation for atrial fibrillation,
- adapt stimulation sequences for heart pacemaker and cardiac resynchronization therapy,
- predict the effect of new pharmaceutical drugs for the treatment of arrhythmias.

Upon using *elastomechanical models*, it will be possible to

- optimize surgical cuts in open heart surgery for therapy of aneurysms and for surgical reconstruction of ventricles,
- evaluate the elastomechanical consequences of an infarction.

Using *fluid dynamical models* of vessels one can

- better understand the etiology of stenoses (plaque in a blood vessel),
- optimize stents to open stenoses and to treat aneurysms.

Using *circulation models* it will be possible to

- evaluate new drugs to treat hypertension (high blood pressure),
- control extracorporal circulation during heart surgery.

For the purpose of further explaining these options, one of them is highlighted in the following.

Planning of RF Ablation at Atrial Fibrillation

RF ablation means the precise destruction of cells (often tumour cells) by feeding a radio frequency current into the tissue. Using an image guided catheter the tissue is heated locally above 42°C so that the proteins are coagulated in an area near to the tip of the catheter. A well-defined scar is created. This therapy is also a preferred choice in case of atrial fibrillation.

Electrophysiological computer models of the heart start with the various ion channels in the membrane of cardiac cells. The dynamics of ion channels can be described by "stiff" ordinary differential equations. Their rate constants are a function of the transmembrane voltage [6]. Single cells lead to a nonlinear system of about 20 coupled differential equations [20]. The spatial coupling of the cells is modelled by partial differential equations, e.g., in the so-called "bidomain model" [7]. As eventually electric potentials in the body are to be determined, basically the equations of electromagnetic field theory have to be applied. Electrophysiological processes in the human body are comparably slow, which is why only the Poisson equation of electrostatic problems—an elliptic partial differential equation—has to be solved. Numerical simulation of these equations requires the discretization of space and time. In biomedical engineering, an explicit Euler discretization is preferred up to now, [6], whereas the mathematical community applies stiff integrators as a standard (see [1] and references therein). Here is space left for improvement via mutual exchange of methods and ideas. Spatial discretization is carried out, e.g., by finite element methods—with uniform or adaptive meshes—or by finite differences. In case of uniform discretization a problem with some million degrees of freedom has to be solved, preferably in the clock pulse of a second. Also the so-called "forward problem", which is used to find the electric signals on the body surface (including the ECG) from the electric source distributions in the heart, leads to a Poisson equation. Figure 7 (top row) shows a simulation of the depolarization of a healthy human atrium: the electric depolarization starts in the sinus node, the natural "pace maker" of the heart, and then spreads out across the right and left atrium.

Next we present details of the modelling of pathologies, here especially of atrial fibrillation. For this application, new parameters have to be assigned to the atrial tissue that are able to describe the pathological case. In this case, the computer model switches into a chaotic state so that patterns of depolarization can be observed in the model which amazingly well resemble the patterns observed in real patients that actually suffer from atrial fibrillation. A more and more applied therapy now is to heat small spots on the atrial tissue in such a way that a scar is formed, which then will not conduct the excitation furtheron (this is the RF ablation, see bottom row of Fig. 7). This procedure can easily be implemented in a computer model.

An important, but still open question is how to find the optimal strategy for RF ablation. What would be the best choice of ablation points and lines in the atrium so that fibrillation is terminated reliably using as little scar as possible and, in addition, protecting the patient from flaring up of the disease? Computer models can indeed answer this question by testing different strategies in the virtual atrium. With a "reset" the virtual atrium can be switched back to the original situation and a new test can be started. This is obviously not possible in real patients. A scar that has been burnt into the atrium cannot be removed again!

The above sketched method for optimization of RF-ablation has not been transferred to clinical practice yet. Validation requires still many more experiments. The biggest challenge will be to adapt the computer model to the individual patient using measurements (images and electrical data) to derive a patient-specific ablation strategy. A bottleneck on the way towards this goal are also algorithms for the computation of cell models and the field equations: Here new mathematical methods are needed to speed up the whole simulation process—so as to "catch up" with the heart dynamics.

4 Mathematics in Therapy and Operation Planning

In recent years, mathematics has gained an increasingly important role in medical planning. Palpable successes have been achieved in the cancer therapy hyperthermia as well as in cranio-maxillo-facial (CMF) surgery, liver surgery, and orthopedic surgery.

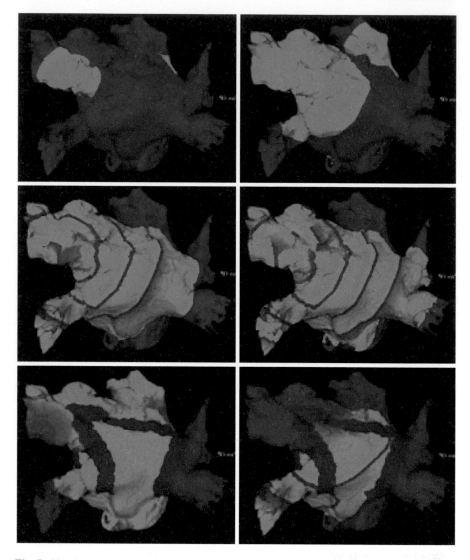

Fig. 7 Simulation of the spread of depolarization in the human atrium. Transmembrane voltages at different time instants of depolarization, see [16]. *Top row:* healthy atrium. *Center row:* atrial flutter. *Bottom row:* test of an ablation therapy

4.1 Success Story: CMF Surgery

Problem Formulation

A "beautiful" face has demonstrably social advantages. But the contrary occurs, when a face has been malformed from birth on or by an accident (cf. Fig. 8). In

Fig. 8 Patients with congenital malformations [4]

CMF surgery, bony structures of the skull are operatively corrected with the aim of both functional and aesthetic reconstruction.

Success by Mathematics

Already since the early 1990s, computer scientists have dealt with computer assisted 3D planning of cranio-maxillo-facial operations on the basis of CT data (see [21] and references therein). Meanwhile mathematics, too, plays a crucial role in the planning of highly complex operations. Its most important tools appeared to be *virtual labs* for therapy and operation planning [2]; this term describes extensive software environments (such as, e.g., Amira™ [19]), within which patient specific geometric models are integrated with partial differential equations for elastomechanics, fluid dynamics, or diffusion as well as with fast algorithms for their numerical solution and visualization. Today, before the surgeon performs his first cut, reliable predictions about the postoperative appearance can be made. For the purpose of illustration, Fig. 8 shows a series of patients whose operation planning has been done in cooperation of clinics and ZIB. In Fig. 9, a comparison of the computed predictions and the actual outcome of two operations is given [21].

4.2 Mathematics as Innovation Factor

In the course of many years of collaboration between mathematics and medicine, a paradigm has crystallized that might be characterized as follows: (see Fig. 10):

- Generation of a "virtual patient" from data of the real patient,
- mathematical therapy and operation planning in a "virtual lab",
- transfer of the results back to the situation of the real patient.

The first step requires the construction of a sufficiently accurate 3D computer model of the patient from medical imaging data. The second step contains the fast numerical solution of partial differential equations over a realistic body geometry of the individual patient. The third step comprises techniques of registration and navigation for an exact implementation of the planning. All steps have the importance of an efficient 3D visualization in common. The corresponding mathematics and

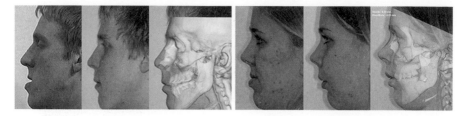

Fig. 9 Comparisons of two patients: before operation, after operation, overlays with predictions [22]

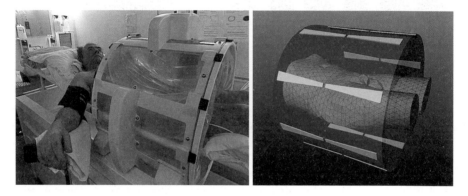

Fig. 10 Hyperthermia: real versus virtual patient in applicator [3]

computer science is not treated here in depth; for intested readers, we refer to the papers [4, 21, 22].

On the way from the real to the virtual patient a sequence of sub-steps has to be taken, which contain a lot of mathematics themselves.

Geometric Modelling [23]

Tomographic imaging (cf. Sect. 2) yields a stack of 2D planar cross sectional images. However, any therapy planning will absolutely require 3D models of the individual anatomy. For this reason, methods need to be provided, which generate a reliable geometric 3D patient model from this kind of 2D information.

As the first task therein, *segmentation* is incurred, whereby the organic substructure is to be derived from the pure density information of CT or MRT images by classification of image subdomains. This substructure is one of the crucial prerequisites of functional patient models. Methods available until a few years ago (such as, e.g., region growing, watershed methods, intelligent scissors, active contours etc.), started from 2D stacks to generate the 3D model and were to their major part *interactive*, i.e. they required the intermittent action of experts. As a consequence, they were too time consuming for clinical practice. A new generation of *fully automatic*

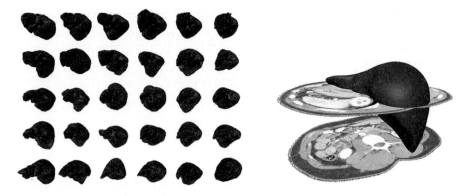

Fig. 11 Statistical 3D shape analysis [10]. *Left:* Test set of known liver data of various patients. *Right:* Segmentation of new liver data (for the purpose of demonstration only two CT cross sectional images)

methods (see, e.g., [10]) uses methods of *statistical 3D shape analysis*: These methods start from some "averaged" 3D model combined with "essential coordinates" established via PCA (principal component analysis) applied to a "training set" of known patient data. With each new data set the training set can be enriched. The explicit use of shape knowledge thus mimicks the process in the mind of the radiologist, who, by education and experience, has a 3D imagination of the situation, into which he can cognitively insert the 2D image under consideration. Some insight into this methodology is given in Fig. 11.

As a result of the segmentation procedure, one obtains *surface meshes* on all tissue interfaces (outer and inner boundaries). In order to achieve a decent balance of a low number of mesh nodes versus a high approximation quality, the surface meshes are coarsened depending on local curvature. These reduced surface meshes then build the basis for the establishment of *volumetric meshes* by tetrahedrons (see [23] and references). Such types of meshes are particularly well-suited for successive refinement—a feature that is crucial in connection with *adaptive multigrid methods* for the fast solution of partial differential equations.

Example: Osteotomy Planning [21]

Osteotomy means the surgical cutting of bone. This part of the planning primarily aims at functional rehabilitation. Whenever several bone segments are involved, which need to be arranged in mutual relationship, or when different therapy variants come into play, then the expected aesthetic outcome will be an important criterion to be taken into account in the planning. A variety of different operation strategies can be planned in the computer with respect to cost efficiency and to surgical safety, in particular for complex bone dislocations. The simulation of the associated soft tissue appearance permits an assessment of relocations of the upper and lower jaws in view of aesthetics. Basic for a reliable prognosis of facial appearance after the intended arrangements of bone structures is, on one hand, an adequate geometric model of the tissue volume with all embedded structures and neighboring bone boundaries, and, on the other hand, a physical deformation model that describes the mechanical properties of biological tissue in acceptable approximation.

Already with the geometric model alone important tasks in medicine can be performed, such as

- the localization of a tumor for the purpose of an exact planning of surgical openings or radiological measures,
- the selection of a patient's hip, knee or tooth implant and its precise implantation into the body,
- a better planning of the relocation of some bone to be performed by computer assistance and navigation techniques.

On the way from a geometrical to a *functional* model some more mathematical steps are necessary, which will be presented in the sequel.

Mathematical Modelling and Simulation

A functional patient model comprises, beyond the geometrical model, additionally a sufficiently accurate mathematical-physical description, mostly via partial differential equations. For illustration purposes, let us mention a few: the Lamé-Navier equations for linear elastomechanics and its nonlinear generalizations (geometry and material properties) in biomechanics, Maxwell's equations and the bio-heat-transfer equation in the cancer therapy hyperthermia, the Navier-Stokes equations for the analysis of fluid motion in the context of plaque building in blood vessels and in aneurysms. Whenever the required answers to the questions from medicine allow, then simpler, so-called reduced models will do. Generally speaking, mathematical models are only useful, if their input parameters have been analyzed with respect to their sensitivity.

A typical feature of medical models is their *multiscale structure*: The mathematical equations express relations between microscopic spatial dimensions and our everyday life dimensions. By intruding sufficiently deep into the mathematical description, we obtain a whole hierarchy of scales to be taken into account, depending on the problems in question. An illustrative example is given by the international project PHYSIOME [8], which spans scales from nm (molecules) via mm (tissue) up to m (organs).

Virtual Labs

The partial differential equations arising in the model must be solved numerically fast and reliably and, in view of clinical application, embedded in a 3D visualization environment, a "virtual lab". As for the simulation of the mathematical models, the aims of mathematics and medicine are in accordance: Both disciplines want the

Fig. 12 Hyperthermia:
adaptive hierarchical spatial
meshes (ZIB)

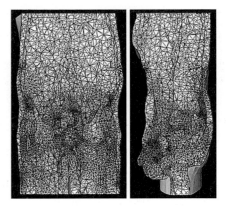

solution in 3D, within short computing times and with reliable accuracy. Only, if
these requirements are met, a mathematical therapy or operation planning can hope
to be accepted in the clinics and serve as a basis for responsible medical decisions.
Among efficient *algorithms* available are: (a) *domain decomposition methods* for
fixed meshes in connection with parallelization or (b) *multigrid methods* in combination with *adaptivity* in space and time.

Example: Hyperthermia [2]

Up to now, research in regional hyperthermia has been restricted for ethical reasons to the treatment of deeply seated, non-operable tumors, i.e. to hitherto hope-
less cases. The aim of this treatment is to heat cancer cells locally thus sensitizing
them to radio or chemotherapy without damaging healthy tissue by too high temperatures. In Fig. 10, the setting of the method is shown. The patient is positioned
in an applicator with (here) eight antennas that emit radio waves into the body. The
heat inside the body is generated non-invasively. By separate control of the individual antennas, the interference field can be adjusted to each individual patient,
to locally apply high thermal energy to the tumor region only. The scientific question is: How should the antennas be tuned (in terms of amplitudes and phases) such
that the tumor is heated within a temperature window between 42.5° and 45°C, but
no healthy tissue. The impact of the different tuning parameters on the therapeu-
tically effective temperature distribution is so complex that optimal therapy plans
can be determined only via numerical simulation. The associated functional patient
model here comprises Maxwell's equations for the description of the electric fields
and the bio-heat-transfer equation which governs the heat distribution inside the
body.

In order to illustrate the importance of spatial adaptivity, we show, in Fig. 12, an
actually generated mesh for the solution of Maxwell's equations: With space adaptivity we obtain about 120 000 nodes, whereas uniform meshes would have required
an estimated number of 16 000 000 nodes. The applied multigrid methods require
computing times proportional to the number of nodes, which implies that adaptive

Fig. 13 Patient preparation by the virtual lab [22]

methods are about a factor of 130 faster—in this medically relevant example and at a comparable accuracy!

Example: Soft Tissue Simulation [4]

In cranio-maxillo-facial surgery the mathematical model consists of the biomechanical differential equations. They are to be solved numerically (by efficient multigrid methods) to permit a reliable prediction of the postoperative facial appearance—assuming the operation went well as planned. The present standard of prediction is represented in Fig. 9 at two examples [21]; the presently achievable prediction quality is already excellent. Apart from the improved operation preparation, this kind of computer assisted operation planning including the soft tissue prognosis leads to an improved patient information and thus eventually to a higher patient motivation, see Fig. 13.

4.3 Perspective: The Virtual Patient

Beyond the described medical applications, mathematics (including computer science) already today provides extensive methods to accompany therapy and operation planning. On an intermediate time scale, it would be reasonable to open more space in public health to mathematics. The following lines of development can be foreseen:

Radiology will more and more move on from mere 2D image interpretation to *3D model reconstruction*. This requires substantial "screening" of individual image data by means of automated segmentation techniques. The corresponding increase of patient specific data will lead to a twofold development: (a) the build-up of centralized medical data bases in large hospitals, and (b) a population-wide introduction of (only personally disposable) individual data carriers (the "electronic patient"). Google-med may be a possible format of storing such data; it will, however, need to be modified due to national differences in health organisations and mentality as well as with respect to its non-guaranteed security of individual data.

Central archives (such as PACS, S-PACS) will, apart from individual data, store general *geometrical anatomy models* on the basis of statistical shape models, i.e. via mean anatomical shapes (averaged over local populations) including essential coordinates that describe the variability of the underlying patient sets. In addition, *mathematical functional models* will be stored, i.e. mathematical descriptions of functions of organs or of moving parts of the skeleton via multiscale models to be differentiated further. Provokingly, the function of an organ will not be fully understood, before it has been expressed by a realistic mathematical model covering both the healthy and the unhealthy case.

Both the legal and the order political frames will have to be clarified. Radiologists will certainly continue to bear the legal responsibility for the correctness of the interpretation of medical image data and the therefrom derived anatomical models. However, in countries like Germany, insurance companies will need to include model assisted planning on the basis of geometrical 3D models and mathematical functional models into their catalogue of financially supported services. Apart from medical indication, the new kind of planning tools is useful in view of an improved patient information as well as of education, documentation, and quality assurance.

5 Vision and Options

This article has shown, at selected examples, how the coaction of mathematics and medicine has pushed a dynamic development towards patient specific models ("virtual heart" in Sect. 3, "virtual patient" in Sect. 4) on the basis of modern medical imaging techniques (Sect. 2). This development will certainly expand in the near future. However, there is still a long way to go, until anatomically correct and medically useful functional models will be available even for the most essential body parts and the most frequent diseases. Within the German funding system, the corresponding research will, on a quite long run, remain dependent on public funding. In any case, political frames in health and research will need to be adjusted in close cooperation with selected medical doctors, engineers, and—mathematicians!

References

1. Colli Franzone, P., Deuflhard, P., Erdmann, B., Lang, J., Pavarino, L.F.: Adaptivity in space and time for reaction-diffusion systems in electrocardiology. SIAM J. Sci. Comput. **28**(3), 942–962 (2006)
2. Deuflhard, P.: Differential equations in technology and medicine: computational concepts, adaptive algorithms, and virtual labs. In: Burkhard, R., Deuflhard, P., Jameson, A., Lions, J.-L., Strang, G. (eds.) Computational Mathematics Driven by Industrial Problems. Springer Lecture Notes in Mathematics, vol. 1739, pp. 70–125. Springer, Berlin (2000)

3. Deuflhard, P., Hochmuth, R.: Multiscale analysis of thermoregulation in the human microvasular system. Math. Methods Appl. Sci. **27**, 971–989 (2004)

4. Deuflhard, P., Weiser, M., Zachow, S.: Mathematics in facial surgery. Not. Am. Math. Soc. **53**(9), 1012–1016 (2006)

5. Dössel, O.: Kausalität bei der Entstehung, der Diagnose und der Therapie von Krankheiten – aus dem Blickwinkel des Ingenieurs. Kausalität in der Technik, Berlin-Brandenburg, Akad. Wiss., pp. 69–80 (2006)

6. Dössel, O., Farina, D., Mohr, M., Reumann, M., Seemann, G.: Modelling and imaging electrophysiology and contraction of the heart. In: Buzug, T.M., Holz, D., Weber, S., Bongartz, J., Kohl-Bareis, M., Hartmann, U. (eds.) Advances in Medical Engineering, pp. 3–16. Springer, Berlin (2007)

7. Henriquez, C.S., Muzikant, A.L., Smoak, C.K.: Anisotropy, fiber curvature, and bath loading effects on activation in thin and thick cardiac tissue preparations: simulations in a three-dimensional bidomain model. J. Cardiovascular Electrophysiol., 424–444 (1996)

8. Hunter, P.J., Borg, T.K.: Integration from proteins to organs: the Physiome project. Nature Rev., Mol. Cell Biol. **4**, 237–243 (2003)

9. Jonas, P., Louis, A.K.: Phase contrast tomography using holographic measurements. Inverse Probl. **20**, 75–102 (2004)

10. Kainmüller, D., Lange, Th., Lamecker, H.: Shape constrained automatic segmentation of the liver based on a heuristic intensity model. In: Heimann, T., Styner, M., van Ginneken, B. (eds.) Proc. MICCAI Workshop 3D Segmentation in the Clinic: A Grand Challenge, pp. 109–116 (2007)

11. Khawaja, A., Dössel, O.: Predicting the QRS complex and detecting small changes using principal component analysis. Biomed. Tech., 11–17 (2007)

12. Louis, A.K.: Medical imaging state of the art and future development. Inverse Probl. **8**, 709–738 (1992)

13. Louis, A.K.: Combining image reconstruction and image analysis with an application to two-dimensional tomography. SIAM J. Imaging Sci. **1**(2), 188–208 (2008)

14. Marr, R.B., Chen, C., Lauterbur, P.C.: On two approaches to 3D reconstruction in NMR zeugmatography. In: Herman, G.T., Natterer, F. (eds.) Mathematical Aspects of Computerized Tomography. Springer LNMI, pp. 225–240. Springer, Berlin (1981)

15. Natterer, F., Wübbeling, F.: Mathematical Methods in Image Reconstruction. SIAM, Philadelphia (2001)

16. Reumann, M., Bohnert, J., Osswald, B., Hagl, S., Dössel, O.: Multiple wavelets, rotos, and snakes in atrial fibrillation—a computer simulation study. J. Electrocardiol., 328–334 (2007)

17. Shepp, L.A., Logan, B.F.: The Fourier reconstruction of a head section. IEEE Trans. Nucl. Sci. **NS-21**, 21–43 (1974)

18. Sörnmo, L., Laguna, P.: Bioelectrical Signal Processing in Cardiac and Neurological Applications. Elsevier, Amsterdam (2005)

19. Stalling, D., Westerhoff, M., Hege, H.-C., et al.: Amira: a highly interactive system for visual data analysis. In: Hansen, C.D., Johnson, C.R. (eds.) The Visualization Handbook, pp. 749–767 (2005). Chap. 38. URL: amira.zib.de

20. ten Tusscher, K.H., Panfilov, A.V.: Alternans and spiral breakup in a human ventricular tissue model. Am. J. Heart Circul. Physiol., 1088–1100 (2006)

21. Zachow, S.: Computergestützte 3D-Osteotomieplanung in der Mund-Kiefer-Gesichtschirurgie unter Berücksichtigung der räumlichen Weichgewebeanordnung. Medizininformatik, Verlag Dr. Hut (2005)

22. Zachow, S., Hege, H.-C., Deuflhard, P.: Computer assisted planning in cranio-maxillofacial surgery. J. Comput. Inform. Technol., 53–64 (2006). Special Issue on Computer-Based Craniofacial Modelling and Reconstruction

23. Zachow, S., Zilske, M., Hege, H.-C.: 3D reconstruction of individual anatomy from medical image data. In: Proc. ANSYS Conference & CADFEM Users' Meeting, Dresden. Siehe auch: Report 07-41, ZIB (2007)

Compounds, Drugs and Mathematical Image Processing

Günter J. Bauer, Dirk A. Lorenz, Peter Maass,
Hartwig Preckel, and Dennis Trede

1 Executive Summary: From the Image to the Drug

The development of new drugs is long-winded and expensive. The first step consists in the search for active pharmaceutical ingredients that are suitable for the treatment of previously difficult to treat diseases. For this purpose, pharmaceutical and biotechnology industry possesses huge substance databases. In these databases, different substances are gathered that can either be synthetically produced or can be extracted from fungi, bacterial cultures or other species, cf. Fig. 1.

What candidate compounds can be used to achieve the desired biochemical or biological mechanism, however, has to be analyzed in laboratories. In this so called screening experiments, millions of interactions are evaluated: currently applicable screening systems analyze over 100 000 compounds per day, each of which brought

G.J. Bauer
Scienion AG, Volmerstr. 7b, 12489 Berlin, Germany
e-mail: bauer@scienion.de

D.A. Lorenz
Institut für Analysis und Algebra, Technische Universität Braunschweig, Pockelsstr. 14, 38118 Braunschweig, Germany
e-mail: d.lorenz@tu-braunschweig.de

P. Maass · D. Trede (✉)
Zentrum für Technomathematik, Universität Bremen, Postfach 330 440, 28334 Bremen, Germany
e-mail: trede@math.uni-bremen.de

P. Maass
e-mail: pmaass@math.uni-bremen.de

H. Preckel
PerkinElmer Cellular Technologies Germany GmbH, Schnackenburgallee 114, 22525 Hamburg, Germany
e-mail: Hartwig.Preckel@perkinelmer.com

M. Grötschel et al. (eds.), *Production Factor Mathematics*,
DOI 10.1007/978-3-642-11248-5_20, © Springer-Verlag Berlin Heidelberg 2010

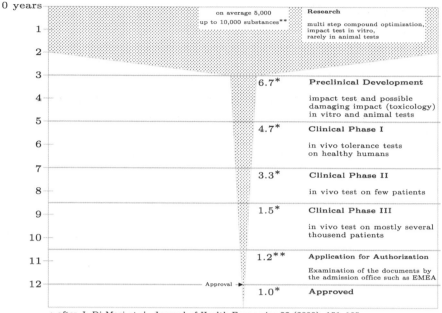

Fig. 1 The stages of development in pharmaceutical research: By far, not every compound is tested on the complex biological systems. From the initially 5000 up to 10 000 substances, on average, only about five remain after the screening experiments for the in vivo testing (clinical phase I), reproduction after [4]

together with several hundred cells that are collected in small reaction tubes (microtiter plates).

The impact of individual substances is here coded in colors (fluorescence marker, polarization properties) and recorded by high resolution systems. The informative value of these experiments strongly depends on the quality of the automated analysis routines for these huge data sets. An additional difficulty arises from the fact that the search for potential candidates leads to the exploitation of increasingly complex biochemical interactions on the cellular level. A standardized evaluation for several experiments is therefore no longer possible, the evaluation routines have to be tailored for every experiment.

These problems require optimized mathematical methods—they are not solvable by pure biotechnological developments. The revolutionary developments in theoretically-mathematical image processing of the last ten years have already led to an enhanced quantitative evaluation of screening experiments in several application scenarios (up to several degrees of magnitude). The potential of mathematical methods in the area of drug discovery is, however, far from being explored. The realization of theoretical results into applicable algorithms is at its outset. The next jump in quality of the screening technology is, therefore, less expected to stem from a technological development but rather from enhanced quantitative methods in mathematical image analysis.

The image processing is thus only at the beginning and is the foundation of the whole development chain, at the end of which a new drug is designed. The main challenges in this area for the next few years are:

- Mathematical modeling of quantitative quality criteria for screening experiments,
- Development of adaptive image processing routines for the determination of critical indicators in complex cell structures.

These challenges must be addressed by teams of biologists, chemists, mathematicians and computer scientists. Mutual projects between scientists and chosen partners from industry are a necessary requirement for future competitiveness of these companies, such as the manufacturer of analysis devices on the chemical, medical and biochemical markets but also companies in the pharmaceutical industry. In Germany and Switzerland, there are e.g. PerkinElmer Cellular Technologies Germany GmbH or the Bruker BioSciences Corporation, and the pharmaceutical side, e.g., Bayer Schering Pharma AG, F. Hoffmann-La Roche AG, Jenapharm GmbH & Co. KG or Schwabe AG.

2 Success Stories

The search for active compounds in the medical chemistry is carried out by screening experiments. These comprise of the automated investigation and testing of molecule activity in test volumes of a few microliters. The analysis of the interaction between the compound and the biosample is based on an optical signal or an image, e.g., an optical absorption, the fluorescence intensity or the polarization. Current screening systems analyze the time evolution of the interaction with so called high-throughput (1000–100 000 substances per day) and ultra-high-throughput systems (more than 100 000 substances). A refinement of the classical (ultra)-high-throughput screening is the so called high-content, high-throughput screening. Here, the optical signals are recorded with a very high cellular resolution and the interaction with subcellular structures are analyzed. Figure 2 shows the high-content,

Fig. 2 Screening experiments are carried out on highly automated platforms for analysis of the interaction of compounds and cells [1, 11]

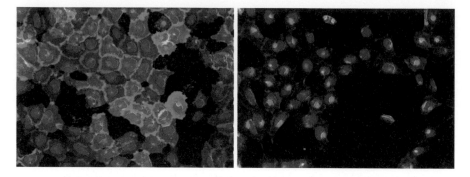

Fig. 3 Translocation of a *green* colored protein. *Left*: unstimulated cells, *right*: stimulated cells

Fig. 4 The analysis of the samples is carried out in microtiter plates with test volumes of a few microliters

high-throughput screening system Opera developed by PerkinElmer Cellular Technologies.

The analysis of high-content data, i.e., the analysis of the interactions in subcellular structures, such as receptors, is usually enabled by the application of fluorescent markers to the investigated structures. The use of different markers allows to simultaneously record different structures. Figure 3 shows cell images of a high-content, high-throughput screening. Here, by a stimulation of the cells, a translocation of colored green protein takes place. While, on the left hand side, it distributes in the cells, on the right hand side, it is concentrated close to the nuclei of the stimulated cell. Nuclei are marked red.

The interaction of the biosamples is analyzed in so called microtiter plates (see Fig. 4), that may contain differently many reaction tubes, so called wells. Along with the identification of the active compounds, i.e., a qualitative statement whether an interaction takes place or not, the realization of the experiment with different doses also provides a quantitative result on the interaction (dose response curve) [3].

In the high-content, high-throughput screening experiments, the biochemical interaction is analyzed for every single cell using image processing methods. For a description of the overall impact of the added substance, one takes the mean over all cells in a well. The central tasks of the evaluation of such screening experiments are:

1. The analysis of the biochemical interaction between the synthetic compound and the biological sample is performed based on an optical signal that is saved as a digital image or as a time-dependent movie. The evaluation of the thus accrued millions of digital images requires the development of reliable adaptive methods in *mathematical image processing*.
2. After having analyzed every image, *statistical methods* are used to verify whether an interaction between the synthetic compound and the biosample exists and, if so, how strong it is. The final statement has to be statistically proved before one can continue with the following steps for the development of new drugs.

For the solution of these tasks, an extensive standardized toolbox of established methods in statistics and mathematics is available, which we illustrate in the following.

2.1 Automated Analysis of Cell Images in the High-Content, High-Throughput Screening

Since the data sets obtained by high-content, high-throughput screening are huge, a manual evaluation of the image data is impossible. For the automated evaluation of cell images various image processing systems are available. There are, e.g., the open-source software CellProfiler [5], or the software Acapella [10] that is used by PerkinElmer for a computer-aided digital image processing. The analysis usually is realized according to the following scheme:

The first step of the processing is the preprocessing. Here, the image is modified such that it is suitable for the subsequent operations. The essential image information shall not be changed. Such an enhancement of the image data can involve, e.g., a reduction of noise or another correction of a non-optimal image record. Figure 5 shows an intensification of the contrast on two example images.

This upgrade of an image is followed by segmentation or object recognition. The goal of this step is the separation of the relevant objects from the remaining image parts. This means, e.g., the separation of cells and background or the decomposition of a cell into small subcellular structures. Such segmentation methods can have highly variable complexity. Figure 6 shows a segmentation of cell nuclei and whole cells.

After the detection of the subcellular structures in the image which are relevant for the analysis, respective properties can be extracted from these (e.g. mean brightness, area or properties of the form). The result is a property vector per cell that forms the basis for a classification: by means of the computed properties it is possible to deduce information on the quantity of interaction for every cell. Taking the mean over all cells of the well, one obtains the mean interaction between the synthetic compound and the biosample and, thus, for each well the following pairs

(dosage of compound, mean stimulation of cells).

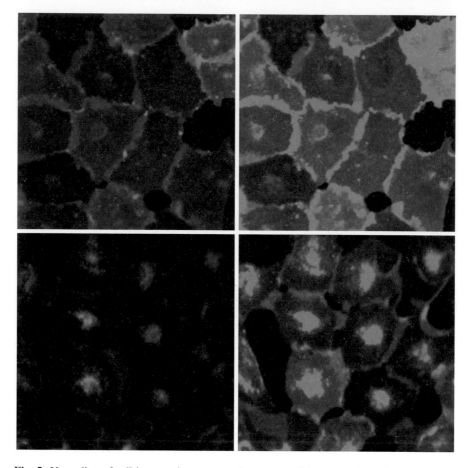

Fig. 5 Upgrading of cell images via contrast enhancement of the *green* channel. *Upper image*: unstimulated cells, *lower image*: stimulated cells. *Left*: original images, *right*: preprocessed images

2.2 Statistical Evaluation of the Data

In order to identify active ingredients, it is sufficient to compare images of cells without the compound and images of cells with very high compound dosage. For a description of the interaction, i.e. a description whether the compound under investigation also reliably works in repeated experiments, in [15], the Z'-value is proposed as a statistical criterion. This value depends on the signals of minimal and maximal intensity and is defined by

$$Z' := 1 - 3 \frac{\sigma_{\text{pos}} + \sigma_{\text{neg}}}{|\mu_{\text{pos}} - \mu_{\text{neg}}|},$$

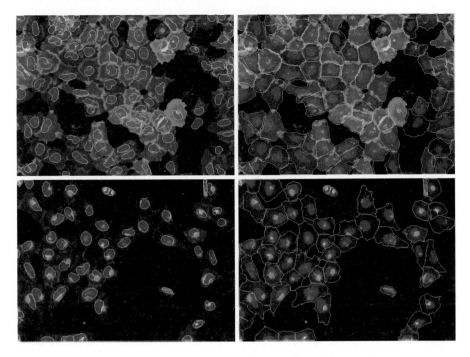

Fig. 6 Segmentation of cell nuclei (*left*) and of the whole cells (*right*)

Fig. 7 Example of a sigmoidal dose response curve

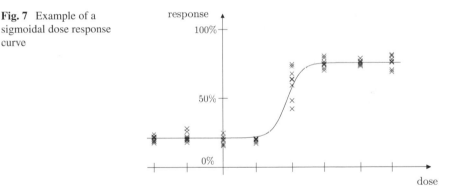

where μ_{pos}, μ_{neg} and σ_{pos}, σ_{neg} are the means and standard deviations over all images with maximal and minimal interaction, respectively. The impact of the compound is larger the closer Z' is to one.

For the development of a new drug, it is necessary to also have a measure for the description of the quantity of the impact. For the computation of a quantitative correlation of dosage and impact, also cell images with medium doses of the substance have to be evaluated. By means of nonlinear regression, it is possible to calculate a dose response curve, see [8, 9, 12]. In Fig. 7, the results of the image analysis are depicted in blue along with a sigmoidal dose response curve in black.

3 Status Quo: Mathematical Methods for the High-Content, High-Throughput Analysis

The mathematical and statistical methods that we have discussed in the previous section are, at least on the academic side, well known and long-established. However, since complex biochemical interactions on the cellular level are used for the development of new drugs, the complexity of the measurable effects is proportionally growing. Increasingly refined and optimized analysis algorithms are therefore needed. This is the domain of mathematical image processing.

The revolutionary research results of the past few years on the mathematical side of image processing, however, have mostly not yet found their way into the life sciences. There are numerous mathematical methods that are literally asking for being used in high-content, high-throughput analysis.

In the following, the successful team play of the newest mathematical methods, biochemical expertise and modern screening technology is demonstrated by two examples.

1. The whole process chain for the evaluation of the screening image material comprises of several steps. Each one of these steps is controlled by its own parameter set. The previously typically separated optimization of these steps is no longer sufficient for handling the increasingly complex tasks. Hence, a global optimization of the whole process chain is necessary. With mathematics, basically a linguistic tool is available that allows to assess the quality of parameters for the whole process chain and to subsequently determine these using *parameter optimization* methods.
2. The condition for the evaluation by means of image processing methods is the error-free recognition of the individual cells—the *cell segmentation*. Usually, most simple methods (thresholding) with a high failure rate are used. Therefore, the available image material can only partially be used. In contrast, modern morphological image processing methods, that are motivated by physical equations, provide results that are more accurate by several degrees of magnitude.

3.1 Parameter Optimization and Inverse Problems

The results of image processing essentially depend on suitably chosen parameters that are specified in accordance on external effects such as the type of the microtiter plate. These are, e.g., thresholds or statistical measures that are part of the description of the activity of a cell. The parameters are usually set manually based on biological experience. Furthermore, there are parameter scanners that searches a predefined finite discrete subset of all possible parameters for a best possible configuration (grid-search). Such a search is obviously very costly and not necessarily optimal.

Recently, in an interdisciplinary cooperation between the Center of Industrial Mathematics of University of Bremen and PerkinElmer Cellular Technologies Germany GmbH, we have developed a method that determines these parameters such

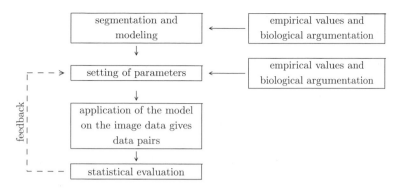

Fig. 8 Processing steps for the high-content, high-throughput analysis and a concept for parameter optimization

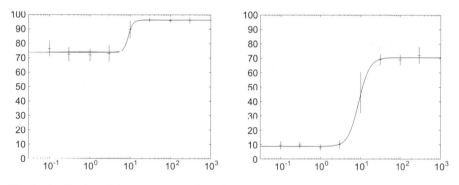

Fig. 9 Application of the parameter optimization to an illustrative test: *in the left* nonoptimized dose response curve, the difference between unstimulated and stimulated cells is not very clear, *in the right* optimized curve the differences are far better distinguishable, from [13]

that the analysis is optimal. The idea of this parameter optimization for the high-content, high-throughput analysis is to iteratively improve the adjustable parameters from the statistical evaluation using a feedback. This can even be done without having a biologically meaningful initial configuration (see Fig. 8). For the optimization, e.g., the Z'-value can be used: the unstimulated and the stimulated cells differ the more, the closer the Z'-value is to one. For the parameter optimization, we consider Z' as a mathematical function of the parameters, whose maximum has to be calculated.

In [13], this parameter optimization was applied to several examples and in all tests it has led to a significant enhancement of the analysis. Figure 9 shows the results of the optimization in an example: in the left nonoptimized dose response curve, the difference between unstimulated and stimulated cells is with approx. 20% not very clear. In the right curve, unstimulated and stimulated cells are with approx. 60% far better distinguishable.

Several possible application areas of parameter optimization have emerged:

1. After modeling the interaction and a rough initial setting of the parameters, the optimization method can be used for fine tuning of the parameters.
2. Based on default parameters, the optimization procedure allows to tune parameters of standard image processing routines. This is even possible if the functioning of the method is not known, and therefore, this is useful for optimizing parameters of black-box methods.
3. Furthermore, the proposed method can be used for adapting parameters of an existing model to other but similar test conditions, such as different microtiter plates or exposure times.

In summary, one can say that the parameter optimization by means of feedback (inverse model) of the statistical evaluation is a powerful auxiliary tool in high-content, high-throughput analysis. The manual tuning of parameters wastes working time and requires extensive experience with analysis of screening data. The application of the presented parameter optimization can safe valuable time in the development phase of new active pharmaceutical ingredients and, thus, new drugs can be brought on the market much faster. Recently, we have filed a patent at the European Patent Office, application number EP08155784.5, entitled "Analysis method for analysing the effectiveness of a test substance on biological and/or biochemical samples".

3.2 Enhanced Cell Segmentation

Most segmentation methods for cells and cell nuclei are based on simple thresholding. Loosely speaking, these are methods that partition the image in segments based on the comparison of the brightness of the pixels. An advantage of thresholding is, among others, the low computational effort. However, the quality of this comparatively simple method is worse compared with more complex methods. One encounters problems, especially for segmentation of whole cells.

At the Center of Industrial Mathematics of University of Bremen, in cooperation with PerkinElmer Cellular Technologies, we have developed and tested a method that employs the comparatively new idea of active contours for segmentation, see [6, 13]. A very similar approach is chosen in [14]. In the method of active contours, the segmentation of cells is performed via the solution of the partial differential equation

$$\frac{\partial u}{\partial t} = g \cdot \left(\text{div}\left(\frac{\nabla u}{\|\nabla u\|} \right) + \rho \right) \|\nabla u\| + \langle \nabla g, \nabla u \rangle,$$

where u implicitly describes the image segments, g is a control function and ρ is a further parameter. This equation is motivated by physical mass transport and is used in fluid mechanics in a very similar manner [2, 7]. For the computation of solutions of the partial differential equation, there exist numerical methods that approximate the solution on a discrete grid.

Fig. 10 Three-dimensional
cell segmentation by means
of active contours

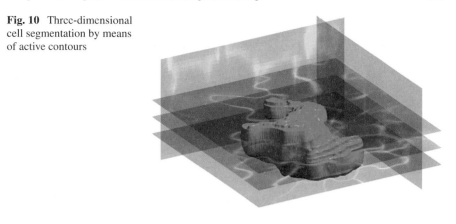

A further advantage of the active contours compared with many other methods
is that it can easily be extended to higher dimensional data. The cell analyzer Opera
has the feature, that images of cut planes can be produced and, hence, a three-
dimensional cell analysis is possible. In Fig. 10, cut planes spaced one above the
other are depicted that were recorded by the device, and also the result of a three-
dimensional segmentation with the developed method.

4 Strengths and Weaknesses Analysis: The Disciplinary Thinking Dilemma

The application of modern mathematical methods in pharmaceutical research and,
in particular, in the evaluation of screening experiments for drug discovery, is still
at the outset. Well established still are classical methods from standard software
packages, that completely ignore the research results of the past ten years. Why is it
so?

The laboratories of the experimentally working research teams in the pharmaceutical industry consist of biologists and chemists, the technical development and software divisions are dominated by physicists and computer scientists. Mathematicians
are a rare species in this environment. At the most, there are some statisticians integrated within bioinformatics teams. However, this one-track thinking power leads
to the fact that solutions to all problems are always searched in the same direction.

The area of image processing, however, pertains to a different, new scientific
area. The healthy self-conception of physicists and computer scientists, who often
claim the mathematical know-how for themselves, does not do the job here. It is
not sufficient to implement some given formulas. The complexity of the application
field as well as the involved methods require a detailed analysis.

The successful future developments in the described application area will cru-
cially depend on the cooperation with mathematicians. In individual highly specialized application scenarios, the advantage of the application of novel mathematical
methods has already been demonstrated. However, there is still a long way until

their application in the full spectrum, from the optimized evaluation of the first cell experiments down to the support of the eventually necessary clinical studies, may become a routine.

The contribution of the mathematician is twofold. On the one hand, mathematics provides a clear and unerring language for the formulation of the objectives for the evaluation of the experimentally obtained data. This evaluation is usually performed within a multi-step process chain. Even if this sounds natural, looking at the optimization of the individual steps of this process chain often veils the overall picture. A consequent mathematical modeling allows a clear description, which at the same time forms the basis for the application of optimization strategies for the overall process. On the other hand, mathematics provides an extensive toolbox of new methods for the evaluation of image data, for modeling of individual process steps and for optimization or parameter identification of the overall process chain.

5 Visions and Policy Recommendations

5.1 Tangible Challenges

The proliferation of mathematical models and methods in drug discovery is the challenge of the next ten years. Mathematical methods can be meaningfully applied in almost all phases of drug discovery. We aim to concentrate our contribution on the first step, the search of active compounds. In particular, the emphasis is on the mathematical modeling of quantitative quality criteria for general high-content, high-throughput screening experiments and the development of adaptive image processing routines for the determination of critical indicators in complex cell structures.

The difficulties for the realization are especially encountered in the continuously growing complexity of modeling biochemical cell processes that have to be mathematically understood, and also, in the proficient evaluation of the many new mathematical methods, which have to be employed for the evaluation of the experimentally obtained data.

The introduction of interdisciplinary degrees is, however, not the solution in this area. The attempt to teach biologists sufficient mathematics or, vice versa, mathematicians sufficient chemistry or biology in interdisciplinary degree programs, and thus, to produce hybrid species of half-biologist-half-mathematician, is doomed to failure. On both sides, the complexity of the scientific landscape has grown in such a way that more than a basic understanding of the other area is no longer possible. It is crucial that the future scientists and developers already gain some basic knowledge in the other disciplines, and, what is more important, take along some openness and curiosity for their results. This also applies to the mathematics degree, where certainly thinkable minors in chemistry or biology are fairly rare, and where the numerical analysis and modeling lectures are mostly limited to discussions of predator-prey-models.

Anyhow, at the end of the day, it is not the lone fighters that are needed here but rather teams of highly professional scientists and developers. The doctoral phase is here the right place for serious cooperations beyond the borders of the own discipline.

The challenge is to create attractive work conditions that allow cooperations of biologists, chemists, computer scientists and mathematicians on the highest level. This is already realized in certain research institutes, however, the establishment of a broad industrial platform for such cooperations is still absent.

5.2 Cooperation

The future lies in the cooperative realization of biomathematically-technical approaches for the optimization of all different process steps in drug design. In future, mathematicians will be just as natural a species in the development teams of the pharmaceutical industry as biologists, chemists and computer scientists.

The potential is obvious, the inhibition threshold between the disciplines, however, is still high. Targeted actions for the initiation of cooperations between industrial development teams and mathematical research groups are necessary. Coordinated support guidelines met by organizations, science support institutions and individual industrial companies, while investing some manageable efforts, would lead to measurable success. On the one hand, in purposefully selected application areas, the contribution of modern mathematics would be demonstrated up to the development of prototype analysis tools, whereas on the other hand, future experts for the industry would be educated with a high multiplier effect.

References

1. Beuthner, A.: Images from the nucleus of a cell, company profit of Evotec Technologies. Frauenhofer Mag. **1** (2005)
2. Caselles, V., Kimmel, R., Sapiro, G.: Geodesic active contours. Int. J. Comput. Vis. **22**(1), 61–79 (1997)
3. Fattinger, C.: High throughput screening. Innov., Mag. Carl Zeiss **12**, 4–5 (2002)
4. German Association of Research-based Pharmaceutical Companies (VFA). In: Labors und Kliniken: Wie entsteht ein neues Medikament? http://www.vfa.de/de/forschung/so-funktioniert-pharmaforschung/wie-ein-medikament-entsteht/amf laborsklinikcn.html (2008)
5. Jones, T.R., Carpenter, A.E., Sabatini, D.M., Golland, P.: Methods for high-content, high-throughput image-based cell screening. In: Proceedings of the First MICCAI Workshop on Microscopic Image Analysis with Applications in Biology, pp. 65–72 (2006)
6. Lorenz, D.A., Maass, P., Preckel, H., Trede, D.: Topology-preserving geodesic active contours for segmentation of high content fluorescent cellular images. In: Proceedings in Applied Mathematics and Mechanics (PAMM), vol. 8 (2008)
7. Malladi, R., Sethian, J.A., Vemuri, B.C.: Shape modeling with front propagation: a level set approach. IEEE Trans. Pattern Anal. Mach. Intell. **17**(2), 158–175 (1995)
8. Malo, N., Hanley, J.A., Cerquozzi, S., Pelletier, J., Nadon, R.: Statistical practice in high-throughput screening data analysis. Nature Biotechnol. **24**(2), 167–175 (2006)

9. Motulsky, H., Christopoulos, A.: Fitting Models to Biological Data Using Linear and Nonlinear Regression: A Practical Guide to Curve Fitting. Oxford University Press, Oxford (2004)
10. PerkinElmer: Acapella brochure. http://www.cellularimaging.com/products/acapella/
11. PerkinElmer: Opera brochure. http://www.cellularimaging.com/products/opera/
12. Ritz, C., Streibig, J.C.: Bioassay analysis using R. J. Stat. Softw. **12**(5) (2005)
13. Trede, D.: Parameteroptimierung und Segmentieren mit aktiven Konturen für die High-Content-Analyse beim Hochdurchsatz-Screening. Diploma thesis, University of Bremen, Center of Industrial Mathematics (September 2007)
14. Yan, P., Zhou, X., Shah, M., Wong, S.: Automatic segmentation of high throughput RNAi fluorescent cellular images. IEEE Trans. Inf. Technol. Biomed. **12**(1), 109–117 (2008)
15. Zhang, J.-H., Chung, T.D.Y., Oldenburg, K.R.: A simple statistical parameter for use in evaluation and validation of high throughput screening assays. J. Biomol. Screen. **4**(2), 67–73 (1999)

Contributors

Prof. Dr.-Ing. **Dirk Abel** is Full Professor at RWTH Aachen University and head of the Institute of Automatic Control which is part of the Faculty of Mechanical Engineering. His research topics are concentrated on advanced control applications in the fields of Automotive, Rail, Industrial Automation and Biomedical Engineering. A methodical focus is—besides others—Model-based Predictive Control, also in connection to Object-oriented Modelling and Simulation of dynamic systems.

Frank Allgöwer is Professor for Systems Theory in Engineering at the University of Stuttgart. His main areas of interest are the development of new methods for systems analysis and control with special emphasis on nonlinear and predictive control. His application interests span a wide range from mechatronics, process control, biomedical engineering and nanotechnology to the new field of systems biology. Frank Allgöwer received several recognitions for his work including the prestigious Gottfried-Wilhelm-Leibniz award of the Deutsche Forschungsgemeinschaft and the State Teaching Award (Landeslehrpreis) of the state of Baden-Württemberg.

Dr.-Ing. **Matthias Baitsch** studied Civil Engineering at the University of Dortmund. For his PhD thesis on the optimization of imperfect slender structures he joined the DFG Post-Graduate Research Programme "Computational Structural Dynamics" at the Ruhr-University Bochum. Current research interests include the p-version of the Finite Element Method and modern software concepts for simulation and optimization.

Dr. **Günter J. Bauer**, PerkinElmer Cellular Technologies Germany GmbH, is supervising scientific and business collaborations to broaden the range of use of existing products and methods and to evaluate new technologies and product ideas.

Prof. Dr. Dr. **Holger Boche** is Full Professor for Mobile Communication Networks at the Technische Universität Berlin and director of the Fraunhofer German-Sino Lab for Mobile Communications and also of the Fraunhofer Institute for Telecommunications (HHI), both Berlin, Germany. His main research areas are communication and information theory as well as applied mathematics. He received the Gott-

M. Grötschel et al. (eds.), *Production Factor Mathematics*,
DOI 10.1007/978-3-642-11248-5, © Springer-Verlag Berlin Heidelberg 2010

fried Wilhelm Leibniz Prize from the Deutsche Forschungsgemeinschaft (German Research Foundation), the "Innovation Award" from the Vodafone Foundation and other awards.

Prof. Dr. Dr. h.c. **Hans Georg Bock** is Professor for Mathematics and Computer Science at the University of Heidelberg, managing director of the Interdisciplinary Center for Scientific Computing (IWR), and director of the Heidelberg Graduate School "Mathematical and Computational Methods for the Sciences". His research focus is on optimization with differential equations, especially realtime methods of optimal control (NMPC), parameter estimation, and optimum experimental design. He has established a large number of scientific cooperations with Southeast Asia, especially with Cambodia and Vietnam. Hans Georg Bock has been awarded the Medal of Honor from the Gottlieb Daimler and Karl Benz Foundation, the Felix Hausdorff Memorial Award and the Microsoft Research Award.

Dr. **Ralf Borndörfer** is deputy head of the optimization department at the Zuse Institute Berlin and associate of the Dres. Löbel, Borndörfer & Weider GbR. His research area is the solution of traffic and transport optimization problems using methods of combinatorial optimization. He was awarded the Joachim-Tiburtius-Prize and the Dissertation Award of the German Operations Research Society.

Prof. Dr. **Wolfgang Dahmen** is Professor of Mathematics at RWTH Aachen. He is a member of the steering committee for the Graduate School "Aachen Institute of Advanced Study in Computational Engineering Science", of the Nordrhein Westfalian Academy of Sciences, of the National Academy of Sciences Leopoldina, of the Senate of the German Research Foundation, and of the panel PE1 for Mathematics of the European Research Council. His research interests cover Applied and Numerical Analysis, Harmonic Analysis and Approximation Theory. They range from foundational studies concerning the development and analysis of adaptive multiscale solution concepts to interdisciplinary applications, for instance, in process engineering and fluid dynamics. In 2002 he was awarded the Leibniz-Prize of the German Research Foundation.

Prof. **Berend Denkena** earned his PhD at Leibniz University Hanover, Germany, in 1992. He than became Manager Standards Engineering at Thyssen Production Systems in Auburn Hills/USA (1993–1995). Back in Germany he served as Manager Mechanical Development at Thyssen Hüller Hille in Ludwigsburg for one more year before he went to Bielefeld as Manager Engineering and Development for Gildemeister Turning Machines. Since 2001 he is a Professor at the Leibniz University Hanover and the head of the Institute of Production Engineering and Machine Tools.

Prof. Dr. Dr. h.c. **Peter Deuflhard** is full professor at the FU Berlin—chair Scientific Computing—, founder and president (since 1987) of the Zuse Institute Berlin (ZIB) as well as co-founder of the DFG Research Center MATHEON. Prof. Deuflhard's interests in mathematical research are within the area of nonlinear systems, ordinary and partial differential equations and Marchov chains, both in numerical analysis and in mathematical modelling. He worked in problems in space theory, chemical and process engineering, medical and biological technology. Prof. Deuflhard has been rewarded among others with the Gerhard Damkoehler Medal for fundamental contributions to chemical engineering and the ICIAM Maxwell prize for originality in applied mathematics.

Prof. Dr. **Moritz Diehl** is Associate Professor with the Electrical Engineering Department, Katholieke Universiteit K.U. Leuven, Leuven, Belgium, and the Principal Investigator of K.U. Leuven's Optimization in Engineering Center OPTEC. His research interests include numerical methods for optimization, in particular for dynamic systems described by differential equations, as well as real-time optimization, state and parameter estimation, and convex optimization.

Prof. Dr. **Olaf Dössel** is member of acatech and head of the Institute of Biomedical Engineering at the Karlsruhe Institute of Technology (KIT). He is member of the executive board of VDE (German Association of Electrical Engineering and Information Technology) and of the board of the German Association of Biomedical Engineering. He is Editor-in-Chief of the journal Biomedizinische Technik/Biomedical Engineering. He also serves as a member of the advisory board of the German Metrology Institute (Physikalisch Technische Bundesanstalt) and of the expert panel on medical technology of the German Ministry of Education and Research (BMBF). He is member of the Berlin-Brandenburg Academy of Science. His research interests are mathematical modeling of the heart, the inverse problem of Electrocardiography (ECG), ECG signal analysis and numerical field calculation in human body. He was one of the two presidents of the World Congress on Medical Physics and Biomedical Engineering 2009 in Munich.

Prof. Dr. **Wolfgang Dreyer** is head of the research group Thermodynamic Modeling and Analysis of Phase Transitions at the Weierstrass Institute for Applied Analysis and Stochastics in Berlin. Since 2008 he is Honorary Professor for Continuum Thermodynamics at the Technical University of Berlin. His research topics are mathematical modelling, phase transitions, thermodynamics, statistical mechanics and kinetic theory.

Jörg Eberspächer is chaired professor for communication networks at the Technische Universität München (TUM). Before joining TUM in 1990, he was with Siemens AG, Munich, where he was responsible for research in high speed communication networks and systems. His current research areas are network architecture and technologies for the Future Internet, 4th Generation Mobile Communication and Network Planning. He is a guest professor at Tongji-University Shanghai, chairing several advisory committees of European research institutes, member of the National Science Academies Leopoldina and acatech and member of the board of the Münchner Kreis.

Professor **Wolfgang Ehlers** holds the Chair of Continuum Mechanics at the University of Stuttgart. At the same time, he acts as the Executive Director of the Cluster of Excellence in Simulation Technology. He was a member of the Senate and the Joint Committee of the Deutsche Forschungsgemeinschaft (DFG) and the Vice Rector for Organisation at his university. His main fields of scientific interest are continuum mechanics, theory of materials, experimental mechanics, and computational mechanics. His specific interest concerns the modelling and simulation of multiphasic and multi-component materials with application to geo- and biomechanical problems. His habilitation thesis on multi-component materials was awarded with the Gottschalk-Diederich-Baedeker prize

Dr. **Andreas Eisenblätter** is co-founder and managing director of atesio GmbH. atesio is a spin-off company of the Zuse Institute Berlin (ZIB), Germany. atesio supports operators of telecommunication network in the short-, mid-, and long-term planning of network infrastructure. His research focus is on the planning and optimization of radio networks (e.g., GSM, UMTS/HSPA, LTE) by means of mathematical optimization.

Dr.-Ing. **Sebastian Engell** is Professor for Process Dynamics and Operations in the Department of Biochemical and Chemical Engineering of Technische Universität Dortmund, Dortmund, Germany. His research interests are control and online optimization of chemical processing plants, hybrid system dynamics and control, logic controller design and planning in scheduling in the process industries. 2002–2006 he was vice-rector of research and international relations of TU Dortmund. He received a Joseph-von Fraunhofer Prize for Applied Research and is a Fellow of IFAC, the International Federation of Automatic Control.

Dr. **Jochen Garcke** is head of a junior research group at Matheon and the TU Berlin. One focus of his research concerns the mathematical study of numerical approaches for applications in data mining and machine learning.

Prof. Dr. **Michael Griebel** is a professor for mathematics at the University of Bonn. There, he is presently the director of the Institute for Numerical Simulation and he is the speaker of the collaborative research center SFB 611 "Singular phenomena and scaling in mathematical models". His research interests comprise numerical simulation and scientific computing for fluid mechanics, material sciences, quantum chemistry, high-dimensional data analysis and data mining.

Prof. Dr. Dr. h.c. mult. **Martin Grötschel** is professor of mathematics at TU Berlin, Vice President of the Zuse Institute Berlin, Secretary of the International Mathematical Union, a Foreign Associate of the US National Academy of Engineering and a member of acatech, BBAW, and Leopoldina. His mathematical research focuses on optimization, discrete mathematics, and operations research; the areas of applications he addresses include telecommunication, traffic, transport, and logistics. He is a Fellow of SIAM and received, among other distinctions, the Leibniz, John von Neumann, Dantzig, and Beckurts Prize.

Prof. Dr. **Lars Grüne** is head of the Chair of Applied Mathematics at the University of Bayreuth. His research interests lie in the area of mathematical systems and control theory with a particular focus on numerical and optimization based methods for nonlinear systems.

Prof. (em.) Dr.-Ing. **Edmund Handschin** held the chair for electric energy systems and energy economics in the faculty of electrical engineering at the University of Technology in Dortmund/Germany from 1974 until 2007. He founded 1992 the EUS GmbH, an engineering company providing solutions at the interface between energy and communication technologies. He was managing director of EUS from 1992 until 2005. He was a member of the EU technology platform Smart Grids. He is a life fellow of IEEE. His research interests concern dispersed generation, information technology in power engineering, statistical methods in power engineering and optimisation techniques for energy management.

Dietrich Hartmann is (active) Professor Emeritus at the Ruhr-University of Bochum, Department of Civil and Environmental Engineering. Here, he held the chair of the Institute for Computational Engineering from 1987 until 2010. From 1982 to 1987, he was Professor for Structural Mechanics and Structural Optimization at the Technical University of Dortmund, intermitted by a one year stay at the University of California at Berkeley in 1983/1984. Prof. Hartmann is a member of the Northrhine-Westfalian Academy of Sciences and the German Academy of Technical Sciences. His research fields include large-scale and multi-level structural optimization, application of agent-based systems, lifetime-oriented structural design, and structural health monitoring and computer-based demolition of complex structures.

Dietmar Hömberg is a Professor of Mathematics at TU Berlin and head of the research group "Nonlinear Optimization and Inverse Problems" at Weierstrass Institute for Applied Analysis and Stochastics in Berlin. His research interests are modeling, simulation and optimal control of technological processes. Applications considered range from modelling of microstructure evolution during steel production and the investigation of production processes such as hardening, milling, and welding to the optimization of complete process chains.

Ulrich Jäger, managing director WSW mobil GmbH (Wuppertal) worked from 2002–2009 in various management functions for Deutsche Bahn Group. He was managing director for various bus companies in North Rhine—Westphalia and manager for the competence center production of DB Stadtverkehr GmbH. The competence center production was responsible for processes/tools for timetabling and vehicle as well as personal scheduling. An important role for and cost reduction programs played the integrated optimization routine and the mathematical method integrated in the planning software used by the 24 bus companies of DB Stadtverkehr.

Dipl.-Ing. **Moritz Kiese** is a member of the research staff at the Institute for Commutation Networks at TU München. His main research areas are the design and planning of backbone and mobile adhoc-networks using advanced linear programming methods. From 2005 to 2008, he worked in the BMBF-project Efficient Integrated Backbone (EIBONE).

Prof. Dr.-Ing. habil. **Rudibert King** holds the Chair of Measurement and Control at the Berlin University of Technology. He is speaker of the DFG-funded collaborative research center SFB 557 "Control of complex turbulent shear flows" and of the interdisciplinary research group "Bio-compatible ceramic foams". His research focuses on modeling, closed-loop control and optimization-based control of nonlinear and uncertain systems. Applications range from problems in fluid dynamics to bioengineering. He received the Dechema-Preis of the Max-Buchner-Forschungstiftung.

Dr. **Jürgen Koehl** is a Distinguished Engineer in the IBM research and development lab in Boeblingen, Germany. He studied mathematics in Bonn and Paris and developed for many years solutions in chip layout and timing optimization. During that time he worked closely with the Research Institute for Discrete Mathematics at Bonn University and supported the tools developed in this cooperation in IBM. From 2001 to 2003 Jürgen Koehl was on assignment to IBM Burlington, Vermont to lead the world wide ASIC design turn-around-time reduction for IBM's ASIC Design centres. Currently he is a member of the IBM Business Analytics and Optimization organization providing optimization solutions for production and supply chain management. He holds over 30 patents and is a member of the IBM Academy of Technology.

acatech-member Professor Dr. rer. nat. Dr. sc. techn. h. c. **Bernhard Korte** is director of the Research Institute for Discrete Mathematics of the University of Bonn. He is a member of several academies and he has received many national and international awards and distinctions. His main area of research is Discrete Optimization and its application to Chip Design.

Prof. Dr. **Günter Leugering** holds a chair for Mathematics at the University of Erlangen-Nuremberg and chairs the Center for Multiscale Modeling and Simulation within the DFG-Cluster of Excellence: "Engineering of Advanced Materials". He is coordinator of a DFG-Priority Program: "Optimization with PDE-constraints" and an International Doctorate Program within the Elite Network of Bavaria.

Prof. Dr. **Dirk Lorenz** is professor for mathematics at TU Braunschweig. His research interests include inverse problems and mathematical methods in image processing. Moreover, he works on applications in life sciences and engineering.

Klaus Lucas has been the Chair of Thermodynamics at RWTH Aachen from 2000 to 2009. He previously held positions at the University of Stuttgart, at the University of Duisburg, and as the Scientific Director at the Institute of Energy and Environmental Technology in Duisburg-Rheinhausen and was a consultant to various industrial companies in the fields of energy and chemical engineering. He is now vice-president of the Berlin-Brandenburg Academy of Sciences and Humanities. His research interests include the molecular modeling of fluids, energy system analysis, and phase and reaction equilibria in fluids. He is the author and coauthor of more than 150 technical articles in scientific journals and is the author of a well-known thermodynamics textbook.

Prof. Dr. **Peter Maass** is Professor of Applied Mathematics and director of the Center for Industrial Mathematics at the University of Bremen. He held positions as Assistant Professor at Tufts University, Boston, and Saarland University, before he was appointed Full Professor for Numerical Analysis at the University of Potsdam in 1993. Since 1999 he is director of the Center for Industrial Mathematics at the University of Bremen. His major research areas are inverse problems and wavelet analysis with an emphasis on applications in signal and image processing. He is inventor of several patents in the field of image processing, and in 1997 he has been awarded a nationwide innovation prize.

Prof. Dr.-Ing. **Wolfgang Marquardt**, a member of acatech, is a professor of process systems engineering at RWTH Aachen University. He is am member of the Steering Committees of the Cluster of Excellence Tailor-made Fuels from Biomass (TMFB) and the Graduate School "Aachen Institute for Adanced Studies in Computational Engineering Science (AICES)". His research interest focuses on systems engineering methods and in particular on the model-based design, operation and control of chemical process systems. He has been awarded the Leibniz-Preis of DFG.

Prof. Dr. **Alexander Martin** is professor for mathematics at the Technische Universität Darmstadt. His research interests are in algorithmic discrete optimization and operations research. He has many practical experiences in areas such as traffic and transport, energy management and finance as well as production planning. He currently is vice president at the TU Darmstadt and received a prize from the state of Hesse for his very innovative and successful cooperation with Hessian companies.

Prof. Dr. **Volker Mehrmann** is professor of Mathematics at TU Berlin and chair of the DFG Research Center MATHEON. His mathematical research areas are Numerical Mathematics, Control Theory, and Matrix theory with applications in flow control, as well as simulation and control of multi-body systems and electrical circuits.

Rolf Möhring is Professor for Applied Mathematics and Computer Science at Berlin University of Technology, head of the research group "Combinatorial Optimization and Graph Algorithms", and a member of MATHEON. His research interests center around combinatorial optimization and industrial applications in traffic, production and logistics. He has served as chair of the Mathematical Programming Society and received the Scientific Award of the German Operations Research Society.

Prof. Dr.-Ing. **Dirk Müller** is the head of Institute for Energy Efficient Buildings and Indoor Climate (EBC), E.ON Energy Research Center at RWTH Aachen University. After completing his studies of mechanical engineering at Aachen University he started his professional career at the co-operate research departments of the companies Bosch and Behr in Stuttgart. Before starting his new Institute at Aachen University he was chair of the Hermann-Rietschel-Institute at the Technical University of Berlin. His main research areas are energy concepts for buildings and communities, room airflow and thermal comfort and advanced air-condition processes.

Prof. Dr.-Ing. **Wolfgang Nitsche** is Professor of Aerodynamics at the Aerospace Department of TU Berlin. His main research activities are focused on flow control of aerodynamic flows.

Van Vin Nyguyen, M. Sc., is a research associate in the Institute for Computational Engineering at the University of Bochum, where he completed the masters course of "Computational Engineering". Current areas of research include agent-based structural design and optimization, Computational Steering and Artificial Intelligence Technologies.

Dr. **Hartwig Preckel**, PerkinElmer Cellular Technologies Germany GmbH, manages customer specific and product development projects for instrument and new method development for drug discovery. The main focus is software for automated image based cell analysis.

Prof. Dr.-Ing. **Christian Rehtanz** is head of chair for power systems and power economics at the TU Dortmund University. From 2000 to 2007 he held various leading positions with ABB in Switzerland and China. His research activities include new technologies for network enhancement and monitoring, control and protection systems. He holds the "World Top 100 Young Innovators Award 2003" of the Massachusetts Institute of Technology (MIT).

Dr. **Sebastian Sager** is head of a junior research group at the Interdisciplinary Center for Scientific Computing (IWR) at the University of Heidelberg. His research interests are the optimization and optimal control of complex processes, with applications mainly in economics, transport, and engineering. He received the dissertation price of the German Society for Operations Research and the Klaus Tschira Award for Achievements in Public Understanding of Science.

Prof. Dr.-Ing. habil. Prof. E.h. Dr. h.c. mult. **Michael Schenk** is the Director of the Fraunhofer Institute for Factory Operation and Automation IFF in Magdeburg and Holder of the Chair for Logistics Systems and Managing Director of the Institute of Logistics and Material Handling Systems at Otto von Guericke University Magdeburg. He holds an honorary Doctorate from Moscow State Automobile and Road Technical University MADI, Chairman of the Saxony-Anhalt State Chapter of the VDI and a member of the German Logistics Association's (BVL) Scientific Advisory Board.

Prof. Dr. **Rüdiger Schultz** is professor of mathematics at the University of Duisburg-Essen. His research interests in mathematics focus on discrete and stochastic optimization. In applied research he has contributed to optimization in energy systems, gas transportation, and resource planning.

Prof. Dr. **Michael Stingl** is a junior professor at the Department of Mathematics of the University of Erlangen-Nuremberg. He is head of the structural optimization group within the Excellence cluster for Engineering of Advanced Materials in Erlangen.

Michael Thess studied mathematics at the Chemnitz University of Technology and the Saint-Petersburg State University. Since 1998 he is R&D manager of the data mining specialist prudsys AG.

Dennis Trede is a research assistant at the Center For Industrial Mathematics at the University of Bremen. Here studied applied mathematics in Dresden and Bremen, and he did PhD studies in Bremen within the BMBF project "INVERS: Deconvolution problems with sparsity constraints in optical nanoscopy and mass spectroscopy". His research areas are mathematical methods in image and signal processing and inverse problems with an emphasis on applications from engineering and life sciences.

Prof. Dr. h. c. Dr.-Ing. **Eckart Uhlmann** is Director of the Fraunhofer IPK and Director of the Chair of Machine Tools and Manufacturing Technology at the IWF of the TU Berlin in the Production Technology Center Berlin. He was Vice President for Research and Development as well as Application Technology at Hermes Schleifmittel GmbH & Co., Hamburg.

Prof. Dr. **Jens Vygen** is Professor of Discrete Mathematics at the University of Bonn. He manages the project "Discrete Mathematics and Applications" of the North-Rhine-Westphalian Academy of Sciences and the research area "Optimization in Large and Complex Networks" of the Hausdorff Center for Mathematics (cluster of excellence). His research areas are combinatorial optimization and chip design.

Dr. **Roland Wessäly** is managing director of atesio GmbH. Together with colleagues of the Zuse Institute Berlin, he founded atesio back in the year 2000. atesio supports telecommunication network operators in decision making in strategic infrastructure projects as well as in the optimization of their networks w.r.t. cost and quality. His research on the cost optimal design of survivable networks has been recognized with the Vodafone Innovations Award. Moreover he is a member of the DFG research unit 894. His research is focused on semidefinite programming and structural optimization, in particular material design with applications in aerospace industry and medicine.

acatech member Prof. Dr. Ing. habil. **Peter Wriggers** is professor for Mechanics in the Department of Mechanical Engineering at the Leibniz Universität Hannover. Focus of his research are discretization and simulation techniques in the area of material theory and contact mechanics. Applications are materials, structures and interfaces and related simulations in Civil and Mechanical Engineering as well as in Biomechanics. He is member of other scientific societies and academies and obtained several national and international prizes.

Dr. **Stefan Zachow** is heading the Medical Planning group at Zuse Institute Berlin (ZIB) and is associated head of the Computational Medicine group. His areas of research are: Computer assisted surgery, visualization, digital image and geometry processing, model-guided therapy planning, and simulation. He holds degrees in computer engineering (Dipl.-Ing.), computer science (Dipl.-Inform.) and completed a postgradual study in Medical Physics. He is a member of the IEEE computer society, the Int. Society for computer aided surgery, and the German societies for computer and robot assisted surgery (curac) as well as the GI section Visual Computing in Medicine.

Printing: Ten Brink, Meppel, The Netherlands
Binding: Stürtz, Würzburg, Germany